Building Stones of Milan and Lombardy

Milan and Lombardy have played an important role in the Italian country since the Roman period. This importance is reflected also by the diffusion of stone architecture: a persisting trait of Milan architecture was the use of different stones in the same building. Milan lies in the middle of the alluvial plain of the Po, far from the stone quarries; some waterways were dug out in order to supply the building stones from the surrounding territories.

The study of stone as a building material was significant at the end of the 19th century, but then it was largely neglected by both architects and geologists. So it is significant to suggest a study about the stones employed to build in Milan (Volume 1) in relationship with a petrographic study about the features of the stones quarried in the whole Lombard territory (Volume 2).

Volume 2 contains descriptions of the features of the stones reported in Volume 1. These features include metamorphic and magmatic rocks of the Alpine area; sedimentary rocks and loose materials of the Prealpine area; sedimentary rocks of the Apennine area; and loose sediments of the Padania plain. Some stones, coming from other northern Italian regions, and used in Lombard architecture, are also described. Each stone is described in a "card" containing commercial and historical names, petrographic classification, macroscopic features, mineralogical composition, microscopic features, geological setting, quarry sites, transport to yards, morphology of dressed elements and surface handworking, use in architecture in the whole Lombard territory and abroad and decay morphologies. A particular investigation is addressed to the stones used during the 20th century; a great part of them were never used before in Milan and Lombardy.

Roberto Bugini was born in Milan, Italy, in 1952. He graduated in Scienze Geologiche (geology), University of Milan, in 1976. Since 1983 he has been a researcher of CNR-ICVBC Istituto Conservazione Beni Culturali. He has been a lecturer in stone materials at the Scuola di Specializzazione in Archeologia – Università Cattolica di Milano, since 2009. His fields of interest are stones and mortars used in historic architecture with particular application to the territory of Lombardy (Italy).

Luisa Folli was born in Lodi in 1956. She graduated in Scienze Naturali (natural sciences), University of Milan, in 1990. Since 1991 Luisa has worked in the field of stone conservation and has been a teacher (Mineralogy and Petrography) at Scuola di Restauro ENAIP – Percorso quinquennale per Restauratori Beni Culturali (Botticino-Brescia) since 1995. Field of interest: scientific analyses on nature and decay of historic building materials. References: about 150 articles on books, journals and proceedings.

Building Stones of Milan and Lombardy

Volume 2: Stones of Lombardy

Roberto Bugini and Luisa Folli

CRC Press
Taylor & Francis Group
Boca Raton London New York

CRC Press is an imprint of the
Taylor & Francis Group, an **informa** business
A BALKEMA BOOK

Designed cover image: Roberto Bugini and Luisa Folli

First published 2023
by CRC Press/Balkema
Schipholweg 107C, 2316 XC Leiden, The Netherlands
e-mail: enquiries@taylorandfrancis.com
www.routledge.com – www.taylorandfrancis.com

CRC Press/Balkema is an imprint of the Taylor & Francis Group, an informa business

ISBN: 978-0-367-34563-1 (Set hbk)
ISBN: 978-1-032-39118-2 (Set pbk)
ISBN: 978-1-032-42056-1 (hbk)
ISBN: 978-1-032-42059-2 (pbk)
ISBN: 978-1-003-36100-8 (ebk)

DOI: 10.1201/9781003361008

Typeset in Times New Roman
by Apex CoVantage, LLC

Contents

Preface to the second volume

In Milan, building stones were supplied with difficulty because of the distance from the mountains, so bricks, made of the clay easily available in the alluvial plain, were the leading building material since Roman times. Roman architects also used blocks of Alpine stones carried out by glaciers and distributed amid the hills of the Prealps; later on, in the Middle Ages, squared blocks were taken by the dismantling of Roman buildings. Over the centuries, many quarries were opened in the Prealps and in the Alpine valleys of the northern Lombardy region, and building stones were carried to Milan through a network that included rivers, lakes, waterways and roads, as already set up by the Romans.

HOW THE VOLUME IS ARRANGED

This volume describes a series of stones, employed in the buildings of Milan reported in Volume 1; the stones mainly come from different provinces of Lombardy, but some Italian regions and a few foreign countries are also involved.

First, some accounts about geology, working and decay were reported: outlines of the geology of Lombardy together with geological chronology and petrographic classification of stone used in Milan's architecture (Chapter 1); methods and tools employed in Lombard quarries and workshops (Chapter 2); list and classification of natural decay phenomena affecting building stones (Chapter 3).

The volume, like the previous one, contains single "cards", grouped on a geographical basis: provinces of Lombardy (Chapters 4–10), regions of Northern and Central Italy (Chapters 11–18), foreign countries (Chapter 19); each card is dedicated to a particular stone.

Personal surveys carried out in the quarries have allowed us to appreciate the different aspects of each building stone, which may show, in the same quarry, some varieties based on grain size, structure, chromatic appearance, etc. affecting recognition and identification.

Each card contains the significant characteristics of the stone: the common name used in commerce and its origin; the petrographic classification; the visual aspect of the surface and the microscopic texture (macroscopic or microscopic features); the essential and accessory minerals (mineralogical composition); the geological formation and its relationships in a regional area (geological setting); the sheet of Carta Geologica d'Italia and the sheet of Progetto CARG connected to each stone (geological reference); the location of the main quarry sites and the commercial varieties (quarry); the category of elements used in building (stonework categories); the most employed among different kinds of finishing (surface finishing); some significant works in which the stone is used, starting from the area around the quarries (use in architecture); and the most significant decay phenomena observed on buildings (decay morphology).

Special attention was paid to the references of architectural treatises, technical books and atlases on building stones, printed from the 17th century to the early 20th century: these specific references allow for better understanding of the role, importance and value of each stone over the centuries.

ADDITIONAL MATERIAL

Eight chapters refer to peculiar architectural cases that cannot be included in the other sections of the text. Three chapters concern some building materials: list and features of ancient coloured marbles coming from the whole Mediterranean basin and used in Roman times and later (Chapter 20), features of mortars and plasters (Chapter 21) and features of artificial stones (Chapter 22). Two chapters concern the features of building elements: masonries (Chapter 23) and pavements (Chapter 24). Three chapters concern the classification of stones employed to make special artefacts as altars in selected churches (Chapter 25), commemorative monuments of the whole city (Chapter 26) and funeral chapels of Cimitero Monumentale (Chapter 27).

ATLAS OF STONE

The atlas shows a representative picture of the actual size of several stones used in Milan architecture in order to facilitate their identification. A microscopic image of each stone is also reported. Stones are grouped according to historical and modern uses and to their petrographic classification.

General references to volume 2 (stones)

Geological maps

ISPRA (Istituto Superiore per la Protezione e la Ricerca Ambientale – Portale Servizio Geologico – Contenuti Informativi – Collane Editoriali e mappe – Cartografia Geologica – Cartografia Geologica del Servizio Geologico d'Italia alla scala 1:100.000 and Cartografia Geologica alla scala 1:50.000 Progetto CARG).

Ancient treatises

Vasari G., The Lives of the most excellent painters, sculptors and architects (L.S. Maclehose ed.), Dent and Company, London, 1907 (1st edition 1568).
Morigia P., La Nobiltà di Milano, Ponzio, Milano, 1595.
Scamozzi, V., L'idea dell'architettura universale, Venezia 1615.
Rondelet, J.B., Traité théorique et pratique de l'art de bâtir, Chez l'Auteur, Paris, 1802.
Cantù, C., Milano e il suo territorio, Pirola, Milano 1844.
Breislak, S., Descrizione geologica della Provincia di Milano, Silvestri, Milano 1845.

Technical books and articles

Curioni, G., Geologia applicata delle Province Lombarde Parte 2, Hoepli, Milano, 1877.
Jervis, G., I tesori sotterranei dell'Italia Parte Quarta, Loescher, Torino, 1889
Salmojraghi, F., I materiali naturali da costruzione, Hoepli, Milano, 1894.
I Marmi italiani (Peverelli, G., Squarzina, F. eds.), Confederazione Fascista Industriali, Roma, 1939.
Fagnani, G., Giacimenti di rocce e minerali utili tra il lago Maggiore ed il lago di Garda, *in* Natura, 47, 1956, p. 3–55.
Pieri M., I marmi d'Italia, Hoepli, Milano, 1964
Rodolico F., Le pietre delle città d'Italia, Le Monnier, Firenze 1965.

Chapter 1

Geology of Lombardy

Geological chronology and petrographical classification of Lombard stones

The territory of the region is broadly subdivided, from north to south, in four geo-morphological zones: mountains (Alps, Swiss border), hills (Prealps), plain (pianura Padana) and other mountains (Apennines, beyond river Po). Obviously, these morphologies are strongly influenced by the regional geology.

The northern part of the territory is occupied by the mountain ranges of the Alps (Alpi Lepontine and Alpi Retiche – Pizzo Bernina, 4050 m above sea level) and Alpi Orobie (Pizzo di Coca, 3052 m asl) followed by mountains and hills of Prealps (Grigna, 2409 m asl). The longitudinal span of the Prealps increases from west (about 25 km – Varese) to east (about 75 km – lake Garda).

The central part of the territory is occupied by the alluvial plain of the river Po and their tributaries called "Pianura Padana": debris, worn away from the Alpine range and carried out by rivers and glaciers, filled up a Pliocene marine gulf spanning from the Prealps to the Apennine. The height above sea level of the plain decreases from 150 m (north-west) to 25 m (south-east).

The southern part of the regional territory, shaped like a narrow triangle, is occupied by the hills and the mountains of the Apennine range (monte Lesima, 1724 m asl).

The Lombard landscape is also characterized by many river courses flowing north to south from the mountains to the plain, towards the course of river Po (flowing west to east). In the same way, other rivers flow south to north from the Apennine to the river Po. The alpine fluvial valleys were run across and deepened by the Quaternary glaciers, so creating some long and narrow lakes (Orta/Cusio, Maggiore/Verbano, Como/Làrio, Iseo/Sebino and Garda/Benaco). A relevant exception is the upper course of river Adda before flowing into Lake Como: the river runs from east to west (about 90 km) in a glacial valley, called Valtellina, between Alpi Retiche (north) and Alpi Orobie (south). Valtellina follows a tectonic line (Linea Insubrica – Linea del Tonale) separating the Alps from the Prealps.

The geological history of Lombardy involves a period of about 270 million years (Permian to Quaternary). The Alps across the border of Switzerland show Penninic and Austroalpine units. Pennidic includes Permian-Jurassic metamorphic covers (calc-silicate rocks, ophiolites, marbles) and Palaeozoic crystalline basement (gneisses, serpentines, ophicalcites). Austroalpine includes Permian-Jurassic sedimentary covers (conglomerates, dolomites, limestones), metamorphites (calc-silicate rocks, quartzites) and Palaeozoic crystalline basement (gneisses, micaschist, phyllites). Some Tertiary magmatic bodies are present along the Linea Insubrica as Val Màsino-Val Bregaglia (granodiorites and diorites) and Adamello (tonalites). Other magmatic bodies (Graniti dei Laghi, Hercynian granites) outcrop on the western border of Lombardy.

DOI: 10.1201/9781003361008-1

The Prealps, also indicated as Southern Alps, include Mesozoic-Tertiary sedimentary covers (conglomerates together with rhyolites; dolomites, limestones together with marls; sandstones) and a Pre-Permian crystalline basement (gneisses, micaschists, phyllites). Mesozoic sedimentary rocks (nodular limestones, soft bioclastic limestones) are present in the eastern part of the Prealps, located in the province of Verona and Vicenza (region of Venetia).

Quaternary includes continental sediments (northbound – conglomerates, morainic and alluvial loose materials) and marine sediments (southbound – sands, clays) of the Pleistocene. The depth of these sediments, measured from the top of Pliocene formations, reaches about 2000 m in the south-eastern part of the Region (Mantua). The "High plain", covering the southern rim of the Prealps, contains coarse-grained sediments (gravel, sand); the "Low plain", around the course of river Po, contains fine-grained sediments (silt and clay). The alluvial deposits are mixed, in the western part of the territory, with the moraines (glaciers of Ticino and Adda) flowing from the Alpine valleys. The central part is free from morainic deposits because of the presence of a "barrier" (Alpi Orobie). Other morainic deposits (Würm – Riss) are widely diffused in the eastern part of the region, where the glacier of Garda spread out in the plain (south of Lake Garda).

The Apennine includes Ligurian Units (argillites, marls, limestones, ophiolites; Cretaceous – Paleocene); autochthonous sedimentary formations (marls, sandstones, gypsum; Oligocene – Miocene); marine and continental deposits (conglomerates, gravels, sands, clays; Pliocene).

The various kind of magmatic, sedimentary and metamorphic rocks produced a large availability of stone materials useful as building stone. The stones were firstly used in the areas surrounding the outcrops and the quarries, but the stones also reached the wealthy cities growing in the plain.

The case of Milan, located in the middle of the plain, is symptomatic: the first building material was the brick made of clay (mainly baked), but the stones came easily to Milan, despite the distance from the quarry sites, using a network of roads, carried out during the Roman rule (starting from the 3rd century BCE) and some artificial waterways (Navigli) dating back to the Middle Ages and connecting rivers and lakes.

The main quarry areas (from west to east) are:

1 Val d'Ossola (ruled by the Duchy of Milan until 1748): gneisses (Serizzo, Beola), granites (Baveno and Montorfano), marbles (Candoglia, Ornavasso, Crèvola)
2 Southern shore of Lake Maggiore: dolostones (dimension stone and lime – Angera)
3 Val Ceresio (between Varese and the Swiss border): clastic limestones (Viggiù, Saltrio)
4 Lake Maggiore area: dolostones to make lime; lime was also produced in the Prealps of Varese around Lecco (lake Como) and on the eastern shore of lake Iseo
5 Lake Como area: black limestones (Varenna, Moltrasio), marble (Musso)
6 Brianza, south of Como and Lecco: sandstone (Viganò) and granodiorites (erratic boulders)
7 Gorges of rivers Adda and Brembo: conglomerates and sandstones (Trezzo, Brembate)
8 Valleys north of Bergamo: veined limestone (Val Brembana) and black limestones (Val Seriana)
9 Lake Iseo and Val Camonica: Quaternary conglomerates, black limestone (Iseo); Permian conglomerate, sandstones, rhyolite (Val Camonica)
10 Hills north-east of Brescia: magnesian limestones (Botticino, Rezzato)
11 Apennines (south of Pavia): sandstones (Oltrepò Pavese, between Casteggio and Broni)

12 Pianura Padana (alluvial plain): clay deposits (High plain, bricks), sands and gravels (mortars)
13 Monti Lessini (Verona) and monti Berici (Vicenza), out of the territory of Lombardy: nodular limestones (Verona) and soft limestones (Verona and Vicenza)

The use in Milan mainly involves the western quarry areas; the use of stones in many Lombard cities is linked to each local quarry areas. The use of loose sediments to make bricks or mortars is widespread in the whole Lombard territory. The Venetian stones were used for their colour (Rosso ammonitico) and for the aptitude to carving (Vicenza). The cities of the plain are linked to the Brescia or Verona areas, but the use of building stone was scant in comparison with the use of brick.

REFERENCE

Geological maps of Italian territory are available at the website of ISPRA (Istituto Superiore per la Protezione e la Ricerca Ambientale – Portale Servizio Geologico – Contenuti Informativi – Collane Editoriali e mappe – Cartografia Geologica – Cartografia Geologica

Table 1.1 Pre-Cambrian and Palaeozoic

Commercial name	*Petrographical classif.*	*Geological formation*	*Chronology*
Marmo di Vezza d'Oglio	Crystalline limestone	Gneiss del Monte Tonale	Pre-Cambrian
Beola	gneiss	Monte Rosa Nappe (Upper Penninic)	
Serizzo Ossola	gneiss	Monte Rosa Nappe (Upper Pennidic)	
Granito di Alzo	granite	Alzo-Roccapietra pluton	
Granito nero di Anzòla	gabbro	Strona-Ceneri (Serie dei Laghi)	
Granito di Baveno	granite	Graniti dei Laghi	
Granito di Montorfano	granite	Graniti dei Laghi	
Grigio del Bòden	silicate marble	Dioritico-kinzigitica (Ivrea-Verbano)	
Marmo di Candoglia	crystalline limestone	Dioritico-kinzigitica (Ivrea-Verbano)	Pre-Permian
Marmo di Ornavasso	crystalline limestone	Dioritico-kinzigitica (Ivrea-Verbano)	
Marmo di Valle Strona	crystalline limestone	Dioritico-kinzigitica (Ivrea-Verbano)	
Marmo di Crevola	crystalline dolostone	Monte Leone Nappe (Lower Penninic)	
Pietra d'Oira	serpentinite peridotite	Dioritico-kinzigitica (Ivrea-Verbano)	
Marmo di Musso	crystalline limestone	Micascisti dei Laghi	
Valcundria	granitoid gneiss	Gneiss del monte Gruf	
Granofiro di Cuasso al Monte	granophyre	Granofiro di Cuasso	
Pietra Simona	sandstone	Cong. Dosso Galli	Permian
Porfido Monumentale	rhyolite	Vulcaniti di Auccia	
Verrucano lombardo	conglomerate	Verrucano lombardo	

del Servizio Geologico d'Italia alla scala 1:100.000 and Cartografia Geologica alla scala 1:50.000 Progetto CARG). These maps contain unabridged geological reports (Note Illustrative) together with full references about each stone described here.

GEOLOGICAL CHRONOLOGY OF STONE MATERIALS USED IN MILAN'S ARCHITECTURE

List of stones quarried in Lombardy according to the geological chronology.

The first column reports the commercial name, the second column reports the petrographical classification, the third column reports the geological formation and the fourth column reports the geological chronology.

List of stones quarried outside Lombardy according to the geological chronology.

The first column reports the commercial name, the second column reports the petrographical classification, the third column reports the geological formation and the fourth column reports the geological chronology.

Table 1.2 Mesozoic

Commercial name	*Petrographical classif.*	*Geological formation*	*Chronology*
Volpinite	anhydrite	Carniola di Bovegno	Lower Triassic
Pietra di Varenna	limestone	Calcari di Perledo e Varenna	Middle Triassic
Pietra di Angera	dolostone	Dolomia principale	Upper Triassic
Nero di Bergamo	limestone	Argilliti di Riva Solto	
Arabescato orobico	limestone	Calcare rosso	
Abbazia	limestone	Calcare di Zu	
Occhiadino	limestone	Calcare metallifero	
Marmo di Zandobbio	limestone	Dolomia di Zandobbio	
Rosso Arzo,	limestone	Broccatello	Lower Jurassic
Macchiavecchia	breccia	Calcari selciferi lombardi	
Pietre di Viggiù e Saltrio	dolomite limestone	Calcari selciferi lombardi	
Pietra di Moltrasio	limestone	Rosso Ammonitico	
Rosso di Entratico	limestone	Corna	
Botticino	dolomite limestone	Corna	
Breccia Aurora	conglomerate	Corso	
Pietra di Rezzato	limestone	Medolo	
Medolo	limestone		
Serpentino	serpentinite	Pietre verdi della Margna	Middle Jurassic
Majolica	limestone	Majolica	Lower Cretaceous
Molera	sandstone	Cretaceous Flysch	Upper Cretaceous
Pietra di Sirone	conglomerate	Conglomerato di Sirone	
Pietra di Sarnico	sandstone	Arenaria di Sarnico	
Pietra di Credaro	sandstone	Flysch di Bergamo	
Pietra di Malnate	sandstone	Arenaria di Malnate	

Table 1.3 Cenozoic

Commercial name	*Petrographical classif.*	*Geological formation*	*Chronology*
Ghiandone	granodiorite	Val Masino – Val Bregaglia Pluton	Paleocene
Serizzo	diorite	Val Masino – Val Bregaglia Pluton	
Granito di Dubino	diorite	Val Masino – Val Bregaglia Pluton	
Granito di S. Fedelino	granite	Val Masino – Val Bregaglia Pluton	
Tonalite	diorite	Adamello Pluton	
Ceppo di Travedona	conglomerate	Formazione Ternate	Eocene
Breccia di Urago	conglomerate	Formazione Ternate	
Molera	sandstone	Arenaria di Malnate	Oligocene
Gonfolite	sandstone/conglomerate	Conglomerati Como	
Arenaria pavese	sandstone	Conglomerati di Cassano Spinola membro Arenarie di Monte Arzolo	Miocene
Alabastro	calcareous alabaster	Pleistocene	Pleistocene
Travertino	travertine	Pleistocene	
Ceppo del Brembo	conglomerate/sandstone	Ceppo del Brembo	
Ceppo Grè, Ceppo Poltragno	breccia	Complesso di Poltragno	

Table 1.4 Palaeozoic

Commercial name	*Petrographical classif.*	*Geological formation*	*Chronology*
Marmo di Lasa	marble	Marmi di Lasa, falda Ortles-Campo	Pre-Permian
Fior di pesco carnico	marble	Paleozoico carnico	
Granito Maddalena	monzogranite	Complesso intrusivo	Carboniferous
Bargiolina	micaceous quartzite	Quarzite di monte Bracco	Permian
Porfido trentino	rhyodacite	Porfiriti e porfidi	
Verde Roja	slate	Scisti gneissici	
Rosso Pantheon	granite	Graniti rossi (Serie Canavese)	

Table 1.5 Mesozoic

Commercial name	*Petrographical classif.*	*Geological formation*	*Chronology*
Marmo apuano	Marble (crystalline limestone)	Autoctono Toscano	Upper Triassic
Rosso di Verona	nodular limestone	Rosso Ammonitico Veronese	Lower Jurassic
Giallo Mori	oolitic limestone	Calcari grigi	
Marbles of Sienna	marble	Marmi della Montagnola	
Moncervetto	marble	Calcescisti	Lower Jurassic
Verde Chatillon, Issorie	ophicalcite	Calcescisti con pietre verdi	Upper Jurassic
Verde Alpi	ophicalcite	Calcescisti con pietre verdi	Upper Jurassic
Rosso / Verde di Levanto	ophicalcite	Breccia di Levanto	Middle Jurassic
Pietra d'Istria	compact limestone	Reef limestone	Cretaceous
Repen	limestone (*Rudistae*)	Calcari di Monrupino	Upper Cretaceous
Pietra di Aurisina	limestone (*Rudistae*)	Calcari di Aurisina	

PETROGRAPHICAL CLASSIFICATION OF BUILDING STONES OF LOMBARDY

The list includes stone materials (igneous, sedimentary, metamorphic) quarried in Lombardy and used in Milan's architecture from the Roman period until the first decades of the 20th century.

The first column reports the lithology, the second column reports the commercial name, the third column reports the quarry site and the fourth column reports the geological chronology.

Stones quarried outside Lombardy and used in Milan's architecture.

Table 1.6 Cenozoic

Commercial name	*Petrographical classif.*	*Geological formation*	*Chronology*
Marmo di Chiampo	nummulitic limestone	Calcari di Chiampo	Eocene
Pietra Gallina, Matòn	nummulitic limestone	Calcari nummulitici	
Pietra di Nanto	nummulitic limestone	Calcari nummulitici	
Pietra del Cardoso	meta-sandstone	Pseudomacigno	Oligocene
Sienite di Balma	sienite	Plutone della valle del Cervo	
Trachite euganea	trachyte	Form. vulcaniche paleogeniche	
Pietra di Vicenza	shell limestone	Calcareniti di Castelgomberto	
Diorite di Vico	diorite	Plutone di Traversella	
Pietra di Finale	shell limestone	Arenaria a Scutella	Oligo-Miocene
Pietra dorata	sandstone	Pietra di Finale	
Granito dell'Elba	granodiorite	Granodiorite di monte Capanne	Pliocene
Travertino romano	travertino	Travertino dei Bagni di Tivoli	Pleistocene

Table 1.7 Igneous rocks

Lithology	*Commercial name*	*Quarry site (province)*	*Chronology*
Granite	Granito di Baveno	Baveno (VCO)	Pre-Permian
	Granito di Montorfano	Montòrfano (VCO)	Pre-Permian
	Granito di San Fedelino	Novate Mezzola (SO)	Paleocene
Granodiorite	Ghiandone	Val Màsino (SO)	Paleocene
Diorite	Serizzo	Val Màsino (SO)	Paleocene
	Granito di Dubino	Valtellina	Paleocene
	Tonalite	Edolo (BS)	Paleocene
Rhyolite Granophyre	Porfido Monumentale	Bienno – Croce Domini (BS)	Permian
	Cuasso al Monte	Cuasso al Monte (VA)	Permian

Table 1.8 Chemical sedimentary rocks

Lithology	*Commercial name*	*Quarry site (province)*	*Chronology*
Compact limestone	Pietra di Varenna	Varenna (Lecco)	Middle Triassic
	Nero di Bergamo	Riva di Solto (Bergamo)	Upper Triassic
	Arabescato orobico	Camerata Cornello (Bergamo)	Upper Triassic
	Abbazia	Abbazia (Bergamo)	Upper Triassic
	Occhiadino	Oltressenda (BG), Cividate (BS)	Upper Triassic
	Rosso Arzo, Macchiavecchia	Arzo (Canton Ticino – CH)	Lower Jurassic
	Pietre di Viggiù e Saltrio	Viggiù Saltrio (Varese)	Lower Jurassic
	Pietra di Moltrasio	Moltrasio (Como)	Lower Jurassic
	Rosso di Entratico	Entratico (Bergamo)	Lower Jurassic
	Botticino	Botticino (Brescia)	Lower Jurassic
	Medolo	S. Eufemia (Brescia)	Lower Jurassic
	Majolica	Cittiglio (VA), Monticelli B. (BS)	Lower Cretaceous
Dolostone	Pietra di Angera	Angera (Varese)	Upper Triassic
	Marmo di Zandobbio	Zandobbio (Bergamo)	Upper Triassic
Gypsum	Volpinite	Costa Volpino (Bergamo)	Lower Triassic
Alabaster and Travertine	Alabastro	Induno O. (VA), Gandino (BG)	Pleistocene
	Travertino	Val Ganna (VA), Civate (LC)	Pleistocene

Table 1.9 Clastic sedimentary rocks

Lithology	*Commercial name*	*Quarry site (province)*	*Chronology*
Sandstone	Pietra Simona	Darfo (Brescia)	Permian
	Molera	Viganò (Lecco)	Upper Cretaceous
	Pietra di Sirone	Sirone (Lecco)	Upper Cretaceous
	Pietra di Sàrnico	Sàrnico (BG), Paràtico (BS)	Upper Cretaceous
	Pietra di Credaro	Credaro (Bergamo)	Upper Cretaceous
	Pietra di Malnate	Malnate (Varese)	Oligocene
	Arenaria pavese	Santa Giuletta (Pavia)	Miocene
Conglomerate and Breccia	Breccia Aurora	Paitone (Brescia)	Lower Jurassic
	Ceppo di Travedona	Travedona (Varese)	Eocene
	Breccia di Urago	Urago (Como)	Eocene
	Gonfolite	Camerlata (Como)	Oligocene
	Ceppo del Brembo	Trezzo (MI), Brembate (BG)	Pleistocene
	Ceppo Gré, Ceppo Poltragno	Castro (Bergamo)	Pleistocene

Table 1.10 Metamorphic rocks

Lithology	*Quarry site (province)*	*Commercial name*	*Chronology*
Gneiss	Beura (VCO)	Beola	Pre-Permian
Gneiss	Valle Antigorio (VCO)	Serizzo Ossola	
Granitoid gneiss	Samolaco (SO)	Valcundria	
Serpentinite	Chiesa Val Malenco	Serpentino	Middle Jurassic
Serp. Peridotite	Nonio – Oira (VCO)	Pietra d'Oira	Pre-Permian
Marble	Vezza d'Oglio	Marmo di Vezza d'Oglio	Pre-Cambrian
	Crevola d'Ossola (VCO)	Marmo di Crevola	Pre-Permian
	Madonna Boden (VCO)	Grigio del Boden	Pre-Permian
	Candoglia (VCO)	Marmo di Candoglia	Pre-Permian
	Ornavasso (VCO)	Marmo di Ornavasso	Pre-Permian
	Sambughetto (VCO)	Marmo di Valle Strona	Pre-Permian
	Musso (CO)	Marmo di Musso	Pre-Permian
Gabbro	Anzola (VCO)	Granito nero di Anzola	Pre-Permian

Table 1.11 Italian territory outside Lombardy

Lithology	*Commercial name*	*Quarry site (province)*	*Chronology*
Granite	Rosso Pantheon	Valperga, Belmonte (TO)	Permian
	Granito Maddalena	Isola S. Stefano (Olbia-Tempio)	Carboniferous
Granodiorite	Granito dell'Elba	Campo (LI)	Pliocene
Sienite	Sienite di Balma	S. Paolo Cervo (Biella)	Oligocene
Diorite	Diorite di Vico	Vico Canavese (Torino)	Oligocene
Rhyodacite	Porfido trentino	Val Cembra (Trento)	Permian
Trachyte	Trachite euganea	Colli Euganei (Padua)	Oligocene
Compact limestone	Rosso di Verona	S. Ambrogio Valpolicella (VR)	Lower Jurassic
	Giallo Mori	Mori (Trento)	Lower Jurassic
	Pietra di Aurisina	Aurisina (Trieste)	Upper Cretaceous
	Repen	Monrupino (Trieste)	Upper Cretaceous
	Marmo di Chiampo	Chiampo (Vicenza)	Eocene
	Pietra d'Istria	Vrsar/Orsera (Croatia)	Cretaceous
Soft limestone	Pietra Gallina, Matòn	Avesa, Lessini (Verona)	Eocene
	Pietra di Nanto	Nanto, Berici (Vicenza)	Eocene
	Pietra di Vicenza	Costozza, Berici (Vicenza)	Oligocene
	Pietra di Finale	Finale Ligure (Savona)	Miocene
Travertine	Travertino	Rapolano (Siena)	Pleistocene
		Guidonia (Roma)	Pleistocene
Marble	Marmo apuano	Alpi Apuane (Massa Carrara)	Lower Jurassic
	Marmo di Lasa	Laas/Lasa (Bolzano)	Palaeozoic
	Moncervetto	Monastero di Vasco (Cuneo)	Upper Triassic
Slate	Verde Roja	San Dalmazzo di Tenda (Fra)	Permian
Ophicalcite	Verde Chatillon, Issorie	Chatillon *et al.* (Aosta)	Jurassic
	Verde Alpi	Cesana (Torino)	Jurassic
	Rosso/Verde Levanto	Levanto, Bonassola (Spezia)	Jurassic-Cretaceous
Quartzite	Bargiolina	Barge (Cuneo)	Permian
Meta-sandstone	Pietra del Cardoso	Seravezza (Lucca)	Oligocene

Chapter 2

Stone quarrying, transporting and dressing in Milan and Lombardy

QUARRYING

The use of stone in Milan's architecture, despite the distance of the quarries, has been documented since the Roman Republican period (1st century BCE): some remains of this period show worked stone artefacts (i.e. drums of column shafts made of Molera sandstone or even capitals made of Vicenza soft limestone, now in exhibition at Museo Archeologico). Nevertheless, if it is possible to identify the stones used for buildings together with their geological formation and the site of provenance, it is impossible to find a witness of Roman quarry works. In fact, the same quarry sites were exploited for centuries, and the marks of the ancient works were completely removed.

Ancient quarries

The Roman craftsmen used the stone materials available on the building site. Stones, such as pebbles and cobbles, came from river courses or from glacial deposits spread in a wide area of the region; in other cases, stones came from rock outcrops inside the town itself (i.e. cities in the Prealpine area, such as Como, Bergamo and Brescia).

The boulders, left by the glaciers on the slopes of the Prealps, represented a particular kind of quarry site. These boulders (igneous or metamorphic rocks) lay many kilometres far from the specific outcrop, but close to the towns (i.e. dozens of boulders of Ghiandone spread around the southern part of Lake Como); they were worked considering their particular features and considering also the advantage to cut a block of stone already detached from the rock outcrop. This was a sign of the skill of the craftsmen appreciating the higher qualities of these rocks in comparison with the qualities of sedimentary rocks (dolostone, limestone) outcropping in the same area.

At the same time the craftsmen, exploring the mountain sites, were able to understand the presence of rocks useful for building through the observation of the rock fragments piled up downhill by rockslides: once the characteristics of a rock were verified, it was possible to find a better place to open a quarry, and the quarries, it is worth it to note, were often located in recessed areas on rough mountain slopes.

On the other hand, to build the towns of the plain, the main source of raw materials were the argillaceous sediments, locally available in great amounts and exploited to make brick and terracotta.

The greater part of the sites was worked as an open-pit quarry, removing the top layers of rock affected by decay and alteration: i.e. dolomite limestone of Botticino shows broad

DOI: 10.1201/9781003361008-2

phenomena of surface erosion by dissolution with the formation of pockets of clayey material, called "terra rossa" (red earth), made of the insoluble silicate fraction of this rock. The ancient methods of excavation were completely hidden by new ones, but these methods probably followed the nature of the stone outcrop: excavation by steps starting from the bottom or excavation by vertical faces.

Massive silicate rocks, such as granite and gneiss, were divided in rough blocks starting from the natural joints, using levers and wedges. Marbles were exploited through the excavation of small pits and trenches around the block, the lower face of the block was detached using levers and wedges. Sedimentary rocks were divided using the natural beds and the cracks perpendicular to the bedding plane; soft stones (bioclastic limestone) were worked by cutting a narrow trench (using a pick or an axe) to obtain a block of appropriate size. In particular, an ancient stone quarry produced blocks of different sizes according to the arrangement of the natural beds or joints or cracks of the outcrop; on the contrary, for a modern quarry, it is compulsory to produce blocks of the same size.

The presence of underground quarries in some sites was also probable, despite the problem of darkness and ground waters; the underground quarries at Viggiù and Saltrio (huge cavities with large stone supports left to bear the load of the ceiling) or at Viganò Brianza or Malnate (small cavities without supports) were exploited in this way because of the specific conditions of the rock outcrop: the presence of hard nodules of chert at Viggiù and Saltrio and the presence of useless thin argillaceous beds at Viganò and Malnate.

Vasari distinguished the "squared ashlar-work" as plain dressing to make straight corners and rectangular shapes, appropriate for masonry; the "carved ashlar-work" to make moulding, frieze, foliage, egg, spindle, dentel and other sorts of carving, appropriate for decoration [Vasari, 1907; ch. 2].

These operations were guided by instruments for measuring and were carried out using a series of tools, each one with an appropriate use (see *infra*).

The splitting of blocks of marble or limestone into regular pieces, was done using a saw (a toothless metallic ribbon) together with sand and water, as reported by Plinius: "The cutting of the marble is effected apparently by iron, but actually by sand, for the saw merely presses the sand upon a very thinly traced line, and then the passage of the instrument, owing to the rapid movement to and fro, is in itself enough to cut the stone" [Plinius, 1962; book 36, 51]. Scamozzi also reports: "The Ancient sawed the marbles using an iron saw, toothless and very thin, sprinkling with water and different kinds of sand" [Scamozzi, 1615; part 2, book 7, ch. 5].

The splitting of blocks of granite or porphyry was done by carving a series of holes and through the percussion of wedges inserted in the holes.

Modern quarries

Ancient quarry methods based on hand-working remained unchanged for centuries. New methods based on controlled explosions were improved in the second half of the 19th century. The mines were useful in the presence of hard rocks without joints, but they were useless in the presence of softer or brecciated stones (marble, limestone, ophicalcite, etc.). Different kinds of explosives were employed (gunpowder, dynamite, etc.) in the aim to release, from the quarry face, a great amount of rock fragments of different sizes, ready to be cut in regular shapes. These explosions caused incidental damage to the surroundings (the stone blocks were thrown at a distance of 500 m) and also an incredible amount of debris along the slopes below each quarry (the so-called ravaneto, clearly visible in the quarries of the Apuanian Alps and a negative environmental impact). Some mines at granite quarries of Baveno (Piedmont), involved 2–3 metric tons of explosives, so moving about 100.000 metric tons of rock.

A lower quantity of explosives was also employed to make the final detachment of a block from the rock outcrop: two faces of a prismatic block were worked through trenches, two remaining faces (base and back of the block) were detached through the explosion of appropriate mines (i.e. granite quarries at Baveno and Montorfano, gneiss quarry at Villadossola, dolostone quarry at Angera).

The use of the wire saw, since the end of the 19th century, produces blocks of regular shape and to dramatically reduce the amount of debris. A twisted, multi-strand wire cuts the rock through the carrying of a slurry of abrasive material (i.e. sand) as used by the ancient sawing methods. The need to cool the metallic wire forced the preparation of very long loops (up to 1 km), regulated by pulleys; in this way the wire was cooled before going back on the cutting line. This method was preferred in carbonate rocks like marbles (Candoglia, Musso) or limestones (Botticino). On the contrary, the cut of igneous (i.e. Baveno granite) or metamorphic rocks (Serizzo Ossola) was made through an improvement of the ancient method: metallic plugs and feathers allowed the division of the rock along the natural joints or along a series of holes carved on the rock surface.

The quarry works were made as open pit or underground, if required by the features of the rock: the quarries of Candoglia marble, particularly the main quarry called "Cava Madre", proceeded inside the mountain; a quarry of Ornavasso marble, on the road to Madonna del Boden, was totally developed underground and the quarry face was reached using long tunnels perfectly covered with shaped stone blocks.

Present-day quarries

Since the middle of the 20th century, after World War II, the Lombard quarries were mostly disused because of the rising cost of labour, the decrease of interest in stone in architecture in favour of other materials (artificial stone, metal, glass, ceramic, wood) and the falling off of the availability of stone material in many sites. Finally, stronger entreaties for environmental conservation really conditioned quarry works in the 21st century, and the restoration of ancient sites took place involving the plantation of different species of trees. Few quarry areas are now working in Lombardy, such as in Botticino (dolomitic limestone), Val Malenco (Serpentinite), Lake Iseo (Ceppo conglomerate), Val Camonica (porphyry) or in the neighbouring Val d'Ossola as Antigorio (gneiss), Crévola (dolomitic marble), Candoglia (calcitic marble), etc. The quarry works try to reduce the environmental impact: a long tunnel was dug in a quarry of Ceppo (Castro, Lake Iseo) in order to avoid the ravage of a lake shore landscape; the tunnel runs parallel to the mountain slope. In many cases, nevertheless, the quarries were neglected after closing down, and they still break the uniformity of the mountain scenery. Despite this decrease of exploitation of local stones, the workshops are still working to saw blocks coming from abroad.

The quarry works in Lombardy are marked today by some operations carried out with different methods according to the rock texture. The primary cut to split a block from the rock outcrop is made by diamond wire or by chainsaw, drill or flame-jet. The large portion of rock, involved by the primary cut, is afterwards divided into large squared blocks by perpendicular cuts; specific devices, such as inflatable splitting cushions (metal or polyester), overturn each block on a "bed" made of rubble, and the overturned block is then cut into smaller blocks of appropriate size.

The diamond wire saw machine was employed in the last decades of the 20th century and was an important improvement in quarry technique, requiring minimal manpower to operate. The wire is abrasive by threading diamond impregnated beads; the cut is lubricated by water. The wire allows a precise cut mainly on carbonate but also on silicate rocks, and different accessories allow other kinds of cut (horizontal, blind, etc.). The use of the diamond wire involves the drilling of two perpendicular holes, the insertion of the wire in the holes and its junction to create a loop; a simple engine moves the wire and simultaneously runs back, on tracks, to maintain the tension needed to cut the rock. As an advantage, the diamond wire is four to five times faster than previous cutting methods, but as a disadvantage, the diamond wire is less durable than a toothed saw blade, and the cutting operations produce a great amount of fine matrix (called "marmettola") made of water and marble dust together with oils, fats, hydrocarbons, etc. Diamond wire cut is made using different kinds of wire according to materials (marble or granite) or purposes (squaring or shaping), and its use produces a certain regularity of the quarry faces, mainly in marble quarries.

The chainsaw or diamond belt arm saw machine is employed to cut soft stones (i.e. bioclastic limestones of Monti Lessini and Monti Berici); the arm saw operates on a vehicle and the cut is allowed in any direction. This system reduces waste and is essential in underground quarries, but on the contrary it requires a great amount of water.

The drill machine, including rods and specific bits, is employed to cut igneous rocks with a granular texture where a low amount of water is needed: a series of holes is made along a definite line using compressed air or hydraulic drifters; coplanarity and parallelism of holes are essential to make a good cut (i.e. granite of Baveno and Montorfano).

The flame-jet includes a blow-pipe generating a flame (2000–2800°C, hydrocarbons fuel) that is moved along the cutting line, and it is used mainly on silicate rocks.

According to different kinds of stones, other kinds of cutting machines were employed, such as: very soft stones (i.e. calcareous tufa of Southern Italy and Sicily) provide a series of prismatic pieces using a combination of three circular saws, with diamond blades arranged at a right angle. Igneous rocks with particular texture (Granofiro from Cuasso al Monte; porphyry from Trentino-Alto Adige/Süd Tirol, etc.) provide more or less regular cubes for road paving using a press split machine. Block cutters with a blade saw supported by a cast iron structure or giant blade saws supported by a steel structure provide slabs of various thickness from marbles or granites.

TRANSPORTATION

The first movement of stones is made in the same quarry to reach the workshop; this transfer involves rollers in the case of horizontal ways or inclines where the stone went down or went up, through ropes and winches.

The handling of rough blocks in ancient quarries was made through cranes, with cables and pulleys; the blocks were held by lifting bosses, protruding from opposite faces and then removed, or using lifting tongs or lewis inserted in apposite lewis-holes carved on a face. The stone pieces were prepared in an intermediate state before the journey towards the yards in order to reduce the weight of each element. Sarcophagi of Aphrodisia (Turkey, Western Anatolia) were worked to the required shape, but some parts were left untouched in order to realize later the final decorations (i.e. garlands of flowers). Different shipwrecks showed groups of four-column shafts almost completely finished except for a narrow strip of stone to maintain a bond among the shafts.

To move rough blocks in a present-day quarry, several devices are employed: derricks/cranes (lifting devices, with cables and pulleys, made of two or more metallic beams fixed in an appropriate site of the quarry), diggers, skid steer loaders, wheel loaders and backhoe loaders, articulated trucks, etc.

The transportation of stones in antiquity, from the quarries to the yards, was a very big challenge. Vasari and Scamozzi, among the ancient authors, provide peculiar information about the transporting in relation to some stones quarried in Lombardy and Venetia, pointing out the importance of waterways.

Vasari reported the travel from Verona to Venice, through the river Adige and the Adriatic Sea (about 150 km), of Valpolicella coloured stones and other kind of stones [Vasari, 1907; ch. 1].

The marbles used to build Milan's cathedral (Candoglia and Ornavasso) coming from the lower Val d'Ossola were carried on specific boats on the river Toce, and they reached Milan sailing the Lake Maggiore, the river Ticino and the Naviglio Grande, a waterway connected to the urban waterways encircling the old town Milan [Scamozzi, 1615; part 2, book 7, ch. 5].

Ceppo del Brembo quarried along the banks of rivers Adda and Brembo was carried to Milan thorough a waterway called "Naviglio della Martesana", linking the river Adda to the urban waterways [Scamozzi, 1615; part 2, book 7, ch. 11]. The black limestones of Val Degagna (north of Salò, Lake Garda) were carried on paved roads, then transferred by boats on Lake Garda to Lazise and then transported to Verona [Scamozzi, 1615; part 2, book 7, ch. 5]. The stones from Viggiù and Saltrio (Val Ceresio, north-east of the province of Varese) were also loaded on carriages to cover about 50 km to reach Milan [Scamozzi, 1615; part 2, book 7, ch. 11] using the network of Roman roads always maintained on duty. Stones quarried outside of Lombardy were also considered: the white marble of the Apuanian Alps, called *marmor Lunense* by the Romans, were embarked at the harbour of *Luna* (Tyrrenian sea), close to the mouth of river Magra, then transported to Rome and other destinations [Scamozzi, 1615; part 2, book 7, ch. 5].

Other information about the stone transportation to Milan's Duomo were reported by Serviliano Latuada: blocks of "white marble" coming from Candoglia quarries, were unloaded from boats in a pond (called Laghetto, this name is still maintained), connected to the Naviglio and located in front of the church of Santo Stefano; then the marbles were loaded to carriages in order to reach the stone workshop, close to the apse of the cathedral (about 200 m) [Latuada, 1738; book 1, #46].

An outstanding case of transportation of very uncommon stone pieces occurred in 1839–42: two obelisks (10.28 m high, more than 22 metric tons in weight), made of granite of Baveno, were translated from the quarry of Val d'Ossola (Lake Maggiore) to the gardens of villa Torlonia (via Nomentana, Rome) as reported by Gasparoni. Firstly, two roughly squared blocks, quarried on monte Camoscio (Baveno) by means of the mine's explosion (1839), were rolled to the Lake Maggiore shore near Feriolo (about 3 km). Then worked blocks were loaded on special boats and taken to Milan through Lake Maggiore, the Ticino River and the Naviglio Grande waterway. The workshop of Antonio Pirovano (located on Naviglio della Martesana) finished the surfaces of the monoliths using toothed hammers. Starting 15 August 1839, the transportation from Milano to Venice was made using new boats and travelling through the Ticino and Po rivers, some canals and the Adige River course; then the boats crossed the Venetian lagoon from Chioggia to the Venice's Arsenale. Here the obelisks embarked together on a special boat to circumnavigate the Italian peninsula from the Adriatic Sea (reaching also the Albanian coast at Durrës) to the Ionian Sea to the strait of Messina between Calabria and Sicily to the Tyrrenian Sea to the mouth of the Tevere river (late 1839). The Nomentano bridge (Sacco-Pastore) in Rome was reached through the Tevere and Aniene river courses; then the boat, still containing the obelisks, was hoisted (January 1840) by winches and ropes to via Nomentana and to the erection site at Villa Torlonia, with a distance of about 2.8 km and a rise of about 30 m from the river level. The journey of about 2900 km lasted six months (1839–40); two years were devoted to the carving of the laudatory inscriptions, written by the friar Luigi Ungarelli using ancient Egyptian hieroglyphics. The obelisks were finally erected on high bases in two distinct days (4 June and 25 July 1842), opposite both façades of the "Casino Nobile" (built by G.B. Caretti, 1823–40). Pope Gregorius XVI led the ceremonies together with Ludwig I, king of Bavaria, and a great crowd of Roman citizens [Gasparoni, 1842].

An efficient network of railways was gradually available beginning in the second half of the 19th century, before the Italian Unification (the first line, connecting Milan to Monza, was inaugurated 17 August 1840), making possible the movement and the transport of stone materials from the quarry sites. The lines from Milan went northward (Como or Varese and the Switzerland), westward (Novara, Turin and France), southward (Genoa and the harbour), eastward (Brescia, Verona and Venice or Austria) and to the south-east (Bologna, Firenze, Rome and the Southern Italy). Other lines went along the lakes (Maggiore and Como) or into the Alpine valleys of Sondrio (Valtellina), Bergamo (Val Seriana and Val Brembana) and Brescia (Val Camonica). In some cases additional railways of a few kilometres were built to link the quarry areas to the trunk lines of the railway network (i.e. the narrow gauge railway from Viggiù to Arcisate).

The network of paved roads, useful for trucks, was greatly enhanced in the 20th century (the first motorway connecting Milan to Varese was inaugurated 21 September 1924). After World War II, the railways were gradually neglected and stones were moved onto trucks, instead of trains, together with other goods.

DRESSING

"The great difficulty of working very hard and compact stones". The words of Vasari [Vasari, 1907; ch. 1] introduce the specific operations to give a shape to a raw block. Stones may be used in architecture in a dressed form (works to obtain a predetermined form) and the nature and the origin of each stone strongly influences the dress works. The original form (without any sort of working) was only used in particular circumstances (availability of loose deposits) or historical period (Middle Ages).

*Cut forms are classified according to the size and the degree of shaping: ashlar (concio), block (blocco), dressed stone (pietra da taglio), decorative stone (pietra per decorazione), slab (lastra). The arrangement of the aforementioned element leads to: course (regular succession of element) and stonework (disposition of the courses).

*Original forms are classified according to the size: sand (sabbia), gravel (ghiaia), pebble (ciottolo), cobble (ciottolo), boulder (scapolo), block (masso) [natural], marble powder (polvere di marmo), crushed brick (cocciopesto), rubble (pietrisco), splinter (scheggia) [made by artificial operations].

An Italian classification [Salmoiraghi, 1892] based on traditional working and building methods, includes the following forms:

Pietra grezza (as comes from the quarry). Rocks suitable in this way are compact rocks, fragile rocks as porphyry or loose sediments for foundations and for paving.
Pietra concia (only one face worked, other faces rough). Rocks suitable in this way are sedimentary rocks for ordinary masonries.
Pietra da taglio (square or rectangular shape, regular working on each face, drafted margins). Rocks suitable in this way are hard materials, such as granite, trachyte, basalt; compact limestones; massive sandstones; massive gneisses for refined masonries.
Lastra (wide, plane piece with two sizes prevailing to the third one). Rocks suitable for slabs are foliated metamorphic rocks (obtained by natural splitting, such as gneiss, serpentine), compact rocks (obtained by cut, such as limestone, marble, granite, syenite, travertine) for floors, staircases, balconies, roofs, cladding and veneer.
Pietra per decorazioni (particular forms). Rocks suitable for decorations are ranged according to the final operation: polishing or carving. The first ones are mainly used as slab and include polychrome stones showing harmonious disposition of veins or spots or patch or dots and a vivid and warm contrast of colours, such as marbles, ophicalcites, alabasters, carbonate rocks (Lumachella, Breccia) and also igneous rocks (granite, syenite). The second ones include soft limestones (monti Berici, Vicenza) and, chiefly, white fine-grained marbles (Apuanian Alps) [Salmoiraghi, 1892].

Tools

To achieve the desired form, stone pieces were worked, since the Greek times, using a series of tools, mainly metallic, with different shapes according to different purposes. These tools have the leading part in the whole process of stone dressing, and the number of tools needed to work a stone piece increases according to the hardness of the material, to the desired form and the degree of finishing of the surface.

Vasari twice reported a list of tools to make a sculpture with a polished surface, starting from a rough block: *subbia* (punch, a pointed and heavy tool) to roughing out the figure; *calcagnuolo* (a flat tool with a notch or tooth in the middle) and *gradina* (a flat tool with two notches) to give the proportions of the figure; *scarpello* (a smooth chisel) and then curved file to remove all marks; slender file and straight rasp to complete the smoothing process; pumice-stone, Tripoli earth, straw to make the surface lustrous and polished. Moreover, a *trapano* (drill) was used to make holes in order not to fracture the stone material [Vasari, 1907; ch. 1 and ch. 9].

A complete list of tools is almost impossible to write; taking into account the Lombard stones, the most important tools are classified as reported here:

*Tools for splitting: plug (wedge) and feathers set in holes drilled parallel to rock joints; thin metallic splitter to make rough slabs in some kind of foliated metamorphic rocks (i.e. Beola from Val d'Ossola).
*Tools with handle useful from rough work to finishing, showing a head with two elements (both similar or different): axe (blade parallel to the handle, soft stone), adze (blade at right angles to the handle, soft stone), pick (long, curved steel head with

point at each end, hard stone), toothed hammer (toothed blade, hard stone finishing), bush hammer (diamond points on orthogonal rows; hard stone finishing).

*Tools for cutting without handle, in sequence from rough work to finishing: pitcher (flat edge, rough work), punch (pointed), claw tool or toothed chisel (toothed blade), chisel (straight blade), bull-nose chisel (rounded blade), gouge (concave blade).

*Tools with handle useful to driving cutting tools: hammer (metallic head, prismatic shape), mallet (wooden head, frustum of cone).

*Tools based on a "back and forth movement": toothed or toothless saw with different shapes used to make slabs; rasp, file and plane used for finishing and for sculpture.

*Tools based on a "rotating movement": auger, drill or lathe, used for sculptures, reliefs or column shafts.

Each tool produces different marks on stone: grooves and curves (toothless saw); almost parallel lines (axe); furrows (punch, fine or coarse); regular series of facets (claw tool, toothed hammer); continuous little facets (chisel); concave facets with short curved lines (bull-nose chisel); fine, long and irregular furrows (rasp); perpendicular holes (drill); honed to polished-rubbed face to reach a reflecting surface (abrasive powder, lead, textile). The surface finishing is identified by the Italian term "pelle" (skin), and some adjectives are added to this term to indicate the degree of finishing: *pelle grossolana* (*subbia grossa* = punch); *pelle rustica* (*subbia fine* = punch, *gradina* = toothed chisel); *pelle mezzana* (*martellina grossa* = toothed hammer, *bocciarda* = bush hammer); *pelle piana* (*martellina fine* = toothed hammer); *pelle liscia* (*scalpello* = chisel); *pelle arrotata* (quartz sand), *pelle levigata* (pumice or lead filings), *pelle lucidata* (tin oxide or bone ash).

Finally, it is worth it to note different instruments for measuring, drawing and verifying: rule, square, compasses, plumb line, level, goniometer.

Sculpture

"The sculpture is an art which by removing all that superfluous from the material under treatment reduces it to that form designed in the artist's mind" [Vasari, 1907; ch. 8]. This art of carving hard materials needs a close examination: the most important factor is the choice of the stone as the homogeneous composition and the regular texture strongly influence the final result; moreover, the homogeneity of the stone gives solidity to the carving work and allows the sculpture to stand upright. According to Vasari, only a particular kind of stone is suitable for sculpture: this stone must be pure white and milky in tone marble, it is easy and soft to work and quite perfect for carving into figures; this kind of marble is more yielding and receives a finer polish than any other marble [Vasari, 1907; ch. 1].

Lombard sculptors, in the Roman period, used a marble showing these features: the marble from the Apuanian Alps (first from Carrara, then also from Massa and Garfagnana). White, fine-grained (average size of calcite crystals 0.2 mm), flawless (almost without fissures, veinlets, silicate or sulphide minerals), subject to a strong polishing, giving a perfect response to the impact of the tools, this marble sailed the sea around the Italian peninsula and climbed over the mountains to be used in the towns of the whole Pianura Padana. Another marble could be used for the sculpture: the marble from Candoglia (Lake Maggiore), but it showed a coarser grain size and some silicate veins with pyrite crystals, and last but not least, this marble was devoted to the construction of the Milan cathedral. Until the late 19th century, it was used for sculpture only in the cathedral and in the church of Certosa di Pavia.

The tools used for carving are the same as those already described, such as punch, claw tool, chisel, bull-nose chisel, rasp and drill, together with polishing materials.

Present-day dressing

The shape and use of each tool were kept alive over the centuries: the modern tools are similar to those used in antiquity as reproduced on archaeological reliefs and paintings. Only some features are changed: the use of harder metal alloys, the use of compressed air instead of hammer plus muscular strength and, especially, the introduction of computer-assisted technologies together with laser technology.

Nevertheless, some machines produce different elements ready for the use without manual works. Three- or five-axis bridge machines make several kinds of slab working, such as cutting, chamfering, edging, profiling, milling, drilling, slotting and rounding together with other particular operations. A multi-axis lathe machine makes column shafts. A water-jet machine (high-pressure water charged with abrasive particles) rapidly makes different kinds of inlay pieces with high precision cuts (tolerances measured by thousandths of an inch), whereas handmade inlay pieces required months of artistic work by skilled craftsmen (i.e. *Opificio delle Pietre Dure*, Florence).

Surface finishing is now done using different tools, according to the different hardness and texture of each stone: sandblasting (a jet of water added with sand and other abrasive materials); bush hammering (percussion with a tool of toothed head or with parallel blades); flaming (an oxy-propane flame passing on the surface) or water finishing (high-pressure water jet); these kinds of finishing involve a roughening of the surface. Brushing is carried out using an abrasive brusher under high pressure, involving more vivid colours of the stone surface. Polishing is carried out using apposite polishing machines featuring a series of rotating heads with a series of rotating spindles also impregnated with diamonds, involving a honed or glossy surface.

It is worth it to note the different appearances of the same stone after the application of different surface finishing (i.e. a strong colour change between a polished and bush hammered or sandblasted surface).

It is also noticeable the addition of synthetic resins to fill the gap on some stone surfaces (i.e. travertine) or the application of a fiberglass mesh to strengthen a stone slab, allowing the use of reduced thickness.

REFERENCES

Gasparoni F., Sugli Obelischi Torlonia nella Villa Nomentana. Ragionamento storico-critico, Salviucci, Roma, 1842.

Latuada S., Descrizione di Milano, Cairoli, Milano, 1738.

Plinius, Naturalis Historia (D.E. Eichholz ed., Natural History), Harward University Press, tenth volume, 1962.

Salmoiraghi F., Materiali naturali da costruzione, Hoepli, Milano, 1892.

Scamozzi S., L'idea dell'architettura universale, Venezia, 1615.

Vasari G., Le vite de' più eccellenti Pittori, Scultori e Architettori – Introduzione (L.S. Maclehose ed., The lives of the most excellent painters, sculptors and architects – On technique), J.M. Dent, London, 1907 (1st ed. 1568).

Chapter 3

Decay phenomena on building stones of Lombardy

The decay is omnipresent on any surface exposed to the atmosphere and to the weathering agents; the stone materials are no exception to physical, chemical and biological alterations. The classification of the decay patterns affecting stone materials has been debated since the time the aim to preserve cultural heritage became crucial in the scientific world. In Italy the interest of scientists in the field of conservation started after the great flood of 1966 caused irreparable damages to the cultural heritage in Florence and Venice. Scientific researchers of chemistry, biology, geology and physics were involved in a sector until the matter of fine arts and rough practices of repair. The "Commissione Normal", promoted by Consiglio Nazionale delle Ricerche (through "Centri di Studio Opere d'Arte" of Milan, Florence and Rome) and by Ministero per i Beni Culturali (through "Istituto Centrale per il Restauro" of Rome) and composed of several scientists of different disciplines, published a series of "Raccomandazioni NORMAL" in 1979 concerning the scientific analyses and observations of stone materials. The first "Raccomandazione" was devoted to the macroscopic alterations of stone materials (Alterazioni macroscopiche dei materiali lapidei, 1980; updated in 1988 and 2006). The aim was to provide a series of accurate definitions considering the decay phenomena visible to the naked eye and describing it without reference to the causes that produced them (Commissione NORMAL, 1980).

The same aim was performed by the International Scientific Committee for Stone of ICOMOS (International Council on Monuments and Sites) in publishing an "Illustrated glossary on stone deterioration patterns", with a French version (ICOMOS, 2008).

Several books or articles in revues or congress proceedings were also available on this argument.

The present text pays attention to the phenomena connected to the intrinsic composition of the material; the phenomena connected to human activities (i.e. mechanical damage) are disregarded.

The definition of each decay phenomenon refers to "Raccomandazione UNI-NORMAL 11182" and to "Illustrated glossary (ISCS-ICOMOS)". The phenomena are grouped according to their effects on stone elements.

1 Without worsening of the characteristics

- *Alterazione cromatica: Variazione naturale, a carico dei componenti della pietra, dei parametri che definiscono il colore. È generalmente estesa a tutto il litotipo interessato.*

DOI: 10.1201/9781003361008-3

- Discolouration: Change of the stone colour in one to three of the colour parameters: hue, value, chroma. It involves generally the entire surface of the stone.
- *Macchia: Pigmentazione localizzata della superficie correlata sia alla presenza di determinati componenti naturali del materiale (concentrazione di pirite nei marmi) sia alla presenza di materiale estraneo (acqua, prodotti di ossidazione di materiali metallici, sostanze organiche, vernici ecc.).*
- Staining: kind of discolouration of limited extent and generally of unattractive appearance. It is linked to certain natural components of the stone (i.e. pyrite in marble) or to the presence of foreign substances (water, oxidation products of metals, organic and painting materials, etc.).
- *Patina: Modificazione naturale della superficie non collegabile a fenomeni particolari e percepibile come una variazione del colore originario del materiale.*
- Patina: Chromatic modification of the material, generally resulting from natural or artificial ageing and not involving in most cases visible surface deterioration.

2 Loss of material from the stone surface

- *Erosione: Asportazione di materiale dalla superficie che nella maggior parte dei casi si presenta compatta.*
- Erosion: Loss of original surface, leading to smoothed shapes.
- *Rugosita: Perdita selettiva di minuscole particelle da una superficie originariamente liscia.*
- Roughening: Selective loss of small particles from an original smooth surface. The substrate is still sound.
- *Erosione differenziale: Messa in risalto dell'eterogeneità di motivi tessiturali o strutturali tipici del materiale lapideo.*
- Differential erosion (differential deterioration): erosion does not proceed at the same rate from one area of the stone to the other. As a result, the stones show an irregularly deteriorated surface.
- *Alveolizzazione: Formazione di cavità di forma e dimensioni variabili, dette alveoli, spesso interconnesse e con distribuzione non uniforme.*
- Alveolization: Formation of cavities (alveoli or small pits), variable shapes and sizes (generally 1 cm), which may be deepened inside and may be interconnected.

3 Loss of the outward form

- *Disgregazione: Decoesione con caduta del materiale sotto forma di polvere o minutissimi frammenti.*
- Disintegration: Detachment of single grains or aggregates of grains.
- *Esfoliazione: Formazione di una o più porzioni laminari, di spessore molto ridotto e subparallele tra loro, dette sfoglie.*
- Exfoliation (delamination): Detachment of multiple thin stone layers (centimetric scale) that are sub-parallel to the stone surface. The layers may bend or twist in a similar way as book pages.
- *Scagliatura: Distacco di parti di forma irregolare e spessore consistente e non uniforme, dette scaglie, spesso in corrispondenza di soluzioni di continuità del materiale originario.*
- Scaling, flaking, contour scaling, spalling: Detachment of stone as scale or a stack of scales, not following any stone structure and detaching like a fish scale or parallel

to the stone surface. The thickness of a scale is generally of millimetric to centimetric scale, and it is negligeable compared to its surface dimension.

- *Distacco: Soluzione di continuità tra strati superficiali del materiale (ad esempio un intonaco), sia tra loro che rispetto al substrato; prelude, in genere, alla caduta degli strati stessi. Nelle pietre le parti distaccate assumono spesso forme specifiche in funzione delle caratteristiche strutturali e tessiturali dando luogo a scagliatura, esfoliazione, crosta.*
- Detachment: Separation and/or swelling and rupturing of an outer stone layer, usually single, either across or parallel to bedding planes.
- *Rigonfiamento: Sollevamento superficiale localizzato del materiale di forma e consistenza variabili.*
- Blistering: Separated air-filled, raised hemispherical elevation of the face of stone resulting from the detachment of an outer stone layer. This detachment is not related to the stone structure.
- *Mancanza: Perdita di elementi tridimensionali (braccio di una statua, ansa di un'anfora, brano di una decorazione a rilievo, ecc.).*
- *Lacuna: Assenza di parti con sviluppo prevalentemente bidimensionale (parte di un intonaco e di un dipinto, porzione di impasto o di rivestimento ceramico, tessere di mosaico, ecc.).*
- Missing part (gap): Empty space, obviously located in the place of some formerly existing stone part. Protruding and particularly exposed parts of sculptures (nose, fingers . . .) are typical locations for material loss resulting in missing parts.

4 Deposit and/or formation of by-products

- *Concrezione o incrostazione: Accrescimento compatto generalmente di estensione limitata, sviluppato sia parallelamente sia perpendicolarmente alla superficie, in quest'ultimo caso può assumere forma stalattitica o stalagmitica.*
- Encrustation (concretion): Compact, hard, mineral outer layer adhering to the stone. Surface morphology and colour are usually different from those of the stone.
- *Deposito superficiale: Accumulo di materiali estranei di varia natura, quali polvere, terriccio, guano, ecc. Ha spessore variabile, generalmente scarsa coerenza e scarsa aderenza al materiale sottostante.*
- Deposit: Accumulation of exogenic material of variable thickness, generally of poor consistency and lack of adhesion to the underlying material.
- *Crosta: Modificazione dello strato superficiale del materiale lapideo. Di spessore variabile, generalmente dura, distinguibile dalle parti sottostanti per le caratteristiche morfologiche e, spesso, per il colore. Può distaccarsi anche spontaneamente dal substrato che, in genere, si presenta disgregato e/o polverulento.*
- Crust (calcium sulphate skin): Generally coherent coat coming from chemical modification by sulfation of materials on the surface. A crust may include exogenic deposits in combination with materials derived from the stone. A crust is frequently dark coloured (black crust). Crust may have a homogeneous thickness and thus replicate the stone surface or have irregular thickness and disturb the reading of the stone details.
- *Efflorescenza: Formazione di sali, generalmente di colore biancastro e di aspetto cristallino o polverulento o filamentoso, sulla superficie.*

- Efflorescence: Generally whitish, powdery or whisker-like crystals on the surface. Efflorescence is generally poorly cohesive and commonly made of soluble salt crystals.
- *Pellicola: Strato superficiale, trasparente o opaco, di sostanze coerenti fra loro ed estranee al materiale lapideo (es.: pellicola pittorica di rifacimento, pellicola protettiva o con funzioni estetiche, pellicola ad ossalati).*
- Film: Thin covering or coating layer generally of organic nature, generally homogeneous, follows the stone surface. A film may be opaque or translucent. Oxalate patina: Orange to brown thin layer enriched in calcium oxalates.
- *Colonizzazione biologica: Presenza di organismi vegetali sul substrato, riconoscibili microscopicamente (alghe, funghi, licheni, muschi, piante superiori).*
- Biological colonization: Colonization of the stone by plants and micro-organisms, such as bacteria, cyanobacteria, algae, fungi, lichens.

5 Loss of mechanical properties

- *Deformazione: Variazione della sagoma o della forma che interessa l'intero spessore del materiale.*
- Deformation: Change in shape without loss of integrity, leading to bending, buckling or twisting of a stone block.
- *Fratturazione o fessurazione: Soluzione di continuità nel materiale che implica lo spostamento reciproco delle parti.*
- Crack: Individual fissure, clearly visible to the naked eye, resulting from separation of one part from another.

DECAY PHENOMENA ASSOCIATED WITH LOMBARD BUILDING STONES

Decay phenomena listed here are linked to the petrographic and mineralogical features; some phenomena (scaling, deformation, crack) can appear in stronger evidence in consequence of the particular kinds of stone working. Some lithotypes include different varieties, and each variety may show different decay phenomena.

1 Without worsening of the characteristics

Discolouration: coloured sedimentary rocks (i.e. Macchiavecchia breccia; Rosso Verona limestone; black Prealp limestone); "Green stones" (i.e. Oira, Serpentinite)

Staining: marbles or granites with iron sulphides (i.e. Candoglia, Ornavasso, Vallestrona marbles; Montorfano granite)

Patina: polished marbles (i.e. Apuanian white marble)

2 Loss of material from the stone surface

Erosion: clastic (conglomerates and sandstones) and chemical (limestones and dolostones) sedimentary rocks; carbonate metamorphic rocks (i.e. marbles)

Differential erosion: nodular sedimentary rocks (i.e. Rosso Verona limestone); fossiliferous sedimentary rocks (i.e. Aurisina limestone); veined marbles (i.e. Candoglia, Ornavasso)

3 **Loss of the outward form**

Disintegration: igneous rocks (i.e. Baveno, Ghiandone, etc.); clastic sedimentary rocks (i.e. Ceppo, Sarnico, Molera, etc.); marbles (i.e. Candoglia, Ornavasso)
Exfoliation: clastic sedimentary rocks (i.e. Sarnico, Ceppo, Molera, etc.); foliated metamorphic rocks (i.e. Beola, Serpentine)
Scaling: igneous rocks (i.e. Baveno, Ghiandone); chemical sedimentary rocks (i.e. massive limestones)
Detachment: igneous rocks (i.e. Baveno, Ghiandone); clastic sedimentary rocks (i.e. Sarnico, Ceppo); foliated metamorphic rocks (i.e Beola, Serpentine)
Blistering: clastic sedimentary rocks (i.e. Sarnico, Ceppo, Molera, etc.)

4 **Deposit and/or formation of by-products**

Crust: carbonate sedimentary and metamorphic rocks (i.e. limestone, dolomite etc.; marbles)
Efflorescence: all types of rock

5 **Loss of mechanical properties**

Deformation: wide slab of marbles
Crack: all types of marble (i.e. the so-called marmo cotto – baked marble, or a marble showing a network of thin cracks)

Causes of the decay

The causes of decay of building stone have been debated on scientific bases since the second half of the 20th century, and dozens of printed texts are available.

In the present text it is sufficient to remember some causes linked to the decay phenomena affecting the stones used in Lombardy.

*Disintegration or erosion of the stone surface is a phenomenon clearly visible in both carbonate sedimentary and metamorphic rocks. This pattern is linked to the solution of calcium carbonate, the main component of limestone and dolostone (chemical sedimentary rocks) and also present as calcite cement in clastic rocks. The dissolution is recognizable through a general recession of the original stone surface with relief of fossil shells (i.e. Aurisina fossiliferous limestone) or with relief of silicate veins (i.e. Candoglia marble) or through a deepening of nodular structure (i.e. Rosso Verona limestone, differential erosion) or the deepening of natural cracks (limestone, marble). The progression of the decay leads to the powdering of the stone, specifically the "sanding" (sandstone, such as Molera or Sarnico) or the "sugaring" (fine-grained marbles, such as the Apuanian white marble).

The dissolution is operated, as a solvent, by water (rain, dampness, etc.) containing carbonic acid [water + CO_2] at the expense of calcium carbonate [$CaCO_3$]. The result is the formation of very soluble calcium bicarbonate [$Ca(HCO_3)_2$].

*Crust or surface skin is a decay phenomenon well developed on carbonate stones in a climate with a high value of humidity, such as the climate of Lombardy.

The release of sulphur, as sulphur dioxide (SO_2), into the atmosphere is caused by the combustion of fossil fuels (coal, oil and natural gas) used by industry, transport and domestic heating. The rate of sulphur dioxide in the atmosphere grew dramatically until the late 20th century, then the reduction of the sulphur dioxide released into the atmosphere became a challenge for the governments of the world. Sulphur dioxide is very soluble in water, and the reaction produces sulphurous acid (H_2SO_3); this acid, reacting with the oxygen of the atmosphere, produces sulphuric acid (H_2SO_4). Carbonate rocks, attacked by the acid, produce calcium sulphate ($CaSO_4$) and water (H_2O), and this water is taken up by calcium sulphate to produce gypsum ($CaSO_4 \cdot 2H_2O$).

The crystallization of gypsum takes place in sheltered areas (i.e. undercut of a cornice or the folds of a cloak of a statue or the projecting parts of a head), producing crusts or sulphate skins; on the contrary, stone areas exposed to the direct washing by rain do not allow the gypsum formation and they conserve an appearance similar to the original one. Particulate pollutants deposited on the crust are bound to it by gypsum, darkening the crust more and more.

*Efflorescence is a manifest decay phenomenon involving the lower part of buildings, where capillary rise of dampness takes place. The dampness, inside the porous structure of a stone material, produces a decay pattern ("moist area") clearly perceptible as a shading of the surface (masonries of stone or brick or coated with plaster). Dampness contains soluble salts (sodium chloride, sodium or magnesium or calcium sulphates, sodium or potassium nitrates) and the evaporation, occurring at the border of the moist area, allows the crystallization as efflorescence arranged as blotches, patches and margins of white salts. Crystal growth may cause the powdering of a clastic or chemical sedimentary rock and also the sanding of igneous rocks, the scaling or powdering of a brick or the crumbling of a plaster.

REFERENCES

Commissione NORMAL, Alterazioni macroscopiche dei materiali lapidei, CNR-ICR, Rome, 1980 (updated 1988, 2006).

ICOMOS-ISCS, Illustrated Glossary on Stone Deterioration Patterns – Glossaire Illustré sur les formes d'altération de la pierre, Monuments and Sites – Monuments et Sites XV, 2008.

Part I

Lombardy

Chapter 4

Stones from province of Bergamo

SEDIMENTARY ROCKS

Abbazia

COMMERCIAL NAME: the name comes from the Abbazia ("Abbey") of San Benedetto in Vallalta (Val Seriana)

REFERENCES

Jervis: Provincia di Bergamo, Circondario di Bergamo; §383 Vallalta (p. 118); §389 Albino (p. 119)

Salmojraghi: §563 Provincia di Bergamo; Calcari; Marmi, macchiati o venati (p. 362)

Marmi italiani: Lombardia; Abbazia (p. 40)

PETROGRAPHIC CLASSIFICATION: chemical sedimentary rock (limestone, mudstone)

MACROSCOPIC FEATURES: brown to grey colour, together with short and irregular white veins; fine grain size

MINERALOGICAL COMPOSITION: calcite

MICROSCOPIC FEATURES: matrix-supported microcrystalline texture, with bioclasts; thin irregular veins filled by calcite spar

GEOLOGICAL SETTING: "Calcare di Zu" formation (Rhaetian – Upper Triassic) of the sedimentary series of the Southern Alps. Multi-metric cycles of marly, dark marly-limestone and dark grey limestone (bottom); bioclastic limestone, oolitic limestone and patch reef limestone (middle); multi-metric cycles of marly limestone sometimes very thin beds (top). The outcrops are located in the Prealps, from the south-eastern tip of Lake Como (Lecco) to the western shore of Lake Iseo (Zu, province of Bergamo)

GEOLOGICAL REFERENCE: CGI, Foglio 33 Bergamo (1954); CARG, Foglio 098 Bergamo (2012)

QUARRY: some sites are located at the bottom of Valle del Lujo (Val Seriana, Bergamo province), a few kilometres from Abbazia (municipality of Albino) on the road to Gaverina Terme

STONEWORK CATEGORIES: block, slab

SURFACE FINISHING: polishing

USE IN ARCHITECTURE: the use has been regular since the Middle Ages near the quarry area, both for building and for decorative purposes (i.e. Abbazia di San Benedetto in Vallalta, 12th century). The stone has been employed in Milan since the end of the 19th century for cladding both outside and inside. Some examples are the moulded base of Banca d'Italia (Piazza Edison, G. Broggi, 1914), the slabs for cladding of the Stazione Centrale

DOI: 10.1201/9781003361008-5

(U. Stacchini, 1931) and the column shafts of the church of Sacro Volto (via Sebenico, O. Cabiati, 1936).

DECAY MORPHOLOGIES: chromatic alteration, surface erosion and sulphate skin formation

Alabaster (travertine)

COMMERCIAL NAME: the name refers to the terms of Greek (*αλάβαστρος*) and Latin (*alabaster*) languages

REFERENCES

Amati: Alabastro di Bergamo (Note – Alcuni altri marmi, book 2, chap. 7, p. 50)

Jervis: Provincia di Bergamo, Circondario di Bergamo; §389 Albino (p. 119)

Salmojraghi: §563 Provincia di Bergamo; Alabastri (p. 361)

Marmi italiani: Lombardia; Onici (p. 44)

PETROGRAPHIC CLASSIFICATION: non-clastic deposit of calcium carbonate from supersaturated waters

MACROSCOPIC FEATURES: dark to light brown colour; typical concretion morphology made of bands of different colours; fine to coarse grain size

MINERALOGICAL COMPOSITION: calcite

MICROSCOPIC FEATURES: convoluted bands of fibrous calcite, showing elongated crystals, presence of irregular cavities

GEOLOGICAL SETTING: this stone pertains to different Quaternary units. The outcrops, always very scanty, are spread in several sites of the Bergamo's Prealps.

GEOLOGICAL REFERENCE: CGI, Foglio 33 Bergamo (1954); CARG, Foglio 098 Bergamo (2012)

QUARRY: the most important site is located close to Albino (Val Seriana); another one is located close to Lovere (Lake Iseo)

STONEWORK CATEGORIES: slab, almost always of small size

SURFACE WORKING: polishing

USE IN ARCHITECTURE: the main use is devoted to decorative purposes; examples are the inlays, together with black limestone or white marble, of Baroque altars

DECAY MORPHOLOGIES: erosion and sulphate skin formation if exposed to weathering

ADDENDUM

Other alabasters were reported by ancient authors in different sites of Lombard territory; they were used locally

Scamozzi: part 2, book 7, chap. 6; Alabastri (p. 193, line 39)

Jervis: Provincia di Como, Circondario di Varese; §279 Induno Olona (p. 99)

Salmojraghi: §560 Provincia di Brescia; Alabastri (p. 356); §563 Provincia di Como; Alabastro (p. 364)

Arabescato orobico

COMMERCIAL NAME: the Italian term "arabescato" (from "arabesque" or "intricated pattern of interlaced lines") generally refers to a white Apuanian marble with branching black veins (Arabescato Corchia, Arabescato Vagli). In this case, the term is referred to a variegated limestone.

REFERENCES

Rondelet: book 1, sec. 2, art. 11; §185 (p. 141)

Amati: Ardese della Valle Brembana (Note – Alcuni altri marmi, book 2, chap. 7, p. 50)

Curioni: chap. 5; Marmi rossi e rosei; §249 Camerata (p. 62)

Jervis: Provincia di Bergamo, Circondario di Bergamo; §369 Camerata Cornello (p. 116)

Salmojraghi: §563 Provincia di Bergamo; Calcari; Marmi rossi (p. 361)

Marmi italiani: Lombardia; Arabescato (p. 39)

PETROGRAPHIC CLASSIFICATION: chemical sedimentary rock (limestone, micrite)

MACROSCOPIC FEATURES: pink ground with veins and variegations of bloody red and white; other varieties show a grey ground with white variegation or a yellow ground with grey variegations; very fine grain size

MINERALOGICAL COMPOSITION: calcite, dolomite, clay minerals

MICROSCOPIC FEATURES: association of calcareous (microcrystalline or fine-grained calcite) bedding structures (te-pee, cavities, fractures) together with cement of calcite spar (infilling the cavities and the fractures) and alternated levels of argillites with iron oxides

GEOLOGICAL SETTING: "Calcare rosso" formation (Middle Triassic, lower Carnian – upper Anisian) of the sedimentary series of the Southern Alps. Well-bedded (beds until 1 m thick) limestones interbedded with paleosoils of "Terra rossa" (red earth). The formation is heteropic with the upper part of the "Calcare di Esino" formation (western Lombardy). Outcrops are present in the central part of the Prealps, between Val Brembana and Val Seriana.

GEOLOGICAL REFERENCE: CGI, Foglio 33 Bergamo (1954); CARG, Foglio 076 Lecco (2012)

QUARRY: the main site is located in Valle Cespedosio, upstream Camerata Cornello (Val Brembana); some quarries are still actively worked and traded: Rosso, Rosato, Grigio rosa, Grigio (Semea di Gervasoni Paolino)

STONEWORK CATEGORIES: block, slab

SURFACE FINISHING: perfect polishing

USE IN ARCHITECTURE: its specific colour involves a use for decorative purposes, mainly for the altars of the 17th and 18th centuries, both locally or in the whole Lombard territory. Examples are diffused in Milan (i.e. high altar, San Giorgio al Palazzo, 1740): column shafts, cornices, mouldings, tiles, inlays. A particular use involved the upper part of balustrades, using small rectangular slabs, with red veins vertically arranged. The most relevant use in the mid-20th century are the columns of the dome of the "Chiesa Nuova" of the Sanctuary of Oropa (north of Biella, Piedmont): the inside columns show a cladding made of curved slabs of Arabescato with red veins. This church was based on a plan of I.A. Galletti (1887), but the construction was carried out in 1960. Another significant use was the floor of San Pietro, the Vatican Basilica: slabs of Arabescato Rosso stand at the centre of some decorative designs.

DECAY MORPHOLOGIES: the use in interior and the polished surface prevent the development of decay phenomena. If the stone is exposed to weathering, the typical decay of coloured limestone arises: chromatic alteration, surface erosion, sulphate skin formation.

Ceppo del Brembo (Ceppo Brembo)

COMMERCIAL NAME: The term comes from Milan's dialect "Cepp" or "Scepp", meaning "stone with pebbles"

The outcrop concerns the whole upper plain of Lombardy where the rivers go out from the Alpine valleys. The outcrops are subdivided along each river course, so the name of the stone may change according to the river (Olona at Lonate Ceppino, Lambro at Inverigo, Adda at Trezzo, Brembo at Brembate). The stone shows similar features in each outcrop, so our description is devoted to the most important one: Ceppo del Brembo.

REFERENCES

Scamozzi: part 2, book 7, chap. 11; Pietre trattabili (p. 208, line 49)

Rondelet: book 1, sec. 2, art. 15; §261, §263 (p. 197); §303 (p. 201)

Breislak: chap. 4; Delle pudinghe; §41 Ceppo dell'Adda, §43 Ceppo di Trezzo, §44 Ceppo di Brembate

Cantù: Geografia fisica; Puddinghe (Ceppo); Trezzo; San Gervasio; Capriate; Brembate

Jervis: Provincia di Bergamo, Circondario di Bergamo; §362 San Gervasio d'Adda, §363 Brembate di Sotto (p. 114); Provincia di Milano, Circondario di Milano; §364 Trezzo sull'Adda (p. 115)

Salmojraghi: §563 Provincia di Bergamo; Ceppi (p. 363)

Marmi italiani: Lombardia; Ceppi di Trezzo d'Adda e di Brembate (p. 46)

PETROGRAPHIC CLASSIFICATION: terrigenous clastic rock (conglomerate, sandstone)

MACROSCOPIC FEATURES: multicoloured (brown, grey, white, purple, black) clasts (pebbles or cobbles) on a yellowish ground. Three varieties are present: coarse grained (conglomerate) with clasts of spherical or elliptical shape, size averaging from 2–3 to 10 cm (maximum size ranging from 25–30 cm to 100 cm); fine grained with few clasts of sub-spherical shape (interbedded sandstone), size averaging from 0.2 to 10 cm; fine grained (sandstone), size averaging from 0.2 to 0.6 cm.

MINERALOGICAL COMPOSITION: clasts of magmatic and metamorphic rocks (range 16–46%), terrigenous clastic rocks (mainly Verrucano Lombardo; range 8–38%), limestones or dolomites; the cement is always made of calcite spar. The fine-grained variety is made of clasts of quartz and silicates (feldspar, micas), limestone, dolomites. All clasts come from the Bergamasc Alps.

MICROSCOPIC FEATURES:

*Coarse grained and fine grained with a few varieties of clasts. Clasts of quartz (sub-rounded polycrystalline aggregates of metamorphic origin, sub-angular to sub-rounded single crystals), thin tabular crystals of muscovite, clasts of fine-grained limestone, chert, sandstone, etc. Calcite crystals of the cement show an increasing size, starting from the surface of the clasts, and it grows to fill the cavities; other cavities are present around the clast boundaries.

*Fine-grained variety. Rounded clasts of quartz (polycrystalline aggregates of metamorphic origin; single crystals); clasts of metamorphic rocks (micaschist), chert, fine-grained limestone, etc. The cement, made of equi-granular calcite crystals, fill the gaps among the clasts; semicircular cavities are regularly spread in the interstices.

GEOLOGICAL SETTING: "Ceppo del Brembo" formation ("Brembo catchment system", Early Pleistocene) of the "Neogene to Quaternary" deposits. Well-cemented, clast-supported conglomerate with arenaceous matrix and well-rounded polygenic clasts from the Brembo and Adda valleys. Outcrops are located on the gorges of river Brembo and river Adda, before and after the confluence, near Trezzo (province of Milan) and Brembate (province of Bergamo). Around the river gorges, conglomerate outcrops are covered by glacio-fluvial sediments ("Bulgarograsso Unit", pertaining to "Besnate Supersynthem", upper Pleistocene).

GEOLOGICAL REFERENCE: CGI, Foglio 46 Treviglio (1969); CARG, Foglio 097 Vimercate (2012)

QUARRY: many small sites distributed on the aforementioned gorges of rivers Brembo and Adda, near Trezzo and Brembate. Blocks reaching 5 m^3 were also worked.

Three varieties were traded according to the different grain sized: "Ceppo rustico" (conglomerate) and "Ceppo mezzano" (sandstone with isolated pebbles) used as block, column drum, lintel and "Ceppo gentile" (sandstone) used as decorative element, statues

These three varieties irregularly follow one another, showing sometimes very sharp limits inside the outcrops.

STONEWORK CATEGORIES: block, shaft, carving, sculpture

SURFACE WORKING: rough or sawn surface

USE IN ARCHITECTURE: the importance of Ceppo on Milan's architecture is well witnessed by the words written by the architect Vincenzo Scamozzi

The use of this stone lasted for a long time: Roman architects employed blocks of Ceppo to build the theatre (1st century), the amphitheatre (2nd century) and the Circus (early 4th century). The reuse of Ceppo blocks was regular in the Middle Ages: churches of San Lorenzo (4th–5th centuries) and Sant'Ambrogio (11th century). The use embraced other towns of Lombardy, such as Pavia, for churches and towers. Then, after a long use of bricks, it was a new exploitation of Ceppo quarries (16th century), as witnessed by Palazzo Marino (1553), Casa degli Omenoni (1565), Porta Romana (1598), Santo Stefano (late 16th century) and San Sebastiano (late 16th century). Ceppo was also the typical building material of the Neo-Classical Renaissance of Milan at the turn of the 18th century. The architect Luigi Canonica planned the Arena Civica (1809) also using the blocks coming from the demolition of the fortifications of Castello Sforzesco. In the early 20th century, Ceppo, together with granites, was the sole stone used for the bases in buildings made of artificial stones: this kind of use was imposed by the rules of the Municipality of Milan. The use of Ceppo in the 20th century was also diffused in Bergamo: Palazzo di Giustizia (M. Piacentini, 1925) and Cimitero Monumentale (E. Pirovano, 1896–1913).

DECAY MORPHOLOGIES: erosion of the calcite cement with detachment of single clasts (finer variety); scaling (coarser variety); growth of organisms in the cavities; sulphate skin formation. The fall of the pebbles leaves round cavities on the surfaces (coarser variety).

ADDENDUM

Other stone materials of the same nature and origin were described by ancient authors.

Cantù: Geografia fisica; Puddinghe; Canonica al Lambro; Inverigo

Jervis: Provincia di Milano, Circondario di Gallarate; §286 Legnano (p. 101); Provincia di Como, Circondario di Como; §297 Inverigo, §298 Villa Romanò (p. 103); Provincia di Milano, Circondario di Monza; §301 Lesmo (p. 103)

Salmojraghi: §564 Provincia di Como (Varese); Ceppi (p. 367)

Ceppo del lago di Iseo (Ceppo Iseo)

COMMERCIAL NAME: the term conserves the same meaning of Ceppo del Brembo, and it also refers to the area around the Prealpine Lake Iseo (Sebino), between the provinces of Bergamo and Brescia. Specific terms are: Ceppo di Grè (a village in the municipality of Solto Collina) or Ceppo Poltragno (a village in the municipality of Castro).

REFERENCES
Salmojraghi: §563 Provincia di Bergamo; Breccia e Brecciola (p. 363)
Marmi italiani: Lombardia; Ceppo di Gré (p. 46)
PETROGRAPHIC CLASSIFICATION: terrigenous clastic rock (diamictite)
MACROSCOPIC FEATURES: grey clasts of irregular shape and angular corners, grey-yellowish ocraceous matrix and cement of calcite spar; coarse grain size
MINERALOGICAL COMPOSITION: dolomite (clast); calcite (cement)
MICROSCOPIC FEATURES: angular and sub-angular clasts of dolomite joined by a cement made of calcite crystals; calcite crystals grow in size starting from the clast surfaces; calcite crystal morphology varies from elongated to equant (euhedral mosaic)
GEOLOGICAL SETTING: "Ceppo di Poltragno" formation ("Oglio-Borlezza amphitheatre", lower Pleistocene) of the "Neogene to Quaternary" continental deposits. Slope and landslide deposits featuring massive matrix supported diamictites with plurimetric angular boulders; clinostratified clast- or matrix-supported conglomerates, with etherometric angular clasts of Norian dolomites (Hauptdolomit). The outcrop is located on the north-western coast of Lake Iseo (monte Clemo) and near the course and the mouth of river Borlezza.
GEOLOGICAL REFERENCE: CGI, Foglio 34 Breno (1971); CARG, Foglio 078 Breno (2012)
QUARRY: some sites are located on the first slopes of monte Clemo, facing Lake Iseo (Gré, municipality of Solto Collina); a tunnel was excavated for the present-day works ("Marini Marmi") to avoid the destruction of the lacustrine landscape. The blocks attain 6 m^3.
STONEWORK CATEGORIES: slab, block
SURFACE FINISHING: sawn or brushed surface
USE IN ARCHITECTURE: the local use involves piers and lintels in rural buildings. The use started in the early 20th century: Banca d'Italia (Bergamo), Camera di Commercio and Cemetery (Mantua). The use in Milan, as in other Lombard towns, spread after World War II, mainly as slabs for façade cladding. Significant examples are easy to find around Milan; each slab was fixed using metallic cramps and mortar.
The Ceppo Iseo is always employed today as witnessed by the addition of Università Bocconi (Grafton Architects, via Roentgen, 2003–08) or by the "Greenway" apartment buildings (Citterio and Giorgi, via Lomazzo 52, 2010). The use in Europe is witnessed in Köln (Domplatte) and Amburg (hotel).
DECAY MORPHOLOGIES: surface erosion on clasts and matrix, soot deposition on cavities, sulphate skin formation

Other conglomerates

Other clast-supported conglomerates with arenaceous matrix and rounded clasts, were quarried from Pleistocene deposits of the province of Bergamo:

- "Conglomerato del Castello di Casnigo" (middle Pleistocene); outcrop along the course of river Serio (Casnigo, val Seriana) and in the nearby val Gandino (eastward). The use was strictly local, such as in the churches of Santa Maria Annunciata (Ponte Nossa), Santa Maria Assunta and their Baptistery (Gandino).
- "Conglomerato di Castione della Presolana" (lower Pleistocene); outcrop around the mountain pass of Presolana, connecting Val Seriana and Valle Camonica

Artificial conglomerate

An artificial variety of Ceppo, called Aggloceppo (Marini Marmi), has been employed as slab for cladding since the second decade of the 21st century; this variety is made of fragments of Ceppo in a binder made of Portland cement (Cemento bianco Italcementi 42.5)
(i.e. German School at via Legnano 24; apartment buildings at piazzale Cimitero Monumentale 23 and at via Bertini 15; corporate building at via Gioia 26)

Pietra di Credaro

COMMERCIAL NAME: the name comes from a village of val Calepio
REFERENCES
The excerpts of the ancient authors do not report the name of "Credaro", but some excerpts referr to outcrops of similar stone, west of Credaro.
Jervis: Provincia di Bergamo, Val Cavallina; §423 Costa di Mezzate; Circondario di Treviglio; §424 Bagnatica (p. 125)
Salmojraghi: §563 Provincia di Bergamo; Arenarie (p. 364)
PETROGRAPHIC CLASSIFICATION: turbiditic sedimentary rock (calcarenite, lutite)
MACROSCOPIC FEATURES: brown to yellowish-brown colour; fine to medium grain size. Three different varieties are available in the quarry: a medium- to fine-grained sandstone, a crystalline limestone, a bioclastic limestone – biomicrite.
MINERALOGICAL COMPOSITION: quartz and muscovite with cement of calcite spar (sandstone); calcite, quartz (crystalline limestone); calcite (biomicrite)
MICROSCOPIC FEATURES: clastic texture with angular quartz crystals and muscovite thin tabular crystals; sparry calcite cement (sandstone); mosaic of inequigranular calcite crystals (crystalline limestone); microcrystalline calcite ground with Foraminifera and thin, irregular veins filled by calcite crystals (biomicrite)
GEOLOGICAL SETTING: "Flysch di Bergamo" formation (Campanian, upper Cretaceous) of the sedimentary series of the Southern Alps. Alternating beds of sandstone, calcarenite and pelite (bed thickness from 1 cm to 100 cm, also massive bedding). The formation is related to the turbiditic sediments (calcareous and arenaceous) of the "Flysch Cretacico Lombardo" (Cretaceous flysch of Lombardy), together with other Lombard sandstones (Arenaria di Sarnico). The outcrops span from Oggiono (Brianza, Lecco province) to the southern tip of Lake Iseo (Bergamo province) closely related to the outcrops of "Arenaria di Sarnico" formation.
GEOLOGICAL REFERENCE: CGI, Foglio 46 Treviglio (1969); CARG, Foglio 098 Bergamo (2012)
QUARRY: different sites are worked in the hilly area south of the village of Credaro; the quarry features beds of different thicknesses. Three different varieties are traded: Berrettino (medium to fine grained sandstone), Médolo (crystalline limestone), Giasòl (bioclastic limestone – biomicrite).
STONEWORK CATEGORIES: block, slab
SURFACE FINISHING: rough
USE IN ARCHITECTURE: the use was mainly local over the centuries, as rough block for masonries made of elements of different size, block for lintel or door jamb. Some examples at Credaro are the church and bell tower of San Giorgio Martire (13th century); the remote church of San Fermo with a tall bell tower (11th–12th centuries) and the columns of the medieval cloister inside Castello di Montecchio. Other examples are the medieval

village of Castel de' Conti (municipality of Castelli Calepio) where a great number of edifices shows masonries made of blocks of this sandstone; the medieval, then transformed in the 17th century, Castello Gonzaga at Grumello del Monte, with a tall tower made of roughly squared blocks. Several rural buildings show masonries made using a dry-stone system with rough blocks and small stone splinters filling the interstices. Starting from the late 19th century, the use was spread in the whole region; some examples are two villas (1907, 1912) and the mausoleum (1907–08) named Faccanoni at Sàrnico (G. Sommaruga); villa Edwige at Adrara San Martino; other villas and houses (villa Leidi, G. Muzio, 1935) at Bergamo with rustication. Finally, the Torre dei Caduti (Bergamo, piazza Vittorio Veneto, M. Piacentini, 1921–24) where the rustication is made of blocks on the corners together with squared elements on the masonry. In Milan, the use was more limited; the most significant example is the masonry of Castello Cova (via Carducci, A. Coppedè, 1910).

Since the last decades of the 20th century, the stone has been employed to make fake stone masonries on the façades of suburban villas, in the area around the quarries: this purpose is reached using thin slabs applied on a reinforced concrete support

DECAY MORPHOLOGIES: erosion of the calcite cement, loss of material, detachment of thin scales

Rosso di Entratico

COMMERCIAL NAME: the name comes from a village of val Cavallina

REFERENCES

Curioni: Chap. 5, Marmi rossi e rosei; §243 Entratico (p. 61)

Jervis: Provincia di Bergamo, Val Cavallina; §420 Entratico (p. 125)

Salmojraghi: §563 Provincia di Bergamo; Calcari rossi (p. 362)

PETROGRAPHIC CLASSIFICATION: chemical sedimentary rock (limestone – biomicrite or wackestone)

MACROSCOPIC FEATURES: red colour with various shades and some white veins; fine grain size

MINERALOGICAL COMPOSITION: calcite, iron oxide

MICROSCOPIC FEATURES: microcrystalline calcite ground with bioclasts (mollusc shells, mainly as cast filled with calcite spar); irregular thin veins of iron oxides, some very thin straight veins of calcite spar

GEOLOGICAL SETTING: "Rosso Ammonitico" formation referred to "Reduced and condensed Jurassic sequences" of the sedimentary series of the Southern Alps. Vari-coloured calcilutites with ammonoids, crinoids and lamellibranches; thin, plane parallel bedded, nodular marly limestones with ammonites; red calcareous marl with cephalopods. The outcrop is located on a narrow strip of Bergamasc Prealps, south of Entratico (eastward of river Serio).

GEOLOGICAL REFERENCE: CGI, Foglio 33 Bergamo (1954); CARG, Foglio 098 Bergamo (2012)

QUARRY: the main site is located in a small valley a few hundred metres south of Entratico (about 20 km east of Bergamo); the road to approach the quarry is named "via delle Cave" (Quarries road)

STONEWORK CATEGORIES: block (the thickness is always corresponding to the bed thickness)

SURFACE FINISHING: smoothing

USE IN ARCHITECTURE: the deep red colour allowed the use of this stone to replace the more expensive "Rosso di Verona" in the quarry area. The "opus sectile" on the façade of Cappella Colleoni in Bergamo (G.A. Amadeo, 1472–76) includes Rosso di Entratico, black Prealp limestone and white Candoglia marble. Other uses in Bergamo and surroundings include blocks for masonry and elements for the portal of the circular church of San Tomè (Almenno San Bartolomeo, early 12th century); the portal of the church of Sant'Annunciata (Entratico) and the masonry of Castello Giovannelli (Luzzana).

DECAY MORPHOLOGIES: chromatic alteration, surface erosion, scaling

Nero di Bergamo

COMMERCIAL NAME: the name comes from the deep monochromatic hue. The term "Pietra da Paragone" was also used in the past.

REFERENCES

Scamozzi: part 2, book 7, chap. 5; Marmi e misti (p. 190, line 3)

Rondelet: book 1, sec. 2, art. 11; §165 (p. 140), §224 (p. 144)

Curioni: chap. 5; Marmi neri; §203 Riva di Solto, §205 Orezzo, §206 Cene (p. 56)

Jervis: Provincia di Bergamo, Circondario di Clusone; §386 Gazzaniga, §387 Cene (p. 118)

Salmojraghi: §563 Provincia di Bergamo; Calcari neri (p. 361)

Marmi italiani: Lombardia; Nero Gazzaniga (p. 42)

PETROGRAPHIC CLASSIFICATION: chemical sedimentary rock (limestone – micrite or mudstone)

MACROSCOPIC FEATURES: deep and uniform black colour; very fine grain size

MINERALOGICAL COMPOSITION: calcite

MICROSCOPIC FEATURES: microcrystalline calcite with some thin alignments of bioclasts; thin, crossing veins of calcite spar (crystal size 0.05 mm)

GEOLOGICAL SETTING: "Argilliti di Riva di Solto" formation (upper Norian – upper Triassic) of the sedimentary series of the Southern Alps. The formation includes cyclic alternations (up to 10 m thick) of black argillites, marls and dark grey marly limestones with bioclast (bivalves and gastropods), argillites and black marls prevail at the bottom. The outcrop is located on the Bergamasc Prealps between Val Seriana and Lake Iseo.

A black limestone ("Calcare di Zorzino" formation, Norian) with very similar features came from the western shore of Lake Iseo (Riva di Solto, Zu).

GEOLOGICAL REFERENCE: CGI Foglio 33 Bergamo (1954); CGI Foglio 34 Breno (1971); CARG, Foglio 077 Clusone (2012); CARG, Foglio 098 Bergamo (2012)

QUARRY: several sites are spread in the upper part of Val Seriana around the villages of Cene and Gazzaniga

STONEWORK CATEGORIES: block, slab

SURFACE FINISHING: smoothing

USE IN ARCHITECTURE: it is worth it to note the very similar features of some black Triassic limestone of the Lombard Prealps (Val Seriana, Varenna, etc.). Starting from the similarity of these stones, it is possible to presume a use based on a geographic criterion: Nero di Bergamo in the eastern part of Lombardy, Varenna stone in the western one. These two parts coincide with the political partition of ancient Lombard territory between the Venetian Republic (east) and the Duchy of Milano (west).

Regular blocks for masonry, cornices, mouldings and elements for portals were used in the quarry area. The Cappella Colleoni in Bergamo (G.A. Amadeo, 1472–76) is the most important building of the Renaissance period: the *opus sectile* veneering the façade is made using the black limestone, employed also in the restoration works of 1875–76 (A. Cattò). Other examples in the Bergamo province are some sculptures of the church of Santa Maria Assunta (Gandino, early 17th century); the portal of the church of Sant'Ippolito (Gazzaniga, early 18th century) and some moulded elements at the Mausoleum of Famiglia Briolini (Gazzaniga, 1905). The black limestone was also used in different towns of eastern Lombardy (Brescia and Cremona), mainly in chromatic alternation with light-coloured stones.

DECAY MORPHOLOGIES: a chromatic alteration affects the black surfaces, leading to a light grey colour completely different from the original deep black. Other decay phenomena are surface erosion and sulphate skin formation.

ADDENDUM

The black limestone of Val Degagna, a site in the mountains about 12 km north of Salò (western shore of Lake Garda, province of Brescia), is also noticeable. The outcrop was reported by ancient authors, but it is impossible to verify the use of this black stone as building and decorative stone.

Scamozzi: part 2, book 7, chap. 5. Marmi e misti (p. 189, line 46)
Jervis: Provincia di Brescia, Circondario di Salò; §453 Degagna (p. 132)
Salmojraghi: §560 Provincia di Brescia; Calcari compatti (p. 357)

"Pietra di Berbenno" is quarried around a village (Berbenno) located in Valle Imagna, a few kilometres north-west of Bergamo. This is a very fine-grained limestone, pertaining to "Argilliti di Riva di Solto" formation (upper Norian). The almost unique use was as slab (5–10 cm thick) to form the roof of the rural buildings of Valle Imagna.

"Porfiroide di Branzi" or "Ardesia di Bergamo" is another dark stone but totally different as origin. This is a blackish, fine-grained slate pertaining to "Formazione di Collio" (lower Permian) of the sedimentary series of the Southern Alps. This formation features argilloschists in alternating with fine-grained grey to green sandstones together with quartz-porphyroid. The outcrops are spread on the higher part of Alpi Orobie (border between the province of Sondrio and Bergamo. Quarry sites are located in the upper val Brembana: Branzi (Porfiroide grigio), Carona, Valleve (Porfiroide grigio scuro, quarry Fontanafredda, elevation 1700 m) and the use was mainly devoted to make thin roof tiles.

Pietra di Sarnico

COMMERCIAL NAME: the name comes from the village on the southern tip of Lake Iseo
REFERENCES
Scamozzi: part 2, book 7, chap. 8; Pietre vive (p. 199, line 26)
Cantù: Geografia fisica; Arenarie; Mapello
Jervis: Provincia di Bergamo, Circondario di Bergamo; §375 Mapello (116), §409 (p. 123); Provincia di Brescia, Circondario di Chiari; §413 Paratico, §414 Capriolo (p. 124)
Salmojraghi: §563 Provincia di Bergamo; Arenarie (p. 364)
Marmi italiani: Lombardia; Arenaria azzurra di Sàrnico (p. 45)
PETROGRAPHIC CLASSIFICATION: turbiditic sedimentary rock (sandstone – greywacke)

MACROSCOPIC FEATURES: grey colour sometimes changing to yellowish-brown; fine grain size

MINERALOGICAL COMPOSITION: quartz, muscovite, K-feldspar; cement of calcite spar

MICROSCOPIC FEATURES: well-sorted crystals of quartz with sharp boundaries together with thin muscovite tabular crystals and rounded clasts of chert; the cement, made of irregular-shaped calcite crystals, wraps the clasts and occupies the interstices; very fine-grained beds show an alignment of quartz crystals

GEOLOGICAL SETTING: "Arenaria di Sàrnico" formation (lower Santonian – Turonian) of the sedimentary series of the Southern Alps. Beds of sandstones (thickness from 1 cm to 100–200 cm) with thin pelite levels. The formation is related to the turbiditic sediments (calcareous and arenaceous) of the "Flysch Cretacico Lombardo" (Cretaceous flysch of Lombardy), together with other Lombard sandstones (pietra Molera). The outcrops span from Oggiono (Brianza, Lecco province) to the southern tip of Lake Iseo (Bergamo province).

GEOLOGICAL REFERENCE: CGI, Foglio 34 Breno (1971); CARG, Foglio 098 Bergamo (2012)

QUARRY: the wide outcrop of the formation allows the presence of several sites. From Oggiono (south of Lecco, near the outcrops of "pietra Molera" sandstone) to Mapello (monte Canto, west of Bergamo) to Castagneta in the valley of torrent Morla (north outskirts of Bergamo), to the area from Bagnatica to Sarnico on the southern tip of Lake Iseo. The quarry site open in the mount north of Sarnico is clearly visible from the lake shore, showing beds from 0.3 to 3 m thick. Another group of quarries were opened on the hills between Paràtico e Capriolo (province of Brescia).

Finally, it is worth it to note a little outcrop in the centre of the old town of Bergamo (Bergamo Alta, piazzetta L. Angelici): the sandstone still pertains to the "Flysch di Bergamo" formation.

STONEWORK CATEGORIES: block, carving

SURFACE FINISHING: smoothing

USE IN ARCHITECTURE: the use in some villages near the quarry area shows squared blocks, moulded elements (bases) and monolithic elements for lintels, portals or window frames. The medieval use in Bergamo was very large, starting from the masonries made of well-squared blocks of the church of Santa Maria Maggiore (second half of the 12th century), of Palazzo della Ragione (late 12th century), of Torre del Gòmbito (12th century), of the church of Sant'Agostino (mid-15th century) or the shafts of the cloister of Santa Maria del Carmine (15th century).

The use was gradually reduced during the centuries; the most important use in the 19th century was for Accademia Carrara (blocks of the first floor) – S. Elia, 1810.

DECAY MORPHOLOGIES: erosion until edge bedding and contour scaling; these last phenomena were also accompanied by the fall of stone pieces

Volpinite (Bardiglio di Bergamo)

COMMERCIAL NAME: the name comes from a village of Val Camonica; the name "Bardiglio di Bergamo" refers to the strong resemblance to grey marble of the Apuanian Alps

REFERENCES

Curioni: chap. 5; Marmi bianchicci; §252 Volpino (p. 62)

Jervis: Provincia di Bergamo, Circondario di Clusone; §406 (p. 122)

Salmojraghi: §563 Provincia di Bergamo; Volpinite (p. 362)
PETROGRAPHIC CLASSIFICATION: chemical sedimentary rock (evaporite)
MACROSCOPIC FEATURES: grey colour with various shades, coarse grain size
MINERALOGICAL COMPOSITION: anhydrite (anhydrous calcium sulphate); gypsum
MICROSCOPIC FEATURES: sub-hedral anhydrite crystals slightly aligned together with gypsum crystals
GEOLOGICAL SETTING: lenticular bodies included in "Carniola di Bovegno" formation of the sedimentary series of the Southern Alps. Massive, cavernous dolomites, tectonic breccias with clasts of dolomite and marly limestone (lower Anisian – upper Werfen, Triassic). The outcrop occupies a narrow area along the western slope of the lower Valle Camonica.
GEOLOGICAL REFERENCE: CGI Foglio 34 Breno (1971); CARG, Foglio 078 Breno (2012)
QUARRY: the main site is located close to the village of Costa Volpino, about 45 km north-east of Bergamo
STONEWORK CATEGORIES: slab, shaft, moulded element
SURFACE FINISHING: polishing
USE IN ARCHITECTURE: the use involves statues, slabs for cladding, shafts as in the southern porch of the church of Santa Maria Maggiore (Bergamo – Giovanni da Campione, 1360)
DECAY MORPHOLOGY: surface erosion

Marmo di Zandobbio

COMMERCIAL NAME: the name comes from a village of Val Cavallina
REFERENCES
Scamozzi: part 2, book 7, chap. 8; Pietre vive (p. 199, line 18)
Rondelet: book 1, sec. 2, art. 15; §265 (p. 197)
Curioni: chap. 5; Marmi rossi e rosei; §241 Trescore (p. 61)
Jervis: Provincia di Bergamo, Val Cavallina; §421 Trescore Balneario, §422 Zandobbio (p. 125)
Salmojraghi: §563 Provincia di Bergamo; Calcari bianchi (p. 362)
Marmi italiani: Lombardia; Marmi di Zandobbio (p. 43)
PETROGRAPHIC CLASSIFICATION: chemical sedimentary rock (dolostone)
MACROSCOPIC FEATURES: pale pink colour tending to whitish with fine cracks featuring a typical network; fine to medium grain size
MINERALOGICAL COMPOSITION: dolomite
MICROSCOPIC FEATURES: mosaic of sub-hedral crystals of dolomite showing perfect rhombohedral cleavage
GEOLOGICAL SETTING: "Dolomia di Zandobbio" formation (Hettangian – Rhaetian) of the sedimentary series of the southern Alps. Massive bedding, including re-crystallized grey to brown dolomites, in heteropy with "Calcari di Sedrina". The outcrops are located on the first reliefs of the Bergamasc Prealps, between the courses of rivers Serio and Cherio (Val Cavallina).
GEOLOGICAL REFERENCE: CGI Foglio 33 Bergamo (1954); CARG, Foglio 098 Bergamo (2012)

QUARRY: the main site is located east of the village of Zandobbio, a second area is located close to the village of Trescore Balneario. Different varieties are traded again: Bianco or Rosa.
STONEWORK CATEGORIES: block, slab
SURFACE FINISHING: polishing
USE IN ARCHITECTURE: the use in Bergamo was remarkable since the Renaissance. Examples are the pavement of the outer part of Cappella Colleoni (G. A. Amadeo, 1472–76); cladding and decorations of Palazzo Nuovo, now seat of the Biblioteca Angelo Maj (V. Scamozzi, early 17th century, but completed in the 20th century); the Fontana Contarini in front of the Palazzo Nuovo (1780); some gates of the city walls (Porta di San Giacomo, 15th century). The use was implemented in the 20th century: the cladding of Palazzo Littorio (Bergonzo, 1936–40) with slabs worked "a punta di diamante", the cladding of Credito Bergamasco and the cladding of many apartment buildings. Moreover, the stone was used in Milan in the 20th century: entrance of villa Romeo-Faccanoni (G. Sommaruga, 1912–13), floor of the crypt of Sacrario dei Caduti Milanesi Caduti in guerra (1928).
DECAY MORPHOLOGIES: surface erosion with widening of the crack network; sulphate skin formation

Loose deposits

COMMERCIAL NAME: these clastic sediment of continental origin (block, cobble, pebble, sand) are named "Gera" (dialects of Lombardy)
REFERENCES
Salmojraghi: §563 Provincia di Bergamo; Erratici, glaciali (p. 363)
PETROGRAPHIC CLASSIFICATION: extremely various, according to the rock outcrops in the catchment area of Alpine rivers. Carbonate sedimentary rocks (Prealps – limestone, dolostone, conglomerate); metamorphic silicate rocks (Alps – gneiss); igneous rocks (Alps – tonalite).
MACROSCOPIC FEATURES: the shape of sediments is linked to the lithologic nature of the source rocks. Spherical shape of igneous or massive sedimentary rocks; tabular or bladed shape of metamorphic foliated rocks and thin-bedded sedimentary rocks. Finally, the roundness of the corners (degree of abrasion) witnesses the transport before the sedimentation. The sizes are eventually fixed by sedimentological point of view: pebble until 6.4 cm, cobble until 25.6 cm, boulder over this size.
MINERALOGICAL COMPOSITION: mainly carbonates (calcite, dolomite), quartz and silicates (feldspar, plagioclase, micas)
GEOLOGICAL SETTING: loose deposits of different origin pertain to "Po Synthem" of "Neogene to Quaternary" deposits (upper Pleistocene – Holocene) and feature angular clast-supported gravels (slope, landslide, debris flow); layers of gravels and sands (alluvial); massive diamicton with boulders (glacial). These deposits occur inside the riverbeds of both Alpine and Prealpine valleys and also in the upper (northern) part of the river Po plain.
GEOLOGICAL REFERENCE: CGI Foglio 33 Bergamo (1954); CARG, Foglio 098 Bergamo (2012)
QUARRY: every riverbed was used to supply these materials; the most important ones were located along the courses of river Brembo (between Bonate and Brembate) and river Serio (between Seriate and Crema)

STONEWORK CATEGORIES: original form without working; in some cases pebble or cobble may be split into two parts. Normally the pebbles, homogeneous in size, were laid in parallel rows with mortar beds.

USE IN ARCHITECTURE: medieval religious buildings, spread in the territory of the province of Bergamo, show masonries made of stone pebbles, sometimes laid in herringbone style; finer deposits (sand) were used to make mortar

DECAY MORPHOLOGIES: different according to the petrographic nature of each clast

Chapter 5

Stones from province of Brescia

IGNEOUS ROCKS

Porfido Monumentale

COMMERCIAL NAME: the name comes from the resemblance with the Egyptian red porphyry (Porfido Rosso Antico)
REFERENCES
Salmojraghi: §560 Provincia di Brescia; Porfiriti (p. 359).
Marmi italiani: Lombardia; Porfido del Gleno (p. 48)
PETROGRAPHIC CLASSIFICATION: igneous rock (rhyolitic ignimbrite)
MACROSCOPIC FEATURES: dark red, purple to grey colour; medium grain size
MINERALOGICAL COMPOSITION: plagioclase, quartz, biotite
MICROSCOPIC FEATURES: rounded or plated phenocrysts of plagioclase, quartz and biotite; micro to cryptocrystalline groundmass
GEOLOGICAL SETTING: "Vulcaniti di Auccia" formation (lower Permian) of the sedimentary series of the Southern Alps; massive quartz-ignimbrite and tuff with crystals at the base. Outcrops are present in a territory between Val Camonica and Val Trompia.
GEOLOGICAL REFERENCE: CGI, Foglio 34 Breno (1971); CARG, Foglio 078 Breno (2012)
QUARRY: the main site is located near the pass of Croce Domini (municipality of Bienno), and it is still working (Porfido Fratelli Pedretti, Esine) at about 2000 m above the sea level
STONEWORK CATEGORIES: roughly squared block, small cube (10–15 cm side), thick slab
SURFACE FINISHING: from quarry face to polishing
USE IN ARCHITECTURE: the first use was occasional, with block carried out by rockfalls or glaciers as witnessed by some Roman artefacts (Museo Archeologico, Cividate Camuno). In the 20th century the quarrying became more organized, and an outstanding example is the Palazzo delle Colonne (Milan, via Verdi, G. Greppi and G. Muzio, 1933–40): the cladding of the porch was made of deeply carved thick slabs to represent the "coats of arms" of the Lombard provinces (sculptor G. Manzù). In the same period the stone was employed for the doorway of Casa Feltrinelli (Milan, via Manin, L. Belgioioso, 1935), for the baseboard and the cladding of the entrance of Banco

DOI: 10.1201/9781003361008-6

Ambrosiano (Milan, via San Dalmazio, P. Portaluppi, 1930); for the spiral staircase of Cassa Assicurazioni (Milan, piazza Missori, M. Piacentini, 1930). After World War II the stone was still employed both for pavement and for cladding: big circles (made of five pieces) on the pavement of the former Chase Manhattan Bank (Milan, piazza Meda, Studio BBPR, 1969). Recently the cladding and the stairs of the stations of the "Passante Ferroviario", the underground railway linking western and eastern parts of the Milan province.

DECAY MORPHOLOGIES: scaling on rough surfaces

Tonalite

COMMERCIAL NAME: the term comes from "Passo del Tonale", an alpine pass leading from Lombardy to Trentino

REFERENCES

Salmojraghi: §560 Provincia di Brescia; Tonalite (p. 359)

Marmi italiani: Lombardia; Diorite della Val Camonica (p. 47)

PETROGRAPHIC CLASSIFICATION: igneous plutonic rock (diorite)

MACROSCOPIC FEATURES: mainly white colour with dark spots, coarse grain size

MINERALOGICAL COMPOSITION: plagioclase, quartz, amphibole, biotite

MICROSCOPIC FEATURES: mainly plagioclase (andesine, labradorite) with zonal structure together with strained quartz, biotite with inclusions and amphibole (hornblende). Plagioclase crystals may contain a mass of very fine mica and clay minerals, due to alteration phenomena.

GEOLOGICAL SETTING: "Batolite dell'Adamello" (Eocene) of late Alpine magmatism; the pluton is intruded in the crystalline basement, and it is bound by two tectonic lines: "Insubrica" (north) and "Giudicarie" (east). Outcrops are present in the eastern side of upper Val Camonica, from Capo di Ponte to Passo del Tonale, together with granodiorites, diorites and gabbros. In particular, the stone material mainly comes from the "Plutone Re di Castello", featuring medium-grained amphibolic-biotitic Tonalite (tonalite, leuco-tonalite and granodiorite).

GEOLOGICAL REFERENCE: CGI, Foglio 19 Tirano (1971); CARG, Foglio 079 Bagolino (2011)

QUARRY: building stones were extracted for centuries from boulders lying on the riverbeds (Oglio and their eastward tributaries) or on the mountain slopes (Val Palobbia, Val Rabbia). A quarry (Pedretti) is still working at Carisolo (val Rendena, province of Trento) exploiting a massive leucocratic variety of biotitic Tonalite (Unità Val d'Avio – Val di Genova)

STONEWORK CATEGORIES: block, slab

SURFACE FINISHING: bush hammering

USE IN ARCHITECTURE: the local use in Val Camonica involved blocks for masonries; in particular it is worth it to note the use as cornerstone. Some examples are San Siro (11th century) at Cemmo; the tower called Federici (12th–14th centuries) at Cividate Camuno; the portal of the church of San Giovanni Battista (1532) at Edolo; the cornerstones of Santa Maria at Esine (1485) together with the pedestals of pilasters of the façade (1776); the cornerstones of Santissima Trinità at Esine (late 15th century); the portal on the right side of San Martino (1786) at Vezza d'Oglio; the piers of Villa Gheza (1934) at Breno.

DECAY MORPHOLOGIES: scaling

SEDIMENTARY ROCKS

Botticino (Corna)

COMMERCIAL NAME: the name comes from the village near the main quarry area (north-east of Brescia). The geological name "Corna" comes from a local term referring to massive rocks forming high vertical faces on the mountains.

REFERENCES

Scamozzi: part 2, book 7, chap. 8; Pietre vive (p. 199, line 29)

Rondelet: book 1, sec. 2, art. 15; §264 (p. 197)

Amati: Botticino di Brescia (Note – Alcuni altri marmi, book 2, chap. 7, p. 50)

Curioni: chap. 5. Marmi giallicci; §238 Botticino Mattina (p. 60), §250 and §253 Virle, §254 Botticino Sera (p. 63)

Jervis: Provincia di Brescia, Circondario di Brescia; §443 Botticino Mattina (p. 128), §444 Botticino Sera (p. 130), §446 Virle Tre Ponti (131), §456 Mazzano (p. 132)

Salmojraghi: §560 Provincia di Brescia; Calcari compatti. Corna (p. 357)

Marmi italiani: Lombardia; Botticino (p. 38)

PETROGRAPHIC CLASSIFICATION: chemical sedimentary rock (limestone, dolomite – intramicrite; mudstone-wackestone or packstone-grainstone)

MACROSCOPIC FEATURES: ivory-white to light brown colour or yellowish to grey; presence of stylolithes and thin transparent veins; fine grain size

MINERALOGICAL COMPOSITION: calcite; locally dolomite may be prevailing

MICROSCOPIC FEATURES: microcrystalline calcite together with euhedral rhomb-shaped crystals of dolomite. Mudstone-wackestone variety contains peloids, intraclasts and bioclasts; packstone-grainstone variety contains peloids, intraclasts, ooids, oncoids and bioclasts of Bivalvia, Brachiopoda, Gasteropoda, etc.

GEOLOGICAL SETTING: "Corna" formation (lower Jurassic – upper Rhetian to lower Sinemurian) of the sedimentary series of the Southern Alps. Outcrops are spread on the hills north-east of Brescia from Botticino to Serle, the mainly calcareous upper part of the formation outcrops in the Botticino area. Massive bedding (about 1 m thick).

GEOLOGICAL REFERENCE: CGI, Foglio 47 Brescia (1970)

QUARRY: the main site is located on the south-western slope of monte Fratta, north of Botticino Mattina; different varieties are traded: Classico, Fiorito, Semi classico. Other sites are spread on "Serle plateau" (Nuvolera), a wide area farther north of Botticino producing the same varieties. Other quarries are present a few kilometres east of Botticino (Mazzano). The area between Virle and Mazzano is also interested by wide quarries devoted to lime and Portland cement production.

STONEWORK CATEGORIES: block, moulding, statue

SURFACE FINISHING: polishing

USE IN ARCHITECTURE: the use was very large in Brescia, starting from the Roman period with the still-standing Capitolium and the Basilica (second half of the 1st century). Since the Renaissance, after a long period (Middle Ages) of re-use of Roman artefacts (i.e. Duomo Vecchio, 11th century), the use of this stone was always high: Santa Maria dei Miracoli (late 15th century), Loggia (1492–1574), ancient Monte di Pietà (late 15th century) and the new one (late 16th century), Duomo nuovo (from 1711), Biblioteca Queriniana (Marchetti, 1747), Santi Nazaro e Celso (Marchetti, 1752–80), Teatro Grande (Turbini, 1782); Cimitero (Vantini, from 1815), San Clemente (Vantini, 1837–40).

The stone was also widely employed in other towns of the Padania plain: Cremona with Duomo, late 13th century, rearranged in the 16th century), Baptistery (second half of the 13th century, rearranged in the middle of the 16th century), Palazzo Comunale (first half of the 14th century), ionic colonnade of Sant'Agata (Voghera, 1835–45); Mantua with San Sebastiano (L.B. Alberti, 1460 – transformed 1925), the façade of San Barnaba (Galli Bibiena, 18th century); Bergamo with the façade of Duomo (1886); Milan with Cimitero Monumentale (Maciachini, 1863–66) or San Camillo de Lellis (Chiappetta, 1900). In the late 19th century, the stone was also used in some civic and religious buildings of Rome as Altare della Patria (Sacconi, from 1885), Synagogue (Armani e Costa, 1899–1904). Finally, in the 20th century the use regarded public and private buildings (i.e. cladding of the new Ospedale Maggiore – Marcovigi and Arata, 1932–39, Milan).

DECAY MORPHOLOGIES: light chromatic alteration, surface erosion, sulphate skin formation

Breccia Aurora

COMMERCIAL NAME: the name comes from the petrographic origin and from the nuance in colour

REFERENCES

Marmi italiani: Lombardia; Breccia Aurora (p. 44)

PETROGRAPIC CLASSIFICATION: clastic sedimentary rock (brecciated limestone)

MACROSCOPIC FEATURES: light brown to greenish ground with dark-brown fine-grained clasts and white veins

MINERALOGICAL COMPOSITION: calcite

MICROSCOPIC FEATURES: coarse, irregular clasts of limestone (oolitic, bioclastic); calcite crystals (circa 0.04–1 mm) fill the space among the clasts

GEOLOGICAL SETTING: "Corna" formation (Hettangian – Lower Jurassic) of the sedimentary series of the Southern Alps; in particular the outcrops in the hilly area between Serle and Vobarno feature strongly brecciated *facies*

GEOLOGICAL REFERENCE: CGI, Foglio 47 Brescia (1970)

QUARRY: the main site is located near the village of Paitone, along the road connecting Brescia to Salò. Different varieties are traded: Breccia Aurora Classico, Breccia Aurora Blu, Laredo Chiaro and Perlato Aurora.

STONEWORK CATEGORIES: block, slab

SURFACE FINISHING: polishing

USE IN ARCHITECTURE: the stone was mainly employed in the 20th century for flooring (i.e. apartment buildings in Milan) and for decorative purposes

DECAY MORPHOLOGIES: surface erosion, sulphate skin formation

Maiolica

COMMERCIAL NAME: the name refers to the aspect (uniform white colour) almost similar to the aspect of "maiolica" (a ceramic body featuring vitreous opaque coating as enamel or smalto)

REFERENCES

Curioni: chap. 5; Marmi bianchicci; §257 Costorio (p. 63)

Jervis: Provincia di Brescia, Circondario di Brescia; §446 Virle Tre Ponti (131); §455 Serle (p. 132)

Salmojraghi: §560 Provincia di Brescia; Calcari. Maiolica (p. 357)

PETROGRAPHIC CLASSIFICATION: chemical sedimentary rock (limestone – micrite; mudstone)

MACROSCOPIC FEATURES: white colour, conchoidal fracture, presence of stylolithes and very thin white veins, strips and nodules of chert; very fine grain size

MINERALOGICAL COMPOSITION: calcite, chert

MICROSCOPIC FEATURES: microcrystalline calcite with some thin rectilinear veins of calcite spar

GEOLOGICAL SETTING: "Maiolica" formation (lower Cretaceous – upper Jurassic) of the sedimentary series of the Southern Alps; well-bedded limestone and marly limestone with chert. Outcrops are spread in the Prealpine area from Lake Garda to Brescia featuring regular beds (from a few centimetres to a few decimetres). Other outcrops are spread near Lake Iseo and Bergamo then following westward to Lecco, Como and Lake Maggiore (*see Como province*).

GEOLOGICAL REFERENCE: CGI, Foglio 34 Breno (1971); CARG, Foglio 099 Iseo (2011)

QUARRY: small sites were located on the hills south of Lake Iseo (Monticelli Brusati – Gaina) and between Concesio and Villa Carcina (a synclinal near the village of Costorio).

STONEWORK CATEGORIES: block

SURFACE FINISHING: polishing

USE IN ARCHITECTURE: mainly near the quarry area as Sant'Andrea (Iseo, 12th century) or Sant'Alessandro in Canzànica (Adrara San Martino, 12th–16th centuries). Other buildings are present in Val Cavallina (east of Bergamo) as San Cassiano, San Vincenzo alla Torre or the tower Suardi at Trescore Balneario. Finally, the stone was employed in different restoration works carried out on the entablature of Cappella Colleoni (Bergamo) at the end of the 19th century.

DECAY MORPHOLOGIES: surface erosion; sulphate skin formation

Médolo

COMMERCIAL NAME: the name comes from a local term referring to a small block of stone, suitable for building, coming from thin bedded rocks

REFERENCES

Jervis: Provincia di Brescia, Circondario di Brescia; §434 Carcina, §438 Gussago (p. 127), §442 Sant'Eufemia della Fonte (p. 128)

Salmojraghi: §560 Provincia di Brescia; Calcari; Lias superior; Medolo (p. 357)

PETROGRAPHIC CLASSIFICATION: chemical sedimentary rock (marly limestone – micrite; wackestone/mudstone)

MACROSCOPIC FEATURES: uniform light brown to greyish colour; fine grain size

MINERALOGICAL COMPOSITION: calcite, chert

MICROSCOPIC FEATURES: microcrystalline calcite with thin argillaceous veins

GEOLOGICAL SETTING: "Calcare del monte Dòmaro" formation (upper Pliensbachian – middle Jurassic) of the sedimentary series of the Southern Alps; this is a part (together with "Calcare di Gardone Val Trompia" formation, lower Pliensbachian – Sinemurian) of the "Medolo group". This group features well-bedded limestone, with strips and nodules of chert, alternating to greenish marls. The outcrops are present in the first Prealpine hills near Brescia (colle Cidneo, Sant'Eufemia).

GEOLOGICAL REFERENCE: CGI, Foglio 47 Brescia (1970)
QUARRY: some sites were located in the municipal area of Brescia, others were located close to the village of Sant'Eufemia (a few kilometres east of Brescia)
STONEWORK CATEGORIES: block, thick slab
SURFACE FINISHING: rough
USE IN ARCHITECTURE: thin beds are easily transformed in building elements ready to set, so this stone was often used for masonries. A particular use regards the "*opus reticulatum*" (pyramidal shape) of Capitolium in Brescia.
DECAY MORPHOLOGIES: light chromatic alteration, surface erosion till granular disintegration, sulphate skin formation

Nero Venato

COMMERCIAL NAME: the name comes from the macroscopic aspect (black with veins)
REFERENCES
Jervis: Provincia di Brescia, Circondario di Breno; §403 Angolo (p. 122)
Salmojraghi: §560 Provincia di Brescia; Calcari; Compatti, triassici (p. 356)
PETROGRAPHIC CLASSIFICATION: chemical sedimentary rock (limestone – micrite; mudstone)
MACROSCOPIC FEATURES: deep black colour with irregular white veins; very fine grain size
MINERALOGICAL COMPOSITION: calcite
MICROSCOPIC FEATURES: microcrystalline calcite with many intersecting veins filled by calcite spar (crystal size 0.05 mm)
GEOLOGICAL SETTING: "Calcare di Angolo" formation (lower Anisian) of the sedimentary series of the southern Alps. The formation includes massive dark grey limestones with calcite veins and fine-bedded black limestones in alternation with blackish clays. The outcrops are spread in the western slope of middle Val Camonica and eastward of the upper part of Lake Iseo.
GEOLOGICAL REFERENCE: CGI, Foglio 34 Breno (1971); CARG, Foglio 078 Breno (2012)
QUARRY: different sites were spread on the southern slopes of Concarena, a mountain of Valle Camonica (Cerveno, Losine, Lozio, Ossimo)
STONEWORK CATEGORIES: block, slab
SURFACE FINISHING: smoothing
USE IN ARCHITECTURE: mainly local both for structural and decorative purposes (i.e. blocks and stoups of Santa Maria at Esine; tiles for floor together with Vezza d'Oglio white marble); therefore it is very difficult to distinguish among other black limestones of the provinces of Bergamo and Brescia
DECAY MORPHOLOGIES: chromatic alteration from the original deep black to light grey; surface erosion and sulphate skin formation

Occhiadino

COMMERCIAL NAME: the Italian term "occhiadino" is generally referred to gneiss (Augen gneiss), but in this case it is referred to the spotted aspect of the stone surface
REFERENCES
Amati: Occhiadino di Valcamonica (Note – Alcuni altri marmi, book 2, chap. 7, p. 50)

Curioni: chap. 5; Marmi grigi; §213 Valle Camonica (p. 57)
Rondelet: book 1, sec. 2, art. 11; §184 (p. 141)
Jervis: Provincia di Brescia, Circondario di Breno; §398 Losine, §400 Cividate Alpino (p. 121)
Salmojraghi: §560 Provincia di Brescia; Calcari; Compatti; Occhiadino (p. 356)
Marmi italiani: Lombardia; Marmo occhiadino (p. 44)
PETROGRAPHIC CLASSIFICATION: chemical sedimentary rock (limestone and dolomite limestone – biosparite; wackestone)
MACROSCOPIC FEATURES: dark grey colour with convolute white veins, algal structures and cavities; fine grain size
MINERALOGICAL COMPOSITION: calcite, dolomite
MICROSCOPIC FEATURES: brecciated texture with microcrystalline calcite ground together with rounded or angular clasts of organic or oolitic limestones; veins of calcite spar
GEOLOGICAL SETTING: "Calcare metallifero" formation (Carnian, upper Triassic) of the sedimentary series of the Southern Alps; well-bedded (50 cm thick) limestone and dolomite limestone. Outcrops in the Prealps from val Brembana to val Camonica.
GEOLOGICAL REFERENCE: CGI, Foglio 34 Breno (1971); CARG, Foglio 078 Breno (2012)
QUARRY: a small site is still visible close to Sommaprada (municipality of Lozio) on the southern slopes of Concarena mountain (mid Valle Camonica). Other sites were probably present westward (i.e. Val Seriana – Valzurio, province of Bergamo).
Ancient authors refer also the "Occhiadino di Varenna" on the Lake Como eastern shore: probably it is a variety of the "Nero di Varenna" (black limestone with white veins).
STONEWORK CATEGORIES: small block, slab
SURFACE FINISHING: polishing
USE IN ARCHITECTURE: the use, starting from the churches of the Val Camonica, spread over the churches of the whole of Lombardy to make decorations of the altars since the late 17th century (balusters, cornices, mouldings, friezes, veneer) in combination with other coloured materials (Macchiavecchia, Rosso di Arzo, Rosso di Francia, etc.). Locally, the stone was employed to make cladding (i.e. San Lorenzo, Sònico, 16th century, San Salvatore, Breno, 17th century).
DECAY MORPHOLOGIES: a perfectly polished surface is well preserved on a stone used in interior; the stone exposed to weathering shows differential erosion and chromatic alteration.

Rezzato (Corso)

COMMERCIAL NAME: the name comes from a village, east of Brescia. The geological name "Corso" comes from the local term referred to a thin bedded rock.
REFERENCES
Curioni: chap. 5; Marmi rossi e rosei; §244 Virle (p. 61)
Jervis: Provincia di Brescia, Circondario di Brescia. §444 Botticino Sera (p. 130), §445 Rezzato (p. 131).
Salmojraghi: §560 Provincia di Brescia; Calcari; Lias medio; Corso (p. 357)
PETROGRAPHIC CLASSIFICATION: chemical sedimentary rock (limestone – intramicrite; wackestone)

MACROSCOPIC FEATURES: whitish to reddish colour with thin white veins; sometimes a nodular aspect is present; fine grain size

MINERALOGICAL COMPOSITION: calcite, dolomite, chert

MICROSCOPIC FEATURES: euhedral rhomb-shaped crystals of dolomite surrounded by microcrystalline calcite

GEOLOGICAL SETTING: "Corso" formation (Domerian-Sinemurian, lower Jurassic) of the sedimentary series of the Southern Alps; compact white limestone, well bedded (about 25 cm), sometimes with chert, and red limestone without chert in alternating with brecciated limestone. Outcrops are located between Brescia and Lake Garda.

GEOLOGICAL REFERENCE: CGI, Foglio 47 Brescia (1970)

QUARRY: the main site is present on a hill very close (north) to the town of Rezzato

STONEWORK CATEGORIES: block, slab

SURFACE FINISHING: polishing

USE IN ARCHITECTURE: mainly local, as slab to make floors, stairs, chimneys, tabletops.

DECAY MORPHOLOGIES: light chromatic alteration, erosion, sulphate skin formation

Scaglia

COMMERCIAL NAME: the Italian term means "chip" because this kind of stone is easily subdivided into chips

REFERENCES

Jervis: Provincia di Brescia, Circondario di Salò; §457 Manerba, §458 San Felice di Scovolo (p. 133)

Salmojraghi: §560 Provincia di Brescia; Calcari; Diversi; Grossolani (p. 357)

PETROGRAPHIC CLASSIFICATION: chemical sedimentary rock (marly limestone – biomicrite; mudstone)

MACROSCOPIC FEATURES: pinkish to reddish or greenish to greyish colour, very fine grain size, scaly fracture

MINERALOGICAL COMPOSITION: calcite, argillaceous material

MICROSCOPIC FEATURES: microcrystalline calcite, presence of some undulose veins of clay minerals

GEOLOGICAL SETTING: "Scaglia lombarda" formation (middle Eocene – lower Cretaceous) of the sedimentary series of the Southern Alps; thin-bedded marly limestone (thickness ranges from a few centimetres to some decimetres), marl, cherty limestone, sandstone. The outcrops are spread in the lower part of the Prealps, from Bergamo to Brescia, to Lake Garda.

GEOLOGICAL REFERENCE: CGI, Foglio 48 Peschiera (1969)

QUARRY: the long peninsula of Sirmione, stretching out into Lake Garda from the southern shore, shows trace of quarrying on the northern tip (Campo delle Noci), in the area of the Roman villa called "Grotte di Catullo". Other sites were located near Manerba and Moniga del Garda, on the western shore of Lake Garda.

STONEWORK CATEGORIES: thick slab with natural splitting

SURFACE FINISHING: rough

USE IN ARCHITECTURE: the use involves small slabs, coming from the natural bedding, as in the Roman villa "Grotte di Catullo" (1st century) at Sirmione (masonries, substructures and fortifications). Thick slabs were used in the masonries of San Pietro in Mavinas (Sirmione, 11th century).

DECAY MORPHOLOGIES: exfoliation

Pietra Simona

COMMERCIAL NAME: the term probably comes from the site called "Simoni di Gorzone" (municipality of Darfo, val Camonica)
REFERENCES
Jervis: Provincia di Brescia, Circondario di Breno; §401 Darfo (p. 121)
Salmojraghi: §560 Provincia di Brescia; Arenarie; Palaeozoiche o triassiche (p. 358)
Marmi italiani: Lombardia; Pietra Simona (p. 45)
PETROGRAPHIC CLASSIFICATION: clastic sedimentary rock (sandstone – greywacke)
MACROSCOPIC FEATURES: purple colour, the action of organisms living in the loose sediment caused the typical convolute texture ("budellatura"), fine grain size
MINERALOGICAL COMPOSITION: quartz and muscovite together with a clay-hematite cement
MICROSCOPIC FEATURES: muscovite thin tabular crystals with curvilinear alignment and single angular crystals or polycrystalline aggregates of quartz
GEOLOGICAL SETTING: member of "Conglomerato del Dosso dei Galli" formation (Permian), a formation positioned at the base of the sedimentary series of the Southern Alps, featuring sandstone and conglomerate. Outcrops in the middle Val Canonica, north of Lake Iseo.
GEOLOGICAL REFERENCE: CGI, Foglio 34 Breno (1971); CARG, Foglio 078 Breno (2012)
QUARRY: the main site is located on the hills north-west of Darfo, towards Gorzone and Angolo Terme, on the valley of river Dezzo.
STONEWORK CATEGORIES: block, slab, moulding
SURFACE FINISHING: smoothing
USE IN ARCHITECTURE: the local use has been well documented since the Roman period (stelae or altars); later on, the use was concentrated on door or window frames (lintel, jamb, etc.). A regional use was based on the chromatic effects allowed by the dark colour contrasting to the light colour of the other stones as already reported about black limestones from Varenna or Bergamo. It is possible to consider the cornices and the mouldings of the lower part of the façade (early 16th century) of the church of Certosa di Pavia, contrasting with the pinkish colour of Candoglia marble or the pieces of the rose window of the façade (A. Amadeo, 1472–76) of Colleoni chapel in Bergamo. The use of this stone started again in the late 19th century in the funerary art as visible at the Cimitero Monumentale in Milan: a meaningful example is the chapel pertaining to Besenzanica family (Riparto VI, spazio 127) where two bronze peasants leading oxen pulling a plough stand on a rocky element made of rough pieces of pietra Simona ("Allegoria del Lavoro", architect C. Malgarini, sculptor E. Butti – 1907/12).
DECAY MORPHOLOGIES: a kind of exfoliation according to the particular texture

METAMORPHIC ROCKS

Marmo di Vezza d'Oglio

COMMERCIAL NAME: the name is referred to the village of upper Val Camonica
REFERENCES
Jervis: Provincia di Brescia, Circondario di Breno; §396 Vezza d'Oglio (p. 121)
Salmojraghi: §560 Provincia di Brescia; Calcari; Saccaroidi (p. 356)

PETROGRAPHIC CLASSIFICATION: metamorphic rock (marble)
MACROSCOPIC FEATURES: whitish to pinkish colour; medium to fine grain size
MINERALOGICAL COMPOSITION: calcite; quartz (occasional)
MICROSCOPIC FEATURES: "mortar structure" with euhedral calcite crystals (circa 2 mm) surrounded by very fine calcite crystals (circa 0.1 mm); rounded quartz crystals (circa 0.2 mm), elongated polycrystalline quartz (2 mm); mosaic structure with anhedral calcite crystals (0.05 to 3 mm), rounded quartz crystals
GEOLOGICAL SETTING: "Falda Ortles – Campo, Unità di Peio" of the metamorphic basement (Pre-Permian); phylladic micaschists with muscovite and garnet, including white to pink marble lenses. The east–west alignment of lenses concerns the upper part of Val Camonica.
GEOLOGICAL REFERENCE: CARG, Foglio 041 Ponte di Legno (2010)
QUARRY: the main site is located on the southern slope of Cima Mattaciul (Rethian Alps), above the village of Vezza d'Oglio (upper val Camònica)
STONEWORK CATEGORIES: slab, moulding
SURFACE FINISHING: polishing
USE IN ARCHITECTURE: the use in Val Camonica is well documented during a long period of time. Some examples are sculptured elements of the portal of San Siro (11th century) at Cemmo; part of the arch of the portal of Sant'Antonio (1334–59) at Breno; portal jambs (1584) and the plinths of the colonnade (1786) on the right side of San Martino at Vezza d'Oglio; some elements in the shafts of the portal of San Salvatore (17th century) at Breno. Finally, the tiles for flooring together with "Nero venato" black limestone, present in several churches of Valle Camonica.
DECAY MORPHOLOGIES: erosion till disaggregation, sulphate skin formation

Chapter 6

Stones from provinces of Como and Lecco

SEDIMENTARY ROCKS

Gonfolite

COMMERCIAL NAME: this term was introduced by Curioni (1844) referring to the Greek noun "γόμφος, -ου" meaning "joint" and to the verb "γομφόω" meaning "to connect" or "to coagulate". This term indicates a stone made of different elements connected each other.

REFERENCES

Breislak: chap. 8. Dei monti della Brianza; §93 Le psammiti di Castel Baradello presso Como

Salmojraghi: §560 Provincia di Como; Arenaria; Miocenica; Molera (p. 367)

PETROGRAPHIC CLASSIFICATION: terrigenous clastic rock (sandstone together with marly and silty shale)

MACROSCOPIC FEATURES: grey to yellowish colour; clasts of crystalline rocks (lower part of the sequence) or of sedimentary rocks (upper part of the sequence); medium to coarse grain size

MINERALOGICAL COMPOSITION: quartz, feldspar, biotite and muscovite, arenaceous matrix; cement made of calcite spar

MICROSCOPIC FEATURES: sub-rounded clasts with different sizes joined by a cement made of calcite spar

GEOLOGICAL SETTING: "Peliti di Prestino" formation (lower Miocene – Aquitanian) of the sedimentary series of the Southern Alps, pertaining to inter-channel and levee deposits of submarine-fan and featuring massive sandstone (several metres thick) and marly to silty shale interbedded (a few centimetres thick). This formation is a part of the group "Gonfolite di Como" (upper Oligocene – lower Miocene), together with the heteropic formations "Formazione di Como" (clast-supported conglomerate; Oligocene, upper Chattian) and "Arenarie della Val Grande" (massive sandstone and conglomerate beds; lower Miocene, Burdigalian). The outcrop occupies a large area south of Lake Como, reaching also the Italian–Swiss border (Chiasso); this area is bounded by the "Sovrascorrimento di Monte Olimpino" (overthrust of the Gonfolite group above the Mesozoic succession).

GEOLOGICAL REFERENCES: CGI, Foglio 32 Como (1937); CARG, Foglio 075 Como (2011)

QUARRY: many sites of different size (Camerlata – Baradello, Rebbio, San Fermo) are spread around a hill ridge (called "Spina Verde") on the western outskirts of Como.

DOI: 10.1201/9781003361008-7

STONEWORK CATEGORIES: rough blocks
SURFACE FINISHING: bush hammering
USE IN ARCHITECTURE: blocks for masonry were used starting from the Roman period in Como and outskirts (i.e. Castello del Baradello, 12th century). The easy working of the stone allowed builders to make carved elements (pediment of the rear portal (San Fedele, 13th century). It is worth it to note the so-called Stele di Prestino (5th century), found in 1966 in the village of Prestino (west of Como): the stele shows a prismatic shape (3.85 m in length and 0.2–0.4 m high) and a chiselled inscription. The use as millstone is also reported.
DECAY MORPHOLOGIES: erosion till exfoliation or crumbling, sulphate skin formation

Maiolica

COMMERCIAL NAME: the term refers to the aspect of the stone almost similar (white colour, uniformity) to the aspect of a "maiolica" or a ceramic body featuring a vitreous opaque coating (enamel or smalto)
REFERENCES
Jervis: Provincia di Como, Circondario di Lecco; §302 Suello (p. 103)
Salmojraghi: §560 Provincia di Como; Maiolica (p. 366)
PETROGRAPHIC CLASSIFICATION: chemical sedimentary rock (limestone – micrite or mudstone)
MACROSCOPIC FEATURES: white colour with some brown shades; conchoidal fracture, stylolithes, stripes and nodules of chert (basal sector of the formation); very fine grain size
MINERALOGICAL COMPOSITION: calcite, chert
MICROSCOPIC FEATURES: homogeneous microcrystalline calcite with some rectilinear thin veins of calcite spar; almost indistinguishable shells of foraminifera
GEOLOGICAL SETTING: "Majolica" formation (lower Cretaceous – upper Jurassic) of the sedimentary series of the Southern Alps, featuring regular beds (from a few centimetres to a few decimetres) of limestone with organic content (foraminifera). Outcrops are spread in the Prealps area between Como and Lecco and westward from Lake Varese to Lake Maggiore (Caravate) or eastward from Bergamo to Brescia and Lake Garda (see "Maiolica", province of Brescia).
GEOLOGICAL REFERENCES: CGI, Foglio 32 Como (1937); CARG, Foglio 075 Como (2011)
QUARRY: located in northern Brianza near the prealpine lakes (Pusiano, Suello); these sites were later transformed by the excavations of the Merone cement factory. Another site near Caravate (north-west of Varese) supplied raw material to make Portland cement.
STONEWORK CATEGORIES: block
SURFACE FINISHING: smoothing
USE IN ARCHITECTURE: the use was occasional in Como and nearby; the most relevant use is the colonnaded porch of Teatro Sociale, featuring six column shafts made by superimposed drums (Como, G. Cusi, 1811–13)
DECAY MORPHOLOGIES: erosion, sulphate skin formation

Pietra di Mandello

COMMERCIAL NAME: the name comes a village on the eastern shore of Lecco south-eastern branch of Lake Como

REFERENCES
Curioni: chap. 5. Marmi grigi; §216 Mandello (p. 57)
Jervis: Provincia di Como, Circondario di Lecco; §349 Mandello del Lario (113)
Salmojraghi: §560 Provincia di Como; Calcari; Diversi; Del Trias (p. 365)
PETROGRAPHIC CLASSIFICATION: chemical sedimentary rock (limestone – micrite/mudstone)
MACROSCOPIC FEATURES: greyish colour with some brown veins; fine grain size
MINERALOGICAL COMPOSITION: calcite
MICROSCOPIC FEATURES: microcrystalline calcite
GEOLOGICAL SETTING: "Calcare di Esino" formation (middle Triassic) of the sedimentary series of the Southern Alps; featuring massive or thick-bedded limestones, dolomites and calcitic dolomites. The outcrops involve a large area of Lombardy, from Lecco to Val Brembana, Val Seriana and Val Camonica; this formation is heteropic with "Calcare di Perledo" in the area east of the Lecco branch of Lake Como (Grigne group).
GEOLOGICAL REFERENCES: CGI, Foglio 32 Como (1937); CARG, Foglio 075 Como (2011)
QUARRY: sites are close to the shore of Lake Como, near Mandello del Lario, 10 km north of Lecco
STONEWORK CATEGORIES: block, column
SURFACE FINISHING: polishing
USE IN ARCHITECTURE: mainly local, but the most significant examples are the tall column shafts of the Santuario del Crocifisso (Como, first half of the 17th century)
DECAY MORPHOLOGIES: erosion, sulphate skin formation

Molera

COMMERCIAL NAME: this name comes from "mœula", a term of the Milanese dialect meaning "millstone"; so "molera" means "sandstone", that is, the stone employed to make millstone
REFERENCES
Breislak: chap. 8. Dei monti della Brianza; §84 Molera di Viganò
Rondelet: book 1, sec. 2, art. 15. §262 (p. 197)
Cantù: Geografia fisica; Arenaria; Viganò, Briosco, Romanò
Jervis: Provincia di Como, Circondario di Lecco; §290 Bosisio (p. 101), §294 Garbagnate Monastero, §295 Bulciago (p. 102); Provincia di Milano, Circondario di Monza; §300 Briosco (p. 103); Provincia di Como, Circondario di Lecco; §304 Sirtori, §305 Viganò (p. 104), §306 Missaglia, §307 Perego, §308 Rovagnate, §309 Santa Maria Hoè (p. 105)
Salmojraghi: §560 Provincia di Como; Arenarie; Brianza; Cretacee; Miocenica; Molera; Cornettone (p. 367)
Marmi italiani: Lombardia; Arenaria di Oggiono (p. 45)
PETROGRAPHIC CLASSIFICATION: terrigenous clastic rock (sandstone from turbidity currents)
MACROSCOPIC FEATURES: grey to yellow colour; medium to fine grain size
MINERALOGICAL COMPOSITION: quartz, muscovite and calcite
GEOLOGICAL SETTING: turbiditic sediments pertaining to "Flysch Cretacico Lombardo" (upper Cretaceous) of the sedimentary series of the Southern Alps, featuring conglomerates, sandstones and marls; these sediments are subdivided in four geologic formations: "Formazione di Pontida", "Arenaria di Sarnico", "Conglomerato di Sirone" and "Flysch

di Bergamo". Two formations supplied dimension stones: "Arenaria di Sarnico" and "Flysch di Bergamo". Conglomerato di Sirone was exploited to make millstone.

Formazione di Pontida: sandstone beds (thin to thick) coupled with marls together with thick beds of conglomerate (middle Turonian – upper Turonian). Outcrops are mainly spread on the east bank of the course of river Adda (from Calolziocorte to Pontida)

Arenaria di Sarnico: grey sandstone beds (thick to very thick) together with shale and marl beds (Coniacian). Outcrops are scattered around Colle Brianza (a few kilometres south of Lake Annone), other outcrops are also present between Bergamo and Lake Iseo.

Conglomerato di Sirone: conglomerate beds (very thick to massive) with sandstone beds at the top (Santionian). Outcrops are located around the Adda river course: south of the Lake Annone and Lake Pusiano (west bank) and near Sotto il Monte and Mapello (east bank) (see *infra*).

Flysch di Bergamo: thin to very thick sandstone beds (fine to medium grain size) together with beds of calcilutites and conglomerates (Campanian). Outcrops are scattered in three parts, from west to east: Montevecchia, monte Canto and Bergamo.

GEOLOGICAL REFERENCES: CGI, Foglio 32 Como (1937); CARG, Foglio 097 Vimercate (2012)

MICROSCOPIC FEATURES: **Arenaria di Sarnico**: clasts with a sort of alignment, made of quartz crystals (circa 0.08–0.2 mm, fine-grained beds; circa 0.08–1.0 mm, coarse-grained beds), subrounded limestone (up to 0.4 mm) and metamorphic rocks (up to 0.5 mm), chert (up to 0.4 mm); calcareous matrix and cement made of calcite spar. **Flysch di Bergamo**: clasts with a sort of alignment, made of sub-angular quartz crystals (circa 0.08–1.0 mm), muscovite tabular crystals, subrounded limestone clasts (circa 0.2–1.2 mm), plagioclase and chert; the cement is made of calcite spar; irregular veins with euhedral calcite crystals (circa 1.8 mm).

QUARRY: **Arenaria di Sarnico**: the sites are located near Oggiono, on the hills close to lake Annone (about 10 km south of Lecco) featuring cemented material, proof against weathering. **Flysch di Bergamo**: the sites were concentrated between Viganò and Montevecchia (about 15 km north of Monza) featuring low cemented material, easily exposed to weathering

STONEWORK CATEGORIES: block, moulded element (i.e. base, shaft and capital for columns), carved elements

SURFACE FINISHING: smoothing

USE IN ARCHITECTURE: a classification of the stones coming from different quarries and from different geological formations is almost impossible without scientific analyses. A geographic method (based on the use of a stone in a territory surrounding the quarry site) seems to be useful to distinguish the different kinds of "Molera".

Starting from the Middle Ages it was diffused in Romanesque churches and chapels of Brianza (Baptistery of Oggiono) both for masonry (block) and for decorative elements. Many buildings of the late 18th century in Milan used the stone for architectural elements (window frame) as the villa Belgiojoso – Reale (L. Pollack, 1790) or the villa Reale of Monza (G. Piermarini, 1777–80) or for masonry (block) as Porta Nuova (G. Zanoja, 1810–13). The scarce durability of the stone involved the replacement in a short period using other more resistant stones (i.e. the decayed window frames made of Molera at villa Belgiojoso – Reale were replaced using the limestone of Viggiù).

DECAY MORPHOLOGIES: erosion till edge bedding and contour scaling (also accompanied by the fall of stone pieces); sulphate skin formation

Pietra di Moltrasio

COMMERCIAL NAME: the term comes from a village on the western shore of Lake Como
REFERENCES
Rondelet: book 1, sec. 2, art. 6. §304 (p. 201)
Cantù: Geografia fisica; Pietra di Moltrasio
Jervis: Provincia di Como, Circondario di Como; §335 Argegno, §337 Carate Lario, §338 Urio, §339 Moltrasio (p. 109)
Salmojraghi: §560 Provincia di Como; Calcari; Liasici (p. 366)
Marmi italiani: Lombardia; Pietra di Moltrasio (p. 45)
PETROGRAPHIC CLASSIFICATION: chemical sedimentary rock (limestone – intramicrite)
MACROSCOPIC FEATURES: dark grey colour, alternately irregular sandstone beds and very thin marly layers, presence of nodules or bands of chert; fine grain size
MINERALOGICAL COMPOSTION: calcite, quartz, chert
MICROSCOPIC FEATURES: microcrystalline calcite together with intraclasts, wavy laminations of argillaceous material; cement made of calcite spar
GEOLOGICAL SETTING: "Calcare di Moltrasio" formation (Sinemurian, lower Jurassic) of the sedimentary series of the Southern Alps; featuring massive limestone with chert (some parts show abundant chert others are chert free) or well-defined beds, often irregular (10– 50 cm thick). The outcrop is spread between the south-western branch of Lake Como until Lake Lugano.
GEOLOGICAL REFERENCES: CGI, Foglio 32 Como (1937); CARG, Foglio 75 Como (2011)
QUARRY: the most important site is located near the village of Moltrasio (about 10 km north of Como) close to the lake shore; vertical faces of the quarry are still visible from the modern road (Strada Statale 340 Regina); other quarries were located further on (Carate Urio). Quarries on the other shore of the lake were located between Torno and Pognana Lario; only one quarry is still working, and it is located west of Faggeto Lario, where the SP 43 (Strada Provinciale) merges into SP 583 Lariana.
STONEWORK CATEGORIES: block, slab; the thickness of blocks follows the different thickness of the rock beds
SURFACE FINISHING: rough face
USE IN ARCHITECTURE: strictly limited to the Como area for masonry (block) or for roofs and stairs (slab). Outstanding examples are the three apses and the piers of San Fedele (since the 10th century); the façade and the columns, made of superimposed and irregular blocks, of the church of Sant'Abbondio (late 11th century); the urban walls with towers, encircling the old town. The low quality, due to the irregular bedding and uneven fracture, prevent decorative use: i.e. the masonries of palazzo Natta (Como, mid-18th century), then coated by plaster, were made of "Pietra di Moltrasio", but the window frames were made of "Pietra di Viggiù". Many religious or civic buildings of Val d'Intelvi (north of Como) were made using blocks of this stone (Pèllio, Ramponio, San Fedele, Scària).
DECAY MORPHOLOGIES: chromatic alteration, erosion till exfoliation, sulphate skin formation.

Puddinga di Sirone

COMMERCIAL NAME: the name comes from a village of Brianza, south of Lecco

REFERENCES

Breislak: chap. 8. Dei monti della Brianza. §88 Sirone, puddinga acconcia per le macine

Cantù: Geografia fisica. Puddinghe. Sirone

Jervis: Provincia di Como, Circondario di Lecco; §291 Nava (p. 101), §292 Cagliano, §293 Sirone (p. 102)

Salmojraghi: §560 Provincia di Como; Pudinghe e Brecciole (p. 367)

PETROGRAPHIC CLASSIFICATION: terrigenous clastic rock (conglomerate)

MACROSCOPIC FEATURES: grey ground with rounded dark clasts, very coarse grain size

MINERALOGICAL COMPOSITION: dolomite clasts together with limestone, chert, quartz, and arenite coming from the Alps (metamorphic and igneous rocks) and the Prealps (carbonate rocks)

MICROSCOPIC FEATURES: rounded clasts made of dolomite, limestone, chert, quartz crystals; calcite matrix; cement of calcite spar filling the interstices among the clasts

GEOLOGICAL SETTING: turbiditic sediments pertaining to "Flysch Cretacico Lombardo" (upper Cretaceous), "Conglomerato di Sirone" formation (Santoniano – upper Cretaceous) of the sedimentary series of the Southern Alps, featuring a massive bedded conglomerate with sandstone beds at the top. The outcrop spreads from eastern Brianza to Lake Iseo

GEOLOGICAL REFERENCES: CGI, Foglio 32 Como (1937); CARG, Foglio 096 Seregno (2010)

QUARRY: the most important site occupied a hill near Sirone (a few kilometres south-west of Lecco)

STONEWORK CATEGORIES: block

SURFACE FINISHING: rough face

USE IN ARCHITECTURE: exclusively local to make architectural elements (lintel, door jamb) and, particularly, for oil presses and mills

DECAY MORPHOLOGIES: erosion of the cement, crumbling

Pietra di Urago

COMMERCIAL NAME: the name comes from a village of the municipality of Montorfano (south-east of Como) near the quarry area

REFERENCES

Cantù; Geografia fisica. Puddinghe; Montorfano

Jervis: Provincia di Como, Circondario di Como; §287 Montorfano (p. 101)

Salmojraghi: §560 Provincia di Como; Pudinghe e Brecciole (p. 367)

PETROGRAPHIC CLASSIFICATION: clastic sedimentary rock (conglomerate, "matrix supported" if fine grained, "clast supported" if coarse grained) linked to bioclastic limestones of shallow marine environment with interbeds of clay; calcareous matrix, cement (from 11% to 33%) made of calcite spar.

MACROSCOPIC FEATURES: Grey ground with black or dark clasts; fine grain size (2–7 mm) or coarse grain size (about 10 cm). Rod-shaped clasts, rounded or angular boundaries according to the lithologies.

MINERALOGICAL COMPOSITION: limestone clasts containing coralline algae, foraminifera and other bioclasts; chert, quartz

MICROSCOPIC FEATURES: rounded clasts of different sizes, mainly made of limestone (oolitic, bioclastic), chert, metamorphic rocks; cement of calcite spar (size 0.2–1.5 mm) filling the space among the clasts

GEOLOGICAL SETTING: "Membro del Montorfano lariano" featuring conglomerates and biocalcarenites, massive beds (up to 7 m thick). The member pertains to "Formazione di Tabiago" (Eocene-Paleocene; lower Lutetian – Danian), of the edimentary series of the Southern Alps, featuring calcareous marls (lower part), pelites and calcarenites (middle part) and conglomerates (upper part).

GEOLOGICAL REFERENCES: CGI, Foglio 32 Como (1937); CARG, Foglio 096 Seregno (2010)

QUARRY: the main site is located in a hill (almost completely dug out) near the village of Urago (municipality of Montorfano) located a few kilometres south-east of Como and close to a small lake.

STONEWORK CATEGORIES: block, moulded element

SURFACE FINISHING: rough face

USE IN ARCHITECTURE: the first use was local using squared blocks for rural buildings or moulded elements to make lintels or jambs or millstones. The stone was massively employed in Milan, from the late 19th century to the first decade of the 20th century, to make the bases of civil buildings: the coarser in the lower parts, the finer in the upper ones: i.e. apartment buildings of Piazza Castello, E. Pirovano, 1890; the colonnaded entrance added to the early Christian basilica of San Lorenzo, architect C. Nava, 1894; the barracks on via Monti, late 19th century. This stone was frequently imitated as artificial stone, using Portland cement and dark pebbles as aggregate.

DECAY MORPHOLOGIES: erosion of the matrix leaving clasts in relief; sulphate skin formation

Nero di Varenna

COMMERCIAL NAME: the name comes from a village on the eastern shore of Lecco south-eastern branch of Lake Como, near the quarry area

REFERENCES

Scamozzi: part 2, book 7, chap. 5, Marmi e misti (p. 189, line 13)

Milizia-Antolini: part 1, book 4, chap. 12, art. 3, Ordine e gradazione dei marmi moderni; Nero – Nero di Como

Amati: Nero di Varenna (Note – Alcuni altri marmi, book 2, chap. 7, p. 50)

Cantù: Geografia fisica; Marmo nero di Varenna

Curioni: chap. 5, Marmi neri; §202 Varenna (p. 56); Marmi grigi; §212, §217 Lierna (p. 57)

Jervis: Provincia di Como, Circondario di Como; §347 Varenna (p. 112), §348 Olcio (p. 113)

Salmojraghi: §560 Provincia di Como; Calcari. Neri o nerastri del Trias (p. 365)

Marmi italiani: Lombardia. Nero di Varenna (p. 39)

PETROGRAPHIC CLASSIFICATION: chemical sedimentary rock (limestone –mudstone/ wackestone)

MACROSCOPIC FEATURES: uniform deep black colour, very fine grain size. A variety, featuring white veins with different thicknesses and arrangements, is called "Grande antico d'Italia" in comparison with the ancient coloured marble called "Bianco e nero antico" or "Grand-Antique" (a brecciated limestone, coming from Pyrenees, used during the Roman period to make column shafts and decorative elements).

MINERALOGICAL COMPOSITION: calcite

MICROSCOPIC FEATURES: very homogeneous microcrystalline calcite, some crossing veins (0.04 mm thick) made of calcite spar.

GEOLOGICAL SETTING: "Calcari di Perledo e Varenna" formation (Ladinian – Middle Triassic) of the sedimentary series of the Southern Alps; featuring black limestone with regular beds (a few centimetres to a few decimetres). The outcrops involve the mountain area (Grigna) between the eastern shore of Lake Como (southern part of Lecco branch) and the Valsàssina (the valley of river Pioverna, running north-west and joining the Lake Como at Bellano, 5 km north of Varenna).

GEOLOGICAL REFERENCES: CGI, Foglio 17 Chiavenna (1941)

QUARRY: different sites on the lake shore between Varenna and Lierna or on the mountain slope above Varenna (along the road to Èsino Làrio)

STONEWORK CATEGORIES: block, moulding, carving

SURFACE FINISHING: polishing

USE IN ARCHITETURE: it is imperative to take into consideration the impossibility to distinguish this stone from other Lombard black stones (see "Nero di Bergamo"), so we used a geographic criterion based on the presence (mid-15th–late 18th centuries) of the border between the Duchy of Milan (western part of Lombardy) and the Republic of Venice (eastern part) running along the Adda (north) and Oglio (south) river courses. Varenna stone was mainly employed in the territory of the Duchy of Milan (Como, Milan, Pavia), the Nero di Bergamo was mainly employed in the territory of Republic of Venice (Bergamo, Brescia). The use in the territory around the quarries is witnessed by blocks to make column shafts of San Giorgio in Varenna (early 14th century) and in other villages along the lake shores: Mandello del Lario (Beata Vergine del Fiume, San Lorenzo); Bellano (Santi Nazaro e Celso). During the Renaissance, the use was mainly focused on the chromatic contrast with white marble: the cladding of Broletto, the floors of the churches at Como (Duomo, San Fedele, Sant'Agostino), the door of Palazzo Rodari. The use in Milan began in the Middle Ages: Loggia degli Osii (Milan – 14th century, renovated in 1904), the façade of the church of Santa Maria di Brera (Milan – 15th century, a church later transformed as a classroom of the Accademia di Belle Arti). The stone was also used for thin cornices of the lower part of the façade of the church of Certosa delle Grazie near Pavia (G.A. Amadeo, late 15th–early 16th century). The "Grande antico" variety was employed, for some column shafts, in the altars made by Pellegrino Tibaldi during the renovation of the Milan cathedral, encouraged by the Counter-Reformation rules; many churches of the western Lombard territory show this limestone employed for column shafts or for balustrades, in the altars of the 17th and 18th centuries.

DECAY MORPHOLOGIES: in addition to the usual decay phenomena affecting limestone (erosion, sulphate skin formation), it is worth it to note the chromatic alteration. This is a progressive bleaching of the stone surface from deep black to grey. Scientific analyses (X-ray diffraction, scanning electron microscopy, FT-IR spectro-photometry, Raman spectroscopy, etc.) carried out on stone samples (Marinoni et al. 2007) do not validate the hypothesis, already proposed, of a chromatic alteration caused by oxidation processes of the organic matter contained in the stone. The decayed surfaces of the black stones show the presence of gypsum, dolomite and organic matter from biological activity. The chemical decay together with the mechanical one (dissolution of calcite and microcracks), produces a surface decohesion and an increase of roughness: the result is a particular diffusion of the incident light.

Verrucano lombardo

COMMERCIAL NAME: the term was firstly employed to indicate a Permo-Triassic series of reddish metamorphic (phyllite, quartzite) and clastic (conglomerate, breccia) rocks forming the group of Monte Pisano, located between Pisa and Lucca in Tuscany. In particular, the "Monte Verruca" occupies the southern part of this group. The same term was chosen to indicate a formation (Permian) made of reddish clastic rock at the base of the sedimentary series of the Southern Alps.

REFERENCES

Marmi italiani: Lombardia. Porfidi di Como e di Bergamo (p. 48)

PETROGRAPHIC CLASSIFICATION: terrigenous clastic rock (conglomerate)

MACROSCOPIC FEATURES: reddish or purple or sometimes green ground with white and pink clasts; very coarse grain size (from 0.5–5.0 cm to 20–30 cm); clast morphology ranges from spherical to rod-shaped

MINERALOGICAL COMPOSITION: fragments of volcanic rocks (quartz porphyries) or igneous rocks (diorite) or schists; matrix containing quartz, feldspar, micas

MICROSCOPIC FEATURES: clasts of porphyry with sub-angular to sub-rounded corners, clasts of metamorphic rocks, polycrystalline quartz; interstitial cement made of silicate minerals

GEOLOGICAL SETTING: "Verrucano Lombardo" formation (upper Permian) of the sedimentary series of the Southern Alps, featuring massive beds (up to 1 m thick) of matrix-supported conglomerates together with sandstones with cross-bedding, micaceous siltites and clayey cement; fluvial environment (braided streams). Outcrops are spread in a long belt of Prealps from Valsàssina (Lake Como, west) to Val Camonica (Brescia, east) and to Valli Giudicarie (Trentino, north).

GEOLOGICAL REFERENCES: CGI, Foglio 32 Como (1937); CARG, Foglio 076 Lecco (2012)

QUARRY: different local sites in the outcrop area, as in Valsàssina from Margno to Introbio

STONEWORK CATEGORIES: rough block

SURFACE FINISHING: bush hammering

USE IN ARCHITECTURE: small altars witness its use during the Roman period. Rural buildings of the Middle Ages, show masonries made of this stone, which was also employed to supply blocks for carriageway or slabs (a few centimetres thick) for stairs. The use strongly increased in the first half of the 20th century: squared blocks were employed for cladding or for bases or piers both locally (present-day Deutsche Bank, Introbio) and in many Lombard towns. Significant Milanese examples are the staircase of the vestibule of the Palazzo delle Colonne (via Verdi, Muzio and Greppi, 1940) or the doorways of some apartment building (via Gonzaga 5–7, Muzio, 1949).

DECAY MORPHOLOGIES: scaling

Loose deposits

COMMERCIAL NAME: there is no specific name for these materials of various sizes (sand, gravel, pebble, cobble, boulder, etc.) and showing lithologies mixed together. A term of the Lombard dialect "gera" was employed to indicate loose materials useful as aggregate to make bedding mortars and plasters.

REFERENCES

Scamozzi: part 2, book 7, chap. 20, Sabbia (p. 235, line 14)

Salmojraghi: §560 Provincia di Como; Erratici; Lago di Lecco, Como, Pian d'Erba, Brianza (p. 366)

PETROGRAPHIC CLASSIFICATION: loose sediments with different lithology, according to the rock outcrops in the catchment areas. Mainly igneous rocks (Alps – granite, granodiorite, diorite, etc.) and metamorphic rocks (Alps – phyllite, gneiss, micaschist, etc.) together with carbonate or silicate sedimentary rocks (Prealps – limestone, dolomite, sandstone, etc.).

MACROSCOPIC FEATURES: rounded pebbles, cobbles and boulders are spherical, tabular, bladed or rod-shaped according to the lithology. In fact, igneous or massive sedimentary rocks make spherical elements; metamorphic foliated rocks and thin-bedded sedimentary rocks make tabular or bladed elements. The size is ordinarily fixed by sedimentological point of view: pebble up to 6.4 cm, cobble up to 25.6 cm, boulder over this size.

MINERALOGICAL COMPOSITION: mainly quartz, silicates (feldspar, plagioclase, micas) and carbonates (calcite, dolomite)

GEOLOGICAL SETTING: Pleistocene deposits of glacio-fluvial origin, mainly occurring where the Alpine valleys go into the plain. Other deposits occur along the present-day courses of rivers (alluvium). Erratic boulders (volume of several cubic metres) are spread along the mountain slopes of the Prealps.

GEOLOGICAL REFERENCES: CGI, Foglio 32 Como (1937); CARG, Foglio 096 Seregno (2010)

QUARRY: every morainic deposit, mainly in the western part of Lombardy, was used to supply stone material; the most important area was located around the southern part of Lake Como and the hilly region south of Como (Brianza). Deposits along river courses were also exploited: the most important was the Adda river course between Trezzo and Lodi. Erratic boulders are spread on the mountain slopes until an elevation of about 700 m above sea level; in many cases they were completely transformed as building material, in other cases the boulders show the marks of the working tools.

STONEWORK CATEGORIES: original form, in some cases pebble or cobble may be split into two parts. Erratic boulders were singularly worked to supply dimension stones (block, slab, etc.).

SURFACE FINISHING: rough

USE IN ARCHITECTURE: medieval religious buildings of western Lombardy show masonries made of pebbles as San Vincenzo (early 11th century) and the Baptistery of San Giovanni (11th century) at Galliano near Cantù (Como); San Carpoforo (11th century) in the southern outskirts of Como; Sant'Eufemia in Erba (11th–12th centuries); the Baptistery of Mariano Comense (11th century); San Pietro al Monte in Civate (Lecco, 11th century); Santi Pietro e Paolo and the Baptistery in Agliate (Milan, late 10th century); the Baptistery of San Giovanni alle Fonti in Milan (4th century). Ghiandone used to make the huge threshold of the church of Santa Tecla (Milan, 5th century), certainly came from an erratic boulder.

DECAY MORPHOLOGIES: different according to the petrographic nature of each pebble

METAMORPHIC ROCKS

Marmo di Musso

COMMERCIAL NAME: the name comes from a village located on the west shore of the upper part of Lake Como

REFERENCES

Scamozzi: part 2, book 7, chap. 5, Marmi e misti (p. 188, line 54)

Amati: Musso; Olgiasca (Note – Marmi moderni; Alcuni altri marmi, book 2, chap. 7, p. 50)

Cantù: Geografia fisica; Marmo d'Olgiasca; Presso Musso, sulla riva opposta

Curioni: chap. 5, Marmi bianchi; §259 Piona, §263 Musso (p. 64)
Jervis: Provincia di Como, Circondario di Como; §332 Musso (p. 108), §343 Colico (p. 110)
Salmojraghi: §560 Provincia di Como; Calcare, Saccaroide talor dolomitico, pre-triassico (p. 365)
Marmi italiani: Lombardia; Bianco di Musso (p. 38)
PETROGRAPHIC CLASSIFICATION: Regional metamorphic rock (marble, crystalline limestone)
MACROSCOPIC FEATURES: the colour varies from grey to yellowish to white to grey with dark veins. The grain size is generally coarse, but in some cases it may be very fine. The marble body is intersected by dark green amphibolite veins (thickness from some centimetres to almost 1 m) with "boudinage" phenomena.
MINERALOGICAL COMPOSITION: calcite crystals and accessory minerals as quartz, silicates (muscovite), amphibole (tremolite), biotite together with pyrite and pyrrhotite. Accessory minerals are more frequent on the sides of the marble lens, near the host rock. The amphibolite veins contain hornblende, phlogopite, andesine, titanite, calcite, pyrite and pyrrhotite.
MICROSCOPIC FEATURES: "mortar structure" with few large anhedral calcite crystals (size circa 0.4–5.0 mm) with curved polysynthetic twinning, surrounded by finer calcite crystals (size circa 0.04–0.15 mm); rounded quartz crystals (0.08–0.4 mm) are sometimes included in larger calcite crystals.
GEOLOGICAL SETTING: marble lens interspersed into gneissic rocks of the South Alpine basement (Hercynian, Dervio-Olgiasca zone, Palaeozoic) and laid out in a west–east direction. The marble outcrop, near the upper tip of Lake Como, is divided into two parts: the most important one on the western shore of Lake Como, near Musso and Dongo; the less important one on the eastern shore of the lake, near Olgiasca and Piona. The western outcrop forms a sub-vertical mountain crest (about 2 km long and 0.5 km thick); this crest, called "Sasso di Musso", reaches the lake shore and separates Musso from Dongo; the upper part of the marble lens is in contact (discordance) with an outcrop of "Dolomia principale" (Hauptdolomit – upper Triassic) with sub-horizontal bedding.
GEOLOGICAL REFERENCES: CGI, Foglio 17 Chiavenna (1941)
QUARRY: the western outcrop shows two amphitheatre quarries, now disused, located at different elevations into the marble lens of "Sasso di Musso". A scarce activity was interested in the outcrop located on the eastern shore of the lake (Olgiasca, Villa Malpensata).
STONEWORK CATEGORIES: block, mouldings, reliefs and sculptures
SURFACE FINISHING: polishing
USE IN ARCHITECTURE: the local use along the shores of Lake Como was significant in the Middle Ages and later on, as noted at Gravedona (Santa Maria del Tiglio, San Vincenzo, Santa Maria delle Grazie); Piona (Abbazia); Corenno Plinio (arched tombs of Conti Andreani); Bellano (Santi Nazaro e Celso); Mandello del Lario (Beata Vergine del Fiume); Lecco (Duomo).
The first important use occurred in Roman times (*Novum Comum*) in colonnaded temples, pediments (one pediment, unfinished, was found in piazza Mazzini and now is located in the gardens near the Mausoleo Voltiano), friezes, sculptures and also in medieval masonries as re-used material. The use in Como was regular throughout the centuries: church of Sant'Abbondio (portal and bases of inner piers); Duomo using marble quarried for this purpose as blocks, decorative elements and sculptures, mainly on the façade and on the side entrances. Another important medieval and subsequent use were the white tiles of

two-colour floors (i.e. Duomo and San Fedele, with hexagonal marble tiles in alternating with hexagonal black limestone tiles).

The use was also significant in the Roman period in Milan: the so-called Colonne di San Lorenzo (3rd century, completely restored in 1951–55) are made of bases, two parts fluted shafts and capitals.

The use of marble was renewed in the 20th century, mainly as veneer slabs on apartment buildings, in Como and other Lombard towns. In Milan it is worth it to note: the Sacrario ai Caduti Milanesi (Muzio et al., 1927–30) with an imposing structure of arches and niches made of well-squared blocks; the cladding of the first floor and the moulded portals of the residential complex Bonaiti (piazza Repubblica, Muzio, 1935); the spiralled funerary chapel of Bernocchi family (Minali, 1936) with sculptured scenes of the Passion of Christ (Castiglioni), located in the centre of Cimitero Monumentale (Necropoli 1A).

DECAY MORPHOLOGIES: erosion till crumbling; sulphate skin formation; the presence of amphibolite veins carrying sulphur minerals produces rust spots with a morphology variable according to the cut

Chapter 7

Stones from province of Pavia

SEDIMENTARY ROCKS

Arenaria dell'Appennino pavese

COMMERCIAL NAME: the term refers to the area of outcrop, there is not a defined name
REFERENCES
Jervis: Provincia di Pavia, Circondario di Voghera; § 694 Oliva Gessi (arenaria), §697 Santa Giulietta (p. 187)
Salmojraghi: §566 Provincia di Pavia; Arenarie; Terziarie (p. 369)
PETROGRAPHIC CLASSIFICATION: clastic sedimentary rock (sandstone – greywacke)
MACROSCOPIC FEATURES: yellowish colour; fine grain size
MINERALOGICAL COMPOSITION: quartz and muscovite; calcite as cement
MICROSCOPIC FEATURES: rounded clasts of composite quartz together with muscovite shreds; the cement of calcite spar fill the interstices among the clasts
GEOLOGICAL SETTING: "Arenarie di Monte Arzolo" member of "Conglomerati di Cassano Spinola" formation (Upper Miocene – Messinian) of the post-Messinian phase Succession (Apennines). The formation includes lenticular beds of clast supported alluvial conglomerate with sandy matrix, sandy lenses and massive badly cemented conglomerate. The member, in the upper part of the formation (Upper Messinian), includes cross-laminated alluvial sandstone together with sand, silt and clay levels (a few centimetres thick).
The member, in the upper part of the formation, includes a sandy-arenaceous facies with cross-bedding and rare clay interbeds. Outcrops are located in a narrow area along the first hills of the Apennines, south of Pavia between Rivanazzano (west) and Stradella (east).
Other formations, pertaining to "Epiligurian Succession", supplied building material for local constructions: "Formazione di Ranzano" (turbiditic formation made of sandstones, conglomerates and pelites; Oligocene – Eocene, Upper Priabonian – Lower Rupelian); "Arenarie di Monte Vallassa" (bioclastic sandstones and biocalcarenites together with conglomerates and sandy marls; Miocene, Upper Burdigalian – Serravallian). Outcrops are located in the more elevated part of the Apennines (southernmost part of the province of Pavia, around Varzi).
GEOLOGICAL REFERENCE: CGI, Foglio 59 Pavia (1965); CARG, Foglio 178 Voghera (2012)
QUARRY: different small sites were located in the area of Santa Giuletta, about 10 km south of Pavia, between Casteggio (west) and Broni (east)

DOI: 10.1201/9781003361008-8

STONEWORK CATEGORIES: block, shaft, carving
SURFACE FINISHING: coarse claw tooling
USE IN ARCHITECTURE: the use was exclusively local in the territory of the province of Pavia; the sandstone was always mixed with some other stones and brickwork: i.e. the Romanesque churches of Pavia as San Michele (first half of the 12th century), San Pietro in Ciel d'Oro (first half of the 12th century) or Santa Maria in Betlem (11th century); Romanesque churches of the Apennines as San Zaccaria (Rocca Susella, 11th century), Sant'Alberto di Butrio (Pizzocorno, 11th century) or San Marcello in Montalino (Stradella, late 11th century). During the centuries, the stone was progressively abandoned because of its poor durability, causing failures in a short lapse of time.
DECAY MORPHOLOGIES: erosion and crumbling with complete loss of the outward form, sulphate skin formation

Loose deposits

COMMERCIAL NAME: there is no name for these materials (block, cobble, pebble)
PETROGRAPHIC CLASSIFICATION: extremely variable according to the rock outcrops in the river catchment basins
MACROSCOPIC FEATURES: pebbles and cobbles are mainly rounded or rod-shaped according to the nature of the stone
MINERALOGICAL COMPOSITION: mainly chemical or clastic sedimentary rocks (limestone, sandstone); minerals: quartz, plagioclase, micas and carbonates (calcite, dolomite)
GEOLOGICAL SETTING: loose deposits (alluvium) mainly occur where the valleys of the Apennine flow into the plain; conglomerates with rounded pebbles and low rate of cementation (Miocene, Messinian) were also involved
GEOLOGICAL REFERENCE: CGI, Foglio 59, Pavia (1965); CARG, Foglio 178 Voghera (2012)
QUARRY: every river course was used to supply the stone materials
STONEWORK CATEGORIES: stone elements mainly maintained the original morphology, but in some cases pebble or cobble were split into two parts. Normally, pebbles of the same size were laid in parallel rows with mortar joints; in some cases, only one lithotype (mainly gneiss) was employed and accurately chosen among the other ones.
SURFACE FINISHING: rough face
USE IN ARCHITECTURE: vernacular architecture of the Apennines (province of Pavia) shows masonries made of pebbles coming from loose deposits
DECAY MORPHOLOGIES: linked to the nature of the rocks forming the sediment

Chapter 8

Stones from province of Sondrio

IGNEOUS ROCKS

Granito di Dubino

COMMERCIAL NAME: the name comes from a village located in the lower part of Valtellina, some kilometres east of the mouth of river Adda into Lake Como.
REFERENCES
Salmojraghi: §562 Provincia di Sondrio; Gneiss (p. 360)
PETROGRAPHIC CLASSIFICATION: plutonic igneous rock (quartz-diorite)
MACROSCOPIC FEATURES: dark grey colour, medium to coarse grain size
MINERALOGICAL COMPOSITION: plagioclase (andesine), hornblende, biotite, quartz
MICROSCOPIC FEATURES: gneissic texture, twin crystals of plagioclase with irregular size together with hornblende and quartz crystals
GEOLOGICAL SETTING: the south-western ramification of "Plutone della Val Màsino e della Val Bregaglia" close to the right bank of river Adda, (lower part of Valtellina) contains the "Quarzodiorite del Monte Bassetta" (Serizzo) with a gneissic structure
GEOLOGICAL REFERENCE: CGI, Foglio 18 Sondrio (1971)
QUARRY: some sites on the mountain close to Dubino, on the right bank of river Adda about 10 km east of Lake Como
STONEWORK CATEGORIES: block, slab
SURFACE FINISHING: polishing
USE IN ARCHITECTURE: this stone went in use in the 1930s; wide slabs were used for cladding (Grattacielo, piazza Repubblica 32, Milan)
DECAY MORPHOLOGIES: scaling

Ghiandone and Serizzo

COMMERCIAL NAME: the name comes from the Milanese term "gianda", meaning "acorn", referring to the rectangular white spots of the rock; the term "giandon" means "coarse granite". The name "serizz-giandon" is a Milanese term to indicate the rocks forming the erratic boulders displaced by the glaciers among the hills north of Milan (Brianza).
REFERENCES
Scamozzi: part 2, book 7, chap. 8, Pietre vive (p. 199, line 7)
Jervis: Provincia di Sondrio, Circondario di Sondrio; §321 Val Màsino (p. 107)
Salmojraghi: §562 Provincia di Sondrio; Graniti; Di varia natura (p. 360)
Marmi italiani: Lombardia; Ghiandone e Sarizzo di Val Màsino (p. 47)

DOI: 10.1201/9781003361008-9

PETROGRAPHIC CLASSIFICATION: intrusive igneous rocks ("Ghiandone" as granodiorite; "Serizzo" as diorite)

MACROSCOPIC FEATURES. "Ghiandone": grey colour with white rectangular spots; coarse grain size; "Serizzo": uniform grey colour with black spots, medium to fine grain size

MINERALOGICAL COMPOSITION. "Ghiandone": potash feldspar, plagioclase (oligoclase-andesine), quartz, biotite, hornblende; "Serizzo": plagioclase (andesine), hornblende, biotite, quartz

MICROSCOPIC FEATURES. "Ghiandone": granular texture with large euhedral crystals of potash feldspar together with twinned crystals of plagioclase and anhedral crystals of quartz; "Serizzo": granular texture with euhedral twinned crystals of plagioclase, hornblende and anhedral crystals of quartz

GEOLOGICAL SETTING: "Plutone della Val Màsino e della Val Bregaglia", a magmatic body outcropping in northern Lombardy, from val Màsino (Valtellina, province of Sondrio) to val Bregaglia-Bergell (Canton Grigioni-Graubünden, Switzerland). This Alpine pluton (Cenozoic) shows a rounded shape (diametre about 10 km) with a narrow western apophysis until Val Chiavenna (Lake Novate Mezzola, province of Sondrio) and Val Leventina (Bellinzona, Canton Ticino, Switzerland). Host rocks are "Orthogneiss" (west) and "Ophiolites" (east) of the Pre-Triassic basement. The "Granodiorite della Val Màsino" occupies the northern part of the pluton (corresponding to the upper parts of both Val Màsino and Val Bregaglia) and it changes gradually towards the south into the "Quarzodiorite del monte Bassetta" (corresponding to the middle part of Val Màsino until the lower part of Val Chiavenna).

GEOLOGICAL REFERENCE: CGI, Foglio 18 Sondrio (1971)

QUARRY: present-day sites are located in the upper Val Màsino, mainly using the fallen boulders and blocks along the mountain slopes. The glacial moraines spread among the slopes of the mountains surrounding the two branches of Lake Como until the hilly region of Brianza, south of Lake Como has been the most important supply area since the Roman times.

STONEWORK CATEGORIES: block, shaft

SURFACE FINISHING: bush hammering, sometimes polishing

USE IN ARCHITECTURE: the excellent characters were already appreciated in the Roman period, as witnessed in many Lombard towns. Blocks of Ghiandone and Serizzo are present in Milan buildings: the Roman amphitheatre; the early Christian basilica of Santa Tecla (a huge prismatic door sill measures 3.5 m in length on a cross section of 0.6 x 0.45 m and a weight of about 2.6 tons); the piers of Loggia dei Mercanti (first half of the 13th century); in the arches of Porta Nuova (late 12th century); the masonry of Romanesque churches (as material pilfered from Roman constructions) like the atrium of the basilica of Sant'Ambrogio. The use decreased in the following centuries together with the increase of granites coming from Lake Maggiore (Baveno, Montorfano). In the 20th century, Ghiandone was employed in the bases of the residential buildings, sometimes with rough finishing (palazzo Castiglioni, corso Venezia, G. Sommaruga, 1904). Noteworthy are the eighteen shafts of the nave of Santa Croce (C. Arpesani, 1917). Polished squared blocks were employed in the funerary monument of the sculptor Adolfo Wildt (1868–1931) made by Giovanni Muzio (Cimitero Monumentale, Circondario di Ponente, 715–16).

DECAY MORPHOLOGIES: scaling

Granito di San Fedelino

COMMERCIAL NAME: the name comes from a village on Val Chiavenna, near the northern shore of the lake Novate Mezzola (north of Lake Como).
REFERENCES
Cantù: Geografia fisica; Granito bianco di San Fedelino (lago di Mezzola)
Jervis: Provincia di Sondrio, Circondario di Sondrio; §330 Samòlaco, §331 Novate Mezzola (p. 108)
Salmojraghi: §562 Provincia di Sondrio; Granito; S. Fedelino (p. 360)
Marmi italiani: Lombardia; Granito di S. Fedelino (p. 46)
PETROGRAPHIC CLASSIFICATION: igneous plutonic rock (granite)
MACROSCOPIC FEATURES: whitish colour with fine black points; roughly aligned arrangement and fine grain size
MINERALOGICAL COMPOSITION: quartz, K-feldspar, plagioclase (oligoclase), muscovite and biotite
MICROSCOPIC FEATURES: xenomorphic texture with slight alignment featuring crystals of quartz and K-feldspar together with tabular crystals of biotite and muscovite laminae
GEOLOGICAL SETTING: this kind of granite (aplite, two micas) features a series of thick veins, and it was intruded on the western side of the Tertiary pluton of Val Màsino – Val Bregaglia. The outcrop is located in the lower part of Val Chiavenna, north of Lake Como and close to the lake of Novate Mezzola; the outcrop is present on both banks of the river Mera. Different acid rocks of similar mineralogical features outcrop in a few kilometres on this area (Granito di San Fedelino – Granito tipo Valcundria – Ghiandone – Granito di Dubino).
GEOLOGICAL REFERENCE: CGI, Foglio 17 Chiavenna (1941)
QUARRY: located on the south-western slope of monte Zucco (left bank of river Mera) close to the village of Novate Mezzola
STONEWORK CATEGORIES: block, thick slab
SURFACE FINISHING: bush hammering
USE IN ARCHITECTURE: the masonry of the oratory of San Fedelino (10th century, close to the western shore of Lake Novate Mezzola) contains some pebbles of this granite together with pebbles of other lithotypes of Val Chiavenna (gneiss, green stones, etc.). The most important use, starting from the early 19th century, was the paving of the urban streets of Milan and other Lombard towns, thanks to its very high resistance to wear: two parallel rows of thick rectangular slabs were disposed to hold the wheels of carts, wagons and coaches in the streets paved with small river pebbles. The use of tramways in the urban streets (since the last decade of 19th century) involved the use of pentagonal blocks set in diagonal rows between the rails. Finally, it is very important to remark the use of prismatic blocks bordering the sidewalks.
DECAY MORPHOLOGIES: scaling

METAMORPHIC ROCKS

Granito Galleggione

COMMERCIAL NAME: the name comes from the mountain of Val Bregaglia, on the right side (north) of river Mera (Maira); the Italian–Swiss border goes through the top

REFERENCES

Salmojraghi: §562 Provincia di Sondrio; Graniti; Gneissico, a grana grossa; Piuro S. Croce (p. 360)

PETROGRAPHIC CLASSIFICATION: regional metamorphic rock (gneiss)

MACROSCOPIC FEATURES: whitish colour featuring "augen" of feldspar and undulose veins of dark silicates

MINERALOGICAL COMPOSITION: quartz, K-feldspar, muscovite

MICROSCOPIC FEATURES: equant crystals of potash feldspar, anhedral crystals of quartz (polycrystalline, undulose extinction), streaks of muscovite, porphyroblasts of garnet

GEOLOGICAL SETTING: "Gneiss di Villa di Chiavenna" pertaining to "Cristallino del Tambò" (System Tambò-Suretta, middle Pennidic, Archaeozoic) and featuring biotitic gneisses, with garnet and staurolite. Outcrops are located along Val Bregaglia in both the Italian and Swiss territories.

GEOLOGICAL REFERENCE: CGI, Foglio 18 Sondrio (1971)

QUARRY: different sites, mainly exploiting the boulders piled along the mountain slopes (Santa Croce, municipality of Piuro)

STONEWORK CATEGORIES: block, slab

SURFACE FINISHING: polishing

USE IN ARCHITECTURE: this stone went in use in the 1930s; G. Muzio chose it for cladding the piers of some buildings (via Albricci 1,3,5 and via Gonzaga 5,7; via Vivaio 1; Galleria Passarella 1, the curved building on piazza San Babila between corso Europa and corso Vittorio Emanuele).

DECAY MORPHOLOGIES: scaling

Serpentinite

COMMERCIAL NAME: the name refers to the aspect of the stone, similar to the skin of a snake

REFERENCES

Scamozzi: part 2, book 7, chap. 5, Marmi e misti (p. 189, line 11)

Jervis: Provincia di Sondrio, Circondario di Sondrio; §320 Chiesa (p. 107)

Salmojraghi: §562 Provincia di Sondrio; Scisti (p. 360)

Marmi italiani: Lombardia; Serpentino della Val Malenco (p. 48)

PETROGRAPHIC CLASSIFICATION: regional metamorphic rocks. Under the Italian term "serpentina" are grouped both serpentinites and serpentine schists. The Italian term "serpentino" refers to the mineral group (serpentine group). Some texts report the term "serpentine" for both rock and mineral group.

MACROSCOPIC FEATURES: dark green colour, fine grain size; serpentinite features a massive texture, serpentine schist features a strong schistosity

MINERALOGICAL COMPOSITION: antigorite; magnetite as accessory

MICROSCOPIC FEATURES: aggregates of fibro-lamellar structure of antigorite

GEOLOGICAL SETTING: Serpentina pertains to the group called "Rocce verdi" (green stones), a Pennidic unit (Pre-Triassic crystalline basement). The outcrops are concentrated in Val Malenco (monte Disgrazia), a tributary of the right side (north) of Valtellina, close to the eastern boundaries of the pluton (Val Masino – Val Bregaglia)

GEOLOGICAL REFERENCE: CGI, Foglio 18 Sondrio (1971)

QUARRY: the stone, in the past, was quarried in Brianza (province of Como), thanks to the erratic boulders carried out by the glaciers (i.e. a famous erratic boulder of Serpentina was cut near Valmadrera, south west of Lecco; about 90 km from Val Malenco). Modern quarries are located in the western part of Val Malenco and different varieties are traded: Serpentino Classico and Serpentino Verde Vittoria. Serpentino Classico (massive) is quarried 100 m from Chiesa in Val Malenco (road to Chiareggio – Nuova Serpentino d'Italia S.p.A.); a quarry of schistose stone (split surface) is opened a kilometre ahead. Serpentino Verde Vittoria is quarried near Lanzada, 3 km east of Chiesa in Val Malenco (road to Franscia).

STONEWORK CATEGORIES: slab of different thickness

SURFACE FINISHING: quarry face (foliated variety), polishing (massive variety)

USE IN ARCHITECTURE: locally was always employed to make roofs thanks to the availability of thin slabs. The use has been spread in many Lombard towns since the 20th century, as cladding of building façades with thick slabs.

DECAY MORPHOLOGIES: scaling of quarry-faced surfaces, mainly in the schistose variety

Valcundria (Granito tipo Valcundria – Samolaco)

COMMERCIAL NAME: the term "Valcundria" indicates a mountain place (elevation 1700 m) above Chiavenna and Prata Camportaccio on the left bank of river Mera. The rock with this term outcrops farther south, close to the village of Samòlaco (a name coming from the Latin expression "In summo laco", meaning "at the tip of the lake").

REFERENCES

This kind of stone seems to be ignored by the ancient authors; the term "Valcundria" is reported in the geological map of Val Chiavenna: "Gneiss biotitici compatti, talora ghiandolari o tipo Valcundria" (CGI, Foglio 17, 1941). Pieri (1964) reports "Granito di Val Cundria" as synonym of "Granito di Samòlaco", quarried in Val Chiavenna, but different from "Granito di San Fedelino".

PETROGRAPHIC CLASSIFICATION: regional metamorphic rock (granitoid gneiss, migmatite)

MACROSCOPIC FEATURES: dark grey to black colour with convoluted white veins of various thickness; fine to medium grain size

MINERALOGICAL COMPOSITION: quartz, K-feldspar, plagioclase together with biotite, amphibole, sillimanite, garnet

MICROSCOPIC FEATURES: alternating bands of leucocratic minerals (quartz, alkali feldspar) and melanocratic ones (biotite, garnet, sillimanite, etc.)

GEOLOGICAL SETTING: "Cristallino dell'Adula" (Archaeozoic) contains two Lower Pennidic Units ("Gneiss del Monte Gruf" and "Gneiss del Monte Provinaccio", both once referred to "Anatessiti"). The outcrop involves the middle part of Val Chiavenna (eastern slope) between Somaggia (south, close to "Granito di San Fedelino") and Prata Camportaccio (north, close to "Meta-ophiolites of Chiavenna Unit"). "Gneiss del monte Gruf" includes biotitic granitoid gneisses rich in potash feldspar; "Gneiss del Monte Provinaccio" includes homogeneous migmatites together with biotitic gneisses.

GEOLOGICAL REFERENCE: CGI, Foglio 17 Chiavenna (1941), Fogli 7–18 Pizzo Bernina-Sondrio (1970)

QUARRY: near Somaggia (municipality of Samòlaco) on the eastern bank of river Mera (val Chiavenna, province of Sondrio)

STONEWORK CATEGORIES: block, shaft, slab

SURFACE FINISHING: polishing

USE IN ARCHITECTURE: this stone went in use, as other coloured ones, in the 1930s, mainly for cladding of façades of apartment buildings. In particular it is worth it to note the use for bases or for portals, always enhancing the dark colour in contrast with whiter stones. It is worth it to note, in Milan, the Sacrario ai Caduti Milanesi (G. Muzio, Alpago Novello, Buzzi, Cabiati, G. Ponti – 1928) with four columns (8 m high), the lintel of the enclosure and the base of the octagonal central building.

DECAY MORPHOLOGIES: scaling

Other metamorphic rocks

Two other metamorphic rocks were quarried: Pietra Ollare and Verde Spluga.

"Pietra Ollare" is a general term referred to a soapstone easy to carve and it is related to Chloriteschists (Clinochlore) and Talcschists (Talc, Magnesite, Lizardite).

Vincenzo Scamozzi reported the pots for cooking of Val Chiavenna; these pots were fabricated using a lathe (part 2, book 7, chap. 11, Pietre trattabili, p. 209, line 3). Rondelet also reported the same characters about this stone (book 1, sec. 2, art. 15, §306, p. 201).

Chloriteschists pertain to "Serpentine della Val Malenco" formation (Mesozoic) to "Pietre verdi del Suretta" and described as serpentinites and serpentine schists with antigorite, olivine and pyroxenes together with ophiolite breccias containing serpentine schists and calcite cement. Quarry sites are located at Val Brutta, about 10 km east of Chiesa in Val Malenco (road to Franscia).

Talcschists pertain to "Pietre Verdi di Val Chiavenna" formation (Mesozoic) referred to Alpine ophiolites of "Falda Tambò" of "Sistema Tambò-Suretta" (middle Pennidic) and described as schists (talc, amphibole, olivine), olivinites and serpetinites (sometimes with orthopyroxenes) together with dark green amphibolithes (hornblende, plagioclase and epidote). Several underground quarry sites are located near Piuro, east of Chiavenna on the left bank of river Mera (road to Maloja Pass). These stones were mainly used to make cooking pots, called "Laveggi", skillfully worked using traditional lathes.

"Verde Spluga" or **"Quarzite verde dello Spluga"** is a commercial name to indicate a quartzite pertaining to Pennidic formations of Splügen and Val Malenco (quartzites and breccias, limestones and dolomite limestones, vuggy dolostones and gypsum rocks). A quarry site is located east of Isola (elevation 1270 m), about 20 km north of Chiavenna, on the western side of Valle San Giacomo (road to Splügen Pass). This stone was mainly used in the second half of the 20th century, as slab (split surface) to make cladding for apartment buildings (i.e. Milan, via Lepetit 19).

Chapter 9

Stones from province of Varese

This list also includes the stones from Arzo, quarried in Canton Ticino (Switzerland), close to Lombard territory.

IGNEOUS ROCKS

Granofiro di Cuasso al Monte (Porfiroide rosso)

COMMERCIAL NAME: the name comes from a village north of Varese. The name "Porfido di Cavagnano" refers to the same kind of stone quarried at Cavagnano, a few kilometres south of Cuasso.
REFERENCES
Jervis: Provincia di Como, Circondario di Varese; §273 Brinzio (p. 98)
Salmojraghi: §564 Provincia di Como; Porfidi. Quarziferi, permiani (p. 367)
Marmi italiani: Lombardia; Porfiroide di Cuasso al Monte (p. 48)
PETROGRAPHIC CLASSIFICATION: hypabyssal igneous rock (Granofiro *Auctorum* – granophyre, alcaline rhyolite)
MACROSCOPIC FEATURES: red-orange colour with some small cavities (geodes) containing well-developed crystals of potash feldspar
MINERALOGICAL COMPOSITION: quartz, potash feldspar, plagioclase
MICROSCOPIC FEATURES: the stone features the so-called granophyric texture, a kind of "micrographic" texture (similar to a cuneiform writing) with radiate intergrowth of quartz and alkaly feldspar (quartz appears as wedges and rods in the feldspar)
GEOLOGICAL SETTING: "Granofiro di Cuasso" formation (Permian) pertains to the post-Hercynian volcanism, as a magmatic body intruded at low depth into the lithosphere. Above the volcanic rocks of "Granofiro di Cuasso", the sandstones of "Servino" formation (Skytian, lower Triassic) mark a clear disconformity (marine transgression), starting the sedimentary series of the Southern Alps. The outcrop occupies a large area between Valganna (west) and Lake Lugano (east) at the core of the "Campo dei Fiori Anticline".
GEOLOGICAL REFERENCES: CGI, Foglio 31 Varese (1932)
QUARRY: the main site was located in Val Ceresio, near the village of Cavagnano, between Cuasso al Monte and Cuasso al Piano.
STONEWORK CATEGORIES: block, paving elements
SURFACE FINISHING: rough face
USE IN ARCHITECTURE: local use as building stone is witnessed, as rough block, in the masonries of the Abbey of San Gemolo (Ganna, 12th century). Far from the quarry area,

DOI: 10.1201/9781003361008-10

the use was mainly in the form of natural pebble coming from glacial deposits (i.e. the masonries of San Giovanni, Castelseprio, 5th–6th century, now in ruins). A large use involved the paving of roads with polygonal blocks, since the late 19th century. In the early 20th century, rough blocks were employed for the base in apartment buildings (via Arimondi 12, Milan) and the stone was also used in some funerary chapels (see Cimitero Monumentale).

DECAY MORPHOLOGIES: scaling

SEDIMENTARY ROCKS

Pietra di Angera

COMMERCIAL NAME: the name comes from a village on the southern tip of Lake Maggiore

REFERENCES

Scamozzi: part 2, book 7, chap. 8, Pietre vive (p. 199, line 12)

Rondelet: book 1, sec. 2, art. 15. §259 (p. 196)

Curioni: chap. 5, Marmi rossi e rosei; §240 Angera (p. 61)

Jervis: Provincia di Novara, Circondario di Novara; §262 Arona (p. 96); Provincia di Como, Circondario di Varese; §270 Angera (p. 97)

Salmojraghi: §564 Provincia di Como; Calcare; Dolomitici, triassici (p. 365)

PETROGRAPHIC CLASSIFICATION: chemical sedimentary rock (dolostone)

MACROSCOPIC FEATURES: colour varies from pink to yellow to white; these changes are not linked to the rock bedding; very fine grain size

MINERALOGICAL COMPOSITION: dolomite

MICROSCOPIC FEATURES: homogeneous microcrystalline dolomite with irregular cavities; many cavities show euhedral crystals (0.02 mm) along the rim (pink variety)

GEOLOGICAL SETTING: the rock outcrop belongs to the sedimentary series of the Southern Alps, featuring massive bedded dolostones. A singular outcrop occupies a hill (called Monte San Quirico, north of Angera) where the medieval Rocca Borromeo is located. The outcrop of dolostones occupies the southern part of the hill; the northern one pertains to "Granofiro di Cuasso" (Permian); the territory around the hill is occupied by loose deposits of glacio-fluvial origin (Quaternary). The outcrop is close to the eastern shore of Lake Maggiore, and it is present on the western side of the lake near Arona. Other Permo – Triassic outcrops are separated and located some kilometres north (Ispra) and north-east (Brinzio). The lack of recent studies compels to report different statements about the geological formation holding the dolomite of Angera.

*"Dolomia Principale – Hauptdolomit" (Norian – upper Triassic) – Carta Geologica d'Italia Foglio 31, 1932 (V. Novarese)

*"Dolomia del Salvatore" (Ladinian – Anisian, middle Triassic) – Carta Geognostico-Geologica della Provincia di Varese, 1932 (L.G. Nangeroni)

*"Dolomia Principale" – Proceedings 3rd Stone Congress Venice, 1979 – Padua, 1983 (A. Fiumara et al.)

*"Dolomia del Campo dei Fiori" (Rhetian – upper Triassic) – Carta Geologica Lombardia, scala 1:250.000, 1980 (A. Montrasio et al.)

QUARRY: the most important site is clearly visible on the western slope of the aforementioned hill, north of Angera; the quarry still shows a tall vertical face. The quarry works ended in the late 17th century, but some underground tests were carried out in the late 19th century. Other quarries were opened near Arona, on the other bank of the Lake Maggiore (province of Novara, Piedmont). The lake was used for the transport to the yards of Milan and Lombardy.

STONEWORK CATEGORIES: block, ashlar, slab, carving

SURFACE FINISHING: claw tool, chisel, polishing

USE IN ARCHITECTURE: the first use is witnessed by some drums of Roman columns (Museo di Varese). In the Middle Ages the use was spread around the quarry site (block for the façade of St Donato church in Sesto Calende), and the use increased in the Renaissance (Collegiata, Arona; Santa Maria Assunta, Angera) until the Baroque period (seminary and San Carlo, Arona; Madonna della Riva, Angera). In the 16th century the stone was also used in Milan: San Fedele (Pellegrini and Bassi – since 1569); San Raffaele; church of Certosa di Garegnano (Alessi and Seregni – early 17th century); great courtyard of the Ospedale Maggiore (Ricchino – early 17th century). The use was important also in Pavia: small shafts and decorative elements of doorways of San Teodoro (12th century); blocks of the façade of San Pietro in Ciel d'Oro (12th century) and the blocks of the façade of Palazzo Ducale (F.M. Ricchino – about 1625) of the Certosa delle Grazie (a few kilometres north of Pavia); the balcony of the Collegio Borromeo (Pellegrini, 1585). The stone also reached Parma (façade of San Giovanni Evangelista, 1604). A long decrease started in the late 17th century, but the stone was still used in Milan: Cimitero Monumentale (Maciachini, 1863); Casa Rossi (corso Magenta, Pestagalli, 1860).

DECAY MORPHOLOGIES: surface erosion leading to a disintegration with loss of material and loss of outward form of carved elements; calcium sulphate skin on recessed sites.

Pietre di Arzo

This is a collective name including three stones of the same origin showing different texture and chromatic features: Broccatello (purple with white spots), Rosso (red with white veins), Macchiavecchia (as "ancient spot" in Italian language, brecciated and multicoloured with red, grey and yellow).

COMMERCIAL NAME: the name comes from a village located in southern Canton Ticino

REFERENCES

Vasari (life of Lione Lioni aretino) described the tomb of Giovanni Iacopo Medici (Milan, Duomo) made of white marble from Carrara together with two columns of "spotted stone like jasper" (Macchiavecchia) and two "black and white" columns (Bianco e Nero Antico). These columns were sent from Rome by Pope Pius IV.

Scamozzi: part 2, book 7, chap. 5, Marmi e misti (p. 189, line 5)

Milizia-Antolini: part 1, book 4, chap. 12, art. 3. Ordine e gradazione dei marmi moderni; Giallo – Breccia detta di macchia vecchia e nuova di Lombardia

Amati: Macchia vecchia Svizzera; Rosso d'Arzo Svizzero (Note – Alcuni altri marmi, book 2, chap. 7, p. 50)

Cantù: Geografia fisica; Broccatello d'Arzo (Stato Svizzero)

Curioni: Chap. 5, Marmi rossi e rosei; §242 Arzo (p. 61)

Salmojraghi: §634 Canton Ticino; Marmi liasici (p. 450)

PETROGRAPHIC CLASSIFICATION: chemical sedimentary rock (limestone, biomicrite – wackestone, sometimes with brecciated structure)

MACROSCOPIC FEATURES

Broccatello: purple-red with white spots and some grey veins, fine grain size
Rosso di Arzo: dark red with irregular white veins and bioclastic fragments, fine grain size
Macchiavecchia: white veined breccia with irregular-shaped clasts that are red, grey or yellow in colour, embedded in a dark red microcrystalline matrix; no description can match the stunning arrangement of the patterns of this breccia
NOTE: this stone may be matched with the ancient coloured marble called "Portasanta", quarried on the island of Chios (Greece) and used by the Romans for column shafts or for wall veneers.
MINERALOGICAL COMPOSITION: mainly calcium carbonate (calcite); calcium-magnesium carbonate (dolomite) is present in some clasts of Macchiavecchia
MICROSCOPIC FEATURES
Broccatello: microcrystalline matrix with abundant skeletal grains of Brachiopods (Rhynconellida and Terebratulida, 1–3 cm wide), crinoidal debris, sponges, shells of bivalves (pectinid and ostreid) and some fragments of coral or bryozoan; sometimes, stratiform accumulations of Brachiopods shells, some of which display tilted geopetal fabrics, are also present.
Rosso di Arzo: typical round shape of the Crinoid articula in a microcrystalline calcite matrix.
Macchiavecchia: very fine-grained microcrystalline calcite forming a matrix among angular fragments of limestone with fossil shells or of fine-grained dolomite.
GEOLOGICAL SETTING: "Broccatello and Breccia Macchia Vecchia" formation of the sedimentary series of the Southern Alps (Sinemurian, lower Jurassic). This formation features: a lower part made of a red biothermal limestone containing Brachiopods and calcareous sponges, as well as small zones of red stromatactis limestone (Biohermal Broccatello, or the quarried variety); an upper part made of a vari-coloured crinoidal grainstone (Crinoidal Broccatello) and a grey bioclastic wackestone. The Broccatello formation features an angular unconformity at the base with the underlying dolomites (upper Triassic); the red marly limestone of the "Morbio" formation and the fossiliferous micritic limestone of Besazio (Pliensbachian) are present at the top. The regional paleogeographic setting of the Arzo area refers to the marginal area of a Jurassic submarine high ("Lugano High"), close to the Lugano–Monte Grona fault scarp separating a higher foot-wall block from the hanging-wall block of the strongly subsiding Monte Generoso basin. The Arzo area is marked by syn-sedimentary tectonic fragmentation along the eastern margin of the "Lugano High"; this tectonic activity resulted in the formation of breccias (Macchiavecchia), featuring a great heterogeneity with decimetric carbonate rock boulders. The breccias are linked to narrow "graben" and contain fragments of the underlying rocks and different phases of overlying sediments. The formation outcrops in a territory across the Italian border (south of Lake Lugano) between Mendrisio and Varese.
GEOLOGICAL REFERENCES: Atlas Géologique de la Suisse au 1:25.000, Numero 152 Mendrisio-Como (2018)
QUARRY: the main site is located in Canton Ticino, the southernmost Canton of the Swiss Confederation, a few kilometres north-east of Arzo on the road to Meride and Serpiano (Lake Lugano), close to the Italian border. The quarry area has been under Swiss rule since the early 16th century, but the commercial relationships with the neighbouring Duchy of Milan were always very strong, thanks to the lack of natural obstacles between the two territories. The site is geologically connected with the Italian quarry site of Brenno, Viggiù and Saltrio, whose brown or grey clastic limestones had been extensively used in architecture since the Roman times, as reported about Viggiù and Saltrio. The main

quarry district includes two distinct sites. The first is located less than 1 km north-east of Arzo and contains one ancient quarry disposed as an amphitheatre with several steps and a modern one with a derrick, both providing "Rosso Arzo" and "Broccatello". The second quarry site is located some 100 m from the first one, closer to the paved road: it is disposed along a cliff including different faces and provided "Macchiavecchia". Any attempt to find on the quarry faces the marks of ancient exploitation methods is frustrated by modern exploitation systems (helical or diamond wire) and by vegetation growth: in fact, the quarry of ancient Rosso has been in disuse for several years, the quarries of Broccatello and Macchiavecchia worked until 2011.

TRANSPORT

The Arzo area was reached by two railway lines in the second half of the 19th century: the line between Milano, Varese and the Lake Lugano in Italy (1865) and the line between Como, Mendrisio and Lugano in Switzerland (1874), both only a few kilometres from Arzo. The stone blocks followed the road to Saltrio and Viggiù (less than 5 km, on the south-west) to reach Milan (about 60 km from Viggiù); transport through Lake Lugano (about 5 km north) was unfeasible because of both the difference in elevation between the quarry sites (490 m above sea level) and Lake Lugano (260 m) and the lack of navigable waterways linking the lake to Lombard rivers.

STONEWORK CATEGORIES: block, slab, carving

SURFACE FINISHING: polishing to enhance the chromatic qualities

USE IN ARCHITECTURE: Arzo stones were at first used locally and in the western part of Lombardy. The use of a stone coming from Canton Ticino (Switzerland) in the Duchy of Milan may be explained by the shared boundaries; moreover, the diocese of the Milan Catholic church included a large part of the territory of Canton Ticino. Later the use went out of the regional borders to reach the whole Italian territory.

The first use for architectural purposes probably dated back to the 13th century (piers of Broletto, Como); the first use for ornamental purpose was probably more recent, dating back to late 15th century (portal of Sacrestia vecchia, Santa Maria delle Grazie, Milan, 1499). These stones were almost entirely employed for ornamentation during the 17th and 18th centuries and were often used together in the same artefact. The geographical distribution of the works made of Arzo stones is very wide, including Como (Baptistery of Duomo, portal of Sant'Agostino, main altar of San Fedele, staircase of Villa Olmo); Varese (San Vittore); Induno Olona (San Giovanni Battista); Busto Arsizio (San Giovanni and Santa Maria di Piazza); Cairate (Monastery of Santa Maria Assunta); Saronno (Sanctuary of Madonna dei Miracoli); Locarno (San Francesco, Sant'Antonio, San Vittore); Brissago (Santi Pietro e Paolo); Lugano (San Lorenzo, San Rocco); Riva San Vitale (Santa Croce); Mendrisio (Parrocchiale); Torino (Carmine, Santi Maurizio e Lazzaro, San Domenico, Duomo, San Lorenzo, Santa Maria di Piazza, San Carlo, Consolata, Santissima Trinità); Varallo Sesia (Sacro Monte), Novara (Duomo, San Gaudenzio); Rome (San Pietro in Vaticano); Naples (San Domenico). In the late 18th century, the decline of the ornamentations due to the rise of the Neo-Classical style, caused a decrease of the fortune of Arzo stones. The use of Macchiavecchia, Broccatello and Rosso di Arzo continued in the 20th century, still with ornamental purpose; the last great work was the floor and the stairs of the new High Altar of the Milan Duomo (1986). In the last decades of the 20th century the quarries provided a scanty production of fireplaces and stairways. The use is different according to the stone variety: Macchiavecchia was more versatile, spanning from monolithic column shaft (about 3 m as maximum length) to moulded baluster for balustrade, from slab for flooring to slab for veneering, from thin slab for inlay to thick slab on tables or consoles sometimes in "open book style". Broccatello was used for handrails or moulded balusters, for veneering or inlay. Rosso was used mainly as slab for flooring or in rail for balustrades. Arzo stones are easy to work thanks to their calcareous composition, and the brightness of polished surfaces enhances the multicoloured ground.

DECAY MORPHOLOGIES: polished surfaces exposed to weathering leads to a strong chromatic alteration, transforming the different colours in a grey-pinkish hue. The presence of calcium carbonate also produces a strong corrosion with roughening and loss of the original surface.

Pietra di Cuvio

COMMERCIAL NAME: the name comes from a village of Valcuvia, on the north side of Campo dei Fiori

REFERENCES

Marmi italiani: Lombardia; Marmo di Cuvio (p. 39)

PETROGRAPHIC CLASSIFICATION: chemical sedimentary rock (limestone)

MACROSCOPIC FEATURES: uniform grey to whitish colour

MINERALOGICAL COMPOSITION: calcite

MICROSCOPIC FEATURES: uniform microcrystalline calcite

GEOLOGICAL SETTING: "Formazione a Conchodon" formation (Rhaetian, upper Triassic) of the sedimentary series of the Southern Alps. Outcrops are present in the area of Monte Campo dei Fiori (north of Varese).

GEOLOGICAL REFERENCES: CGI, Foglio 31 Varese (1932)

QUARRY: a small site is still visible on the provincial road between Castello Cabiaglio and Cuvio (100 m west to the bridge over the river). The outcrop includes a paleo-ground, called "Terra Rossa", a witness of the uplift (Rhaetian) before the marine transgression of Sinemurian (Calcari Selciferi Lombardi formation).

STONEWORK CATEGORIES: block

SURFACE FINISHING: polishing

USE IN ARCHITECTURE: the small size of the quarry does not allow it to supply large quantities of stone, so it is very difficult to recognize the use of this stone in architecture outside the local area

DECAY MORPHOLOGIES: erosion

Pietra di Malnate (Molera)

COMMERCIAL NAME: the name comes from a village south of Varese, but the stone is also known as "Molera" (a term of Milanese dialect meaning rock apt to make millstones)

REFERENCES

Breislak: chap. 8, Dei monti della Brianza; §93 Le psammiti di Malnate presso Varese

Salmojraghi: §564 Provincia di Como; Arenaria; Miocenica; Molera (p. 367)

PETROGRAPHIC CLASSIFICATION: clastic sedimentary rock (sandstone – arkose)

MACROSCOPIC FEATURES: brown to grey colour with bluish shades, fine to medium grain size

MINERALOGICAL COMPOSITION: quartz, plagioclase, potash feldspar, muscovite, amphibole; clay matrix and cement of calcite spar

MICROSCOPIC FEATURES: sub-angular clasts (according to the different levels, size varies from 0.2 to 2.0 mm or from 0.2 to 0.6 mm, some levels show clasts of a few centimetres) made of quartz, plagioclase, potash feldspar, muscovite and biotite, metamorphic rock fragments (polycrystalline quartz and muscovite); argillaceous cement fills the spaces among clasts

GEOLOGICAL SETTING: "Arenaria di Malnate" formation of the sedimentary series of the Southern Alps, pertaining to a turbiditic succession of Caenozoic (middle Miocene – upper Oligocene). The formation includes marly limestones in the lower part (Formazione di Chiasso); conglomerates and sandstones (Conglomerato di Como, Burdigalian – Chattian and Conglomerato di Lucino, Langhian – Burdigalian) in the upper one. The Arenaria di Malnate is positioned between the conglomerate of Como and the conglomerate of Lucino. The bed thickness varies from low to medium, thin clay interbeds separate the rock beds. The outcrop occupies a narrow area from Ròdero (north-east) towards Gornate Olona (south-west) close to the Italian–Swiss border.

GEOLOGICAL REFERENCES: CGI, Foglio 31 Varese (1932)

QUARRY: the main sites are located a few kilometres north of Malnate, along the valley of river Lanza; huge cavities were dug, without piers, in the rock outcrop

STONEWORK CATEGORIES: block, carving

SURFACE FINISHING: rough

USE IN ARCHITECTURE: the local use was pointed out, during the centuries, by piers, door jambs, windowsills, brackets, fireplaces and also millstones. The stone was employed at Castelseprio (pebbles in the ruined church of San Giovanni, 6th–7th centuries, and in the adjacent baptistery, 11th century) and Castiglione Olona (blocks and mouldings in some Renaissance buildings promoted by cardinal Branda Castiglioni in the early 15th century, Palazzo Branda, etc.). Molera matches the features of the Pietra Serena, the grey sandstone (Cretaceous) quarried around Florence (Tuscany) and used in the Renaissance architecture of Brunelleschi (Cappella dei Pazzi, San Lorenzo and Santo Spirito).

DECAY MORPHOLOGIES: erosion of the calcite cement, detachment of single clasts, disaggregation or scaling

Ceppo di Travedona

COMMERCIAL NAME: the name comes from a village south of Varese; the Milanese word "Ceppo" means "rock made of pebbles"

REFERENCES

Jervis: Provincia di Como, Circondario di Varese; §281 Travedona (p. 99)

Salmojraghi: §564 Provincia di Como; Calcari; Eocenici brecciformi (p. 366)

PETROGRAPHIC CLASSIFICATION: clastic sedimentary rock (clast-supported conglomerate)

MACROSCOPIC FEATURES: generally whitish colour with some dark spots, coarse grained

MINERALOGICAL COMPOSITION: calcite, dolomite, chert, quartz, plagioclase

MICROSCOPIC FEATURES: clast of sedimentary rocks (limestone, chert, dolostone), clasts of igneous rocks (granophyre, porphyryte), metamorphic rock (gneiss); calcite spar cement adheres to the clast surfaces filling the interstices among the clasts)

GEOLOGICAL SETTING: "Calcari a nummuliti" (nummulitic limestone) of the "Formazione di Ternate" (Eocene), including conglomerate and marly levels; this is the last term of the sedimentary series of the Southern Alps in western Lombardy. Outcrops are located in the southern hills of the Varese province in a territory featuring lakes (Monate and Comabbio) and glacial deposits.

GEOLOGICAL REFERENCES: CGI, Foglio 31 Varese (1932)

QUARRY: the main site is located at the northern tip of the outcrop area, east of Travedona. In particular, this formation was quarried (a wide site between Ternate and Travedona) to produce Portland cement thanks to the proper percentage of clay (marly interbedding) in the limestone beds.

STONEWORK CATEGORIES: block

SURFACE FINISHING: rough face

USE IN ARCHITECTURE: strictly local in masonries or as lintel and cornerstone

DECAY MORPHOLOGIES: erosion, sulphate skin formation

Tufo – Travertino (alabaster – travertine)

COMMERCIAL NAME: the name "tufo" is properly referred to volcaniclastic sediments; in this case the term refers to the high porosity of this calcareous deposit. The name Travertino refers to the stone of the same origin, quarried around Tivoli (east of Rome) and used by the Romans (*lapis Tiburtinus*).

REFERENCES

Salmojraghi: §564 Provincia di Como; Tufi; Calcarei (p. 366)

Marmi italiani: Lombardia; Travertino (p. 46)

PETROGRAPHIC CLASSIFICATION: non-clastic deposit of calcium carbonate precipitated from supersaturated water

MACROSCOPIC FEATURES. brownish colour, spongy appearance

MINERALOGICAL COMPOSITION: calcite

MICROSCOPIC FEATURES: elongated calcite crystals, plenty of irregular cavities

GEOLOGICAL SETTING: very small Quaternary deposit linked to water sources and concerning the southern part of Val Ganna, a few kilometres north of Varese, where Triassic dolomites outcrop

GEOLOGICAL REFERENCES: CGI, Foglio 31 Varese (1932)

QUARRY: the main site is located at "Grotte di Valganna" (Induno Olona), where several sources created a thick deposit of travertine together with a particular waterfall

STONEWORK CATEGORIES: blocks of different sizes

SURFACE FINISHING: rough face

USE IN ARCHITECTURE: the use features the medieval architecture with ashlars and voussoirs for windows and arches, thanks to the scarce weight. Some examples are still visible in the windows of the apse of San Giovanni in Castelseprio (5th–6th centuries) and in other similar buildings.

DECAY MORPHOLOGIES: disaggregation and erosion

NOTE: Alabaster deposits are connected to the travertine outcrops. A kind of alabaster was quarried at Frascarolo, a former castle west of Induno Olona, according to Jervis ("§279. Induno Olona. Alabastro orientale lionato, bianco latteo nuvolato ed onice biancastro e giallo scuro – calcite stalattitica. Trovasi presso il casale di Frascarolo. . . . Offre una bellissima pietra da ornamento") and Salmojraghi ("Alabastro. Induno Olona, Fraschirolo, monte Martica. Calcareo, giallo o bianco, variegate; giacimenti stalattitici"). It is now impossible to locate this site exactly, but it is worth it to note the so-called Bus de l'Alabaster (Cave of Alabaster), a karst cave in a dolomite outcrop (middle Triassic) located a few hundred metres north of the travertine outcrop of Grotte di Valganna (still in the territory of Induno Olona).

ADDENDUM

Other similar travertines, coming from different Lombard sites, were described by ancient authors.

Jervis: Provincia di Como, Circondario di Como; §271 Claino con Osteno (p. 98), §297 Inverigo (p. 103); Circondario di Lecco; §303 Civate (p. 104), §358 Maggianico (p. 114); Provincia di Bergamo, Circondario di Bergamo; §408 Predore (p. 123)

Salmojraghi: §564 Provincia di Como; Tufi; Calcarei (p. 366)

Pietre di Viggiù, Saltrio and Brenno

COMMERCIAL NAME: the name comes from the village north of Varese, a village of stone cutters, sculptors and architects. Other names come from nearby villages, as Saltrio (pietra di Saltrio) or Brenno Useria (pietra di Brenno).

REFERENCES

Scamozzi: part 2, book 7, chap. 8, Pietre vive (p. 199, line 14)

Amati: Nero di Saltrio; Pietra granita di Viggiù (Note – Alcuni altri marmi, book 2, chap. 7, p. 50)

Rondelet: book 1, sec. 2, art. 15. §260 (p. 197)

Cantù: Geografia fisica; Pietra di Viggiù; Pietra di Saltrio

Curioni: chap. 5, Marmi mandorlati e brecciati, §228, §229, §230 Viggiù (p. 59); Marmi grigi, §209 Saltrio (p. 57); Marmi giallicci, §232 Saltrio (p. 59); Marmi bianchicci. §255 Brenno (p. 63).

Jervis: Provincia di Como, Circondario di Varese; §282 Saltrio (p. 99), §283 Viggiù, §284 Brenno Useria (p. 100)

Salmojraghi: §564 Provincia di Como; Calcari; Compatti, del Lias inferiore (p. 365)

Marmi italiani: Lombardia; Pietre di Viggiù e di Saltrio (p. 44)

PETROGRAPHIC CLASSIFICATION: chemical sedimentary rock (dolomitic limestone; intraoosparite – grainstone)

MACROSCOPIC FEATURES: each group of quarries supplies different varieties of stones with different macroscopic and microscopic features

Viggiù: "Granitello" (calcarenite, grey colour and coarse grain size), "Comune" (calcarenite, grey colour, medium grain size), "Rossetto" (calcarenite, variegated grey and reddish, medium grain size)

Brenno: "Nocciola" (brownish colour), "Grigio" (grey colour), medium to fine grain size

Saltrio: "Grigio" (bottom; limestone grey colour, fine grain size, laminations with marly levels); "Rossetto" (calcarenite, compact, fine to medium grain size); "Piombino" (limestone, black or dark grey colour, fine grain size); "Cappello" (top; cherty limestone, yellowish colour, fine grain size, thin laminations with marly levels, cracks; unsuitable as building stone)

MINERALOGICAL COMPOSITION: calcite, dolomite, quartz, chert

MICROSCOPIC FEATURES

Viggiù (Granitello): packstone texture with clasts of irregular shape and interstitial cement; the grain size varies according to the different beds quarried. In particular: intraclasts (micrite); lithoclasts featuring dolomitic limestone with oolites sometimes broken (concentric structure and small nucleus of sparry calcite, diameter until 1 mm) pertaining to a previous formation called "Dolomia del Campo dei Fiori" (upper Triassic), eroded during the marine transgression at the origin of "Calcari selciferi" formation (lower Jurassic); peloids; rare bioclasts (echinoderm plates); quartz crystals; fine-grained iron oxides. Cement shows a drusy mosaic of euhedral sparry calcite (calcite crystals wrap the clasts and increase in size towards the centres of the interstices); irregular cavities as a result of the growth of the drusy mosaic of calcite cement.

Viggiù (*Comune*): similar to the Granitello and featuring small chert nodules, referred to the top of the calcarenite beds of "Calcari selciferi" formation

Brenno Useria ("Nocciola" and "Grigio" show the same features, but the Grigio variety is finer): packstone texture with clasts of irregular shape; calcite matrix and interstitial cement; intraclasts (micrite); lithoclasts featuring dolomitic limestone with oolites sometimes broken (round or elliptical shape, diameter until 0.6 mm). Cement shows a drusy mosaic of euhedral sparry calcite crystals.

Saltrio: a more compact texture in comparison with varieties of "pietra di Viggiù". "Grigio": intraclasts, chert, bioclasts. "Rossetto": intraclasts, chert, bioclasts, reworked oolites. "Piombino": intraclasts, chert, bioclasts; "Cappellaccio": intraclasts, chert.

GEOLOGICAL SETTING: "Calcari selciferi lombardi" formation (Sinemurian, lower Jurassic) of the sedimentary series of the Southern Alps, including calcarenites and cherty limestones. This formation marks a marine transgression over the Rhaetian sediments (Dolomia del Campo dei Fiori, oolitic limestone): the lower part of the formation (calcarenite beds) contains fragmentary oolites. The outcrop is spread along the Prealps, on the central territory of the province of Varese (Campo de Fiori), and it continues to the east until Lake Como and beyond (Triangolo lariano).

GEOLOGICAL REFERENCES: CGI, Foglio 31 Varese (1932)

QUARRY: many sites supplying slightly different stone materials were located in val Ceresio on the north-eastern side of province of Varese, near the Swiss border (Viggiù, Saltrio, Brenno Useria). Both in Viggiù and Saltrio there are many underground quarries (pits) following one another along the mountain slope; in Brenno this is a single underground quarry, now in a private garden. It is worth it to note the destruction of some ancient quarries at Saltrio, quarries located at the higher level on the mountain slope; the destruction is operated by a modern plant crushing the rock to make the gravel used in road foundation.

All quarries were exploited as pit quarries due to the small thickness of the calcarenite beds, over-bedded by 100 m of cherty limestone unable to make building stone. Therefore, the most important feature of these quarries are the huge pillars of rock left in place to support the roof; the front of excavation stands always at a lower level in comparison with the opening of the quarry pit because of the immersion of the rock bed. Therefore, taking away the stone material was very difficult, and it was aided by small wagons and a winch moved by dint of arms. Moreover, it was a priority to remove the water from the excavation front: now, at the end of the works, many quarries are completely flooded by ground water.

STONEWORK CATEGORIES: block, shaft, carving

SURFACE FINISHING: smoothing

USE IN ARCHITECTURE: the use spread from the quarry area to the nearby territory. The most significant example from the Middle Ages is visible in Varese: the partially decorated octagonal monolithic font (7th–8th centuries) in the Baptistery of San Giovanni

(12th–13th centuries). Since the Renaissance the stone has been extensively employed in western Lombardy: the façade of San Lorenzo in Lugano (early 16th century), the façade of the sanctuary dedicated to the Madonna dei Miracoli in Saronno (Buzzi, late 16th century); the porch and the statues of the sanctuary dedicated to the Madonnina in Prato (1678–86) together with several monolithic shafts of the church of San Vittore (late 17th century) both in Varese. The mixed use with Moltrasio limestone (coming from the same geological formation) is witnessed in Como: decorative elements and window frames of palazzo Natta (1735–44) together with masonries made of rough blocks of Moltrasio.

Use in Milan started at the end of the 15th century for the small columns surrounding the dome and the columns of the small cloister of Santa Maria delle Grazie. This little cloister is still recognizable, despite the strong chromatic alteration, the arrangement of a light-coloured shaft together with a dark-coloured base and capital or of a dark-coloured shaft together with a light-coloured base and capital. This alternating use of different coloured varieties of stone is also present in some house façades, built during the Renaissance and then dismantled and recomposed by the architect Luca Beltrami (late 19th century) in the north-eastern side of the main court (piazza d'Armi) of Castello Sforzesco. The use in Milan was increased throughout the 17th and the 18th centuries, mainly for the architectural elements associated with the plastered masonry, on the façades of private, public or religious buildings: San Paolo Converso (G.B. Crespi, 1613), palazzo Annoni (F. Ricchino, 1631), Santa Maria alla Porta (F.M. Ricchino, 1653), Trivulzio (G. Ruggeri, 1707–13), palazzo Litta (B. Bolli, 1752–63), theatre alla Scala (G. Piermarini, 1774–76), Santa Maria della Consolazione (G.B. Chiappa, 1836), San Francesco di Paola (E. Alemagna, 1891). In the second half of the 19th century the stone was also employed for portals, window frames and balustrades of balconies in dozens of apartment buildings. Its use ceased rapidly at the end of the 19th century, in coincidence with the diffusion of the artificial stone: a new material, made of Portland cement together with appropriately coloured sand or crushed stone, allowing architects to obtain complicated decorative elements at a lower price. In particular, it is worth it to note the resemblance (colour, texture) of the artificial stone with the natural one coming from Viggiù. The disuse of the quarry works originated with the emigration of stone cutters to North America. In particular, the craftsmen set in Vermont, where they contributed to the exploitation of the grey granite in the Barre area (Barre granite). One of the last architectural works was the Galleria Vittorio Emanuele II (G. Mengoni) in Milan: the stone was employed in the inner façades (pilasters, 1867) and in the monumental front on piazza del Duomo (cornices and decorative elements, 1878).

DECAY MORPHOLOGY: erosion till crumbling, sulphate skin formation, chromatic alteration

Loose deposits

COMMERCIAL NAME: there is no name for these materials (block, cobble, pebble). The term "gera" of the Lombard dialect was employed to indicate sand or gravel used to make mortar.

REFERENCES

Salmojraghi: §564 Provincia di Como. Erratici; Lago Maggiore, Val Tresa, Val Travaglia, Val Ganna, Val Cuvia, Varesotto (p. 366)

PETROGRAPHIC CLASSIFICATION: extremely variable according to the rock outcrops in the glacial basins

MACROSCOPIC FEATURES: the shape of pebbles and cobbles is linked to the lithological nature of the source rocks: igneous or massive sedimentary rocks make almost equant elements; metamorphic foliated rocks and thin-bedded sedimentary rocks make tabular or bladed elements. Finally, the roundness of the corners (degree of abrasion) witnesses the transport before the sedimentation. The sizes are conventionally fixed by sedimentary point of view: pebble up to 6.4 cm, cobble up to 25.6 cm, boulder beyond this dimension.

MINERALOGICAL COMPOSITION: metamorphic rocks (Alps – gneiss, paragneiss, micaschist, phyllite, green stone); igneous rocks (Alps – granite, diorite, ryolite) and carbonate or silicate sedimentary rocks (Prealps – limestone, dolomite, sandstone); minerals: quartz, silicates (feldspar, plagioclase, micas) and carbonates (calcite, dolomite)

GEOLOGICAL SETTING: Pleistocene deposits, mainly occurring where the Alpine valleys go into the plain, and erratic boulders left by glaciers on the mountain slopes or Olocene deposits of present-day rivers (alluvium)

GEOLOGICAL REFERENCES: CGI, Foglio 31 Varese (1932)

QUARRY: every river course was used to supply loose materials; these materials were dug from alluvial deposits of the current riverbeds (Ticino, between Somma Lombardo and Boffalora Ticino), into the water table (i.e. sites scattered around the Milan urban area). Erratic boulders present in lower parts of the hill slopes were cut on site to obtain dimension stones of suitable size.

STONEWORK CATEGORIES: original form without working, in some cases pebble or cobble were split in two parts. Normally, pebbles of the same size were laid in parallel rows with mortar beds. In some cases, only one lithotype (mainly gneiss) was employed and accurately chosen among the other ones.

SURFACE FINISHING: no finishing

USE IN ARCHITECTURE: Medieval religious buildings of western Lombardy show masonries made of pebbles, such as San Vittore (Arsago Seprio, 9th–11th centuries), baptistery (Arsago Seprio, 12th century – gneiss), Santi Pietro e Paolo (Brebbia, late 12th century – gneiss); San Giovanni (Castelseprio, 5th–6th centuries), baptistery (Castelseprio, 5th century), San Paolo (Castelseprio, 11th century), Santa Maria foris Portas (Castelseprio, 7th–8th centuries), San Michele (Gornate inferiore, 5th–6th centuries); Santi Primo e Feliciano (Leggiuno, 11th century), San Matteo (Malnate, 11th century and restored in the 18th century), baptistery (Riva San Vitale, Canton Ticino, 5th–9th centuries), San Donato (Sesto Calende, 12th century). During the Roman period, loose sediments were used to make Roman cement useful for building foundations: pebbles with homogeneous size and very thick mortar joints thanks to the round shape of pebbles. Some examples are visible in Milan: the amphitheatre (2nd century, via De Amicis), the Herculean baths (4th century, corso Europa) and the Imperial Palace (3rd–4th centuries, via Brisa).

DECAY MORPHOLOGIES: linked to the nature of the rocks forming the sediment; some pebbles or cobbles were already decayed through a natural process (feldspar alteration, called "ferrettizzazione") before setting in masonries

NOTE: Sand, gravel and clay were also quarried to make aggregates for mortars from the same kind of deposits. The clastic materials were selected (washing, sieving) in order to obtain the proper aggregate (filler) for different kinds of mortar. Geologically, three defined sub-surface units (Quaternary) are present in the river Po plain; the total thickness increases moving southward, from the Prealps to the plain. Gravels are predominant in the upper part of the plain sediments (middle Pleistocene – Holocene; glacio-fluvial deposits; about 50 m thick); sands are predominant in the middle part (Calabrian – Middle Pleistocene; fluvial deposits, about 50 m thick); clays are predominant in the deeper part (Calabrian; marine, transitional and continental deposits, more than 50 m thick).

Chapter 10

Stones from province of Verbano-Cusio-Ossola

Ossola, the alpine valley of river Toce, pertained to Duchy of Milan until the late 18th century, before the annexation to the Kingdom of Sardinia, but the trade of the building stones to Milan and Lombardy lasts till now.

IGNEOUS ROCKS

Granito di Alzo

COMMERCIAL NAME: The name comes from the village on the western shore of Lake Orta (Piedmont)

REFERENCES

Jervis: Provincia di Novara, Circondario di Novara; §252 Boletto, §253 Pella (p. 92)

Salmojraghi: §568 Provincia di Novara; Graniti; Verbano e Cusio (p. 371)

Marmi italiani: Piemonte; Granito bianco di Alzo (p. 21)

PETROGRAPHIC CLASSIFICATION: plutonic igneous rock (granite)

MACROSCOPIC FEATURES: white colour with black spots

MINERALOGICAL COMPOSITION: quartz, potash feldspar, plagioclase, biotite

MICROSCOPIC FEATURES: granular texture with anhedral crystals of quartz showing non-uniform extinction, potash feldspar with micro-perthitic texture or partial sericite alteration, zoned plagioclase crystals with polysynthetic twinning, tabular crystals of biotite

GEOLOGICAL SETTING: "Alzo – Roccapietra" pluton pertaining to "Graniti dei Laghi" (Hercynian magmatism), the host rocks are gneisses and micaschists of "Serie dei Laghi". The outcrop is located between the lower part of Val Sesia (west) and the lake Orta (east).

GEOLOGICAL REFERENCES: CGI, Foglio 30 Varallo (1927)

QUARRY: a big site was opened (mid-19th century) near Arzo (municipality of Pella), in a mountain overhanging the western shore of Lake Orta

STONEWORK CATEGORIES: block, slab

SURFACE FINISHING: bush hammering to polishing

USE IN ARCHITECTURE: this granite was firstly used to make wharfs for boat docking, piers for bridges, etc.; nevertheless, the architect A. Antonelli made forty-eight shafts for the high dome of San Gaudenzio (Novara, 1888). Finally, the use in Milan's architecture flourished in the 1930s: i.e. column shafts of the atrium of Palazzo dell'Arte (G. Muzio, 1932), eight big shafts in the public room of Palazzo delle Colonne (via Verdi, G. Greppi and G. Muzio, 1940).

DOI: 10.1201/9781003361008-11

DECAY MORPHOLOGIES: scaling; in particular, this granite is totally free from the rust spot phenomena caused by the oxidation of accessory minerals (iron sulphides)

Granito nero di Anzola

COMMERCIAL NAME: the name refers to a village of val d'Ossola and includes an erroneous petrographical classification (Granito Nero di Anzola)
REFERENCES
Marmi italiani: Piemonte; Diorite di Anzola (p. 24)
PETROGRAPHIC CLASSIFICATION: igneous rocks (amphibole gabbro)
MACROSCOPIC FEATURES: black colour with scattered whitish spots; medium grain size
MINERALOGICAL COMPOSITION: plagioclase, hornblende, olivine
MICROSCOPIC FEATURES: pleochroic prismatic crystals of hornblende with some pseudo-hexagonal cross-sections; crystals of plagioclase (labradorite) with polysynthetic twinning and curvilinear boundaries; anhedral crystals of olivine with irregular fractures
GEOLOGICAL SETTING: "Dioritico – kinzigitica" formation, featuring medium- to high-grade metapelites, including mafic layered intrusions (gabbro) and pertaining to the Ivrea-Verbano zone (Hercynian age). Outcrops from Valle Strona (Piedmont, south-west) to Canton Ticino (Switzerland, north-east).
GEOLOGICAL REFERENCES: CGI, Foglio 30 Varallo (1927)
QUARRY: the main site is located in the middle Val d'Ossola, on the right bank of the Toce River, 1 km south of the village of Anzola d'Ossola and close to Cuzzago, on the left bank
STONEWORK CATEGORIES: block, slab, carving, tombstone
SURFACE FINISHING: polishing
USE IN ARCHITECTURE: mainly thick slabs for cladding and for paving or moulded elements for portals since the last years of the 19th century. The most significant work is the precinct of the Sacrario ai Caduti Milanesi (Milano, piazza Sant'Ambrogio, 1922–29) with squared blocks, voussoirs and drums to make walls, arches and column shafts. Another important use was as tombstones in cemeteries.
DECAY MORPHOLOGIES: scaling

Granito di Baveno

COMMERCIAL NAME: the name comes from a village of the western shore of Lake Maggiore. The ancient term "Migliarolo" or "Miaròlo rosso" comes from the term "miarœu" of Milanese dialect, meaning "granite".
REFERENCES
Scamozzi: part 2, book 7, chap. 8, Pietre vive (p. 199, line 2)
Milizia-Antolini: part 1, book 4, chap. 12, art. 3, Ordine e gradazione dei marmi moderni, Rosso – Granito rosso del lago Maggiore
Rondelet: book 1, sec. 2, art. 7, "Migliarolo" (p. 87)
Amati: Granito (Note – Marmi moderni, book 2, chap. 7, p. 49)
Cantù: Geografia fisica; Granito rosso di Baveno (Stato Sardo)
Jervis: Provincia di Novara, Circondario di Pallanza; §261 Baveno (p. 93)
Salmojraghi: §568 Provincia di Novara; Graniti; Verbano e Cusio (p. 371)
Marmi italiani: Piemonte; Graniti roseo e rosso di Baveno (p. 22)
PETROGRAPHIC CLASSIFICATION: igneous plutonic rock (granite)

MACROSCOPIC FEATURES: pink colour with black spots; other varieties show white or dark red colour (see Addendum); medium grain size

MINERALOGICAL COMPOSITION: quartz, potash feldspar, plagioclase (oligoclase), biotite

MICROSCOPIC FEATURES: granular texture with anhedral crystals of quartz showing non-uniform extinction; potash feldspar with micro-perthitic texture or partial sericite alteration; plagioclase crystals with polysynthetic twinning and multiple zoning; crystals of biotite with tabular habit.

GEOLOGICAL SETTING: Hercynian pluton intruded into metamorphic rocks (gneisses and micaschists) of the "Serie dei Laghi". The pluton pertains to "Graniti dei Laghi". Outcrops in the north-westerm part (monte Camoscio, Baveno) of the Mottarone, a long ridge located in eastern Piedmont, between Lake Orta (west) and Lake Maggiore (east).

GEOLOGICAL REFERENCES: CGI, Foglio 30 Varallo (1927) and Foglio 31 Varese (1932)

QUARRY: the most important site, clearly visible at a distance, is located at "Pradera" between Baveno and Feriolo and close to the western shore of Lake Maggiore. The works started in the first decade of the 16th century and the quarries are still working. It is worth it to note the use of mines in quarry works: starting from the mid-19th century, an increasing number of bigger mines were exploded to obtain rapidly a great quantity of stone material proper to use. The system produced many damages to the rock body and to the nearby villages: according to Salmojraghi (1894) in 1885–86 four mines (total 20 metric tons of explosives) and a fifth mine containing 17 metric tons of explosives were employed to detach an amount of about 180.000 cubic metres of rock with single a boulder ranging from 1500 to 3000 cubic metres.

The pieces of stone reached the Lake Maggiore shore to the embankment; the ships, following the lake to the south and the course of river Ticino, went to Milan and Pavia through the artificial waterways (Naviglio Grande and Naviglio Pavese).

STONEWORK CATEGORIES: block, slab, monolithic shaft, moulding

SURFACE FINISHING: the traditional bush hammering for outside use was replaced, in modern times, by a perfect polishing

USE IN ARCHITECTURE: local use is witnessed in some buildings near the quarry area (Baveno, Santi Gervasio e Protasio, late 12th century). The granite of Baveno was ignored in Roman times: in this period, to make monolithic shafts were used granodiorites of Valtellina (Ghiandone) and granites coming from Mediterranean countries as the red granite of Aswan in southern Egypt (a shaft is visible in the central court of Castello Sforzesco, near the north-eastern wall) or as the "Granito violetto" from the Troad peninsula on the Aegean coast of Turkey (a shaft is still present on the right side of the façade of the church of Sant'Angelo). From the Middle Ages to the end of the 15th century, monolithic shafts were made using gneisses of the upper val d'Ossola (Serizzo, first building of the former Ospedale Maggiore in Milan) or the limestones as pietra di Viggiù (little cloister of Santa Maria delle Grazie, late 15th century). Other materials were: Ceppo conglomerate (piers made of superimposed drums in the court of palazzo Arcivescovile – Pellegrini, 1572–1604); pietra d'Angera (pilasters and columns made of superimposed drums in the façade of San Fedele – Pellegrini and Bassi, since 1569).

The use of granite started from the early 16th century, almost always coupling in the same building the pink granite (Baveno) with the white one (Montorfano). The use was spread in Milan, in Lombardy, in Northern Italy and in a lot of the Italian territory. To understand a larger employ in Lombardy than in Piedmont, it is necessary to consider the Val d'Ossola as belonging to the Duchy of Milan until the mid-18th century (Treaty of Aachen, 1748). Some examples of Milanese architecture are: Lazzaretto (porch – G. Palazzi, early 16th century and demolished in 1880); Orfanotrofio delle Stelline (porch of the central court – F. Mangone, early 17th century); Seminario Arcivescovile (porch and loggia – A. Trezzi and F. Mangone, 1602–10); Accademia di Brera (porch and loggia – F. Ricchino and others, 1627–65); Senato (porch and loggia – F. Mangone and F. Ricchino, 1608–35), Litta (porch – Ricchino, 1648). Baveno granite was also used in religious buildings, such as Santa Maria alla Porta (shafts of two orders – F. Ricchino, 1652–70); Sant'Alessandro (bases of pilaster and portal – Binago, Ricchino and others, from 1601 to 1710); Santa Maria della Passione (jambs and lintel of the main portal, about 1729).

The possibility to obtain firm monoliths spurred the research of enormous artefacts: two columns on the inner façade of Duomo (10.7 m high or 18 "braccia milanesi", 1637–45) or the never-made shafts planned by Pellegrini (1628) for the façade of Duomo (ten shafts 19.6 m or 33 "braccia" long, six shafts 15.5 m or 26 "braccia" long). The use of granite in Milan, during the 18th and 19th centuries, was mainly directed on the bases of buildings and on the architectural elements in combination with plastered masonries. Examples are Litta (façade – Merlo, 1752–63), Teatro alla Scala (Piermarini, 1778). Monolithic elements were still employed in the façade of San Fedele (column shafts – 1835).

Thick slabs of granite as a carriageway of city streets (early 19th century) were rapidly disused because of the scarce hardwearing; from then the more resistant Granito di San Fedelino was used on the carriageways.

Starting from the second half of the 19th century, Baveno granite was employed for the base of the buildings and, when the artificial stone spread on the façades (late 19th–early 20th century), the granite was one of few stone materials employed in the construction. Two remarkable exceptions are present in Milan: Galleria Vittorio Emanuele II (G. Mengoni, 1865/78 – the bases and the shafts on the two main fronts); Cimitero Monumentale (Maciachini, 1863/66 – piers of the loggias and the pilasters of the Famedio).

The use in Lombardy was crucial in Pavia and the province, as witnessed by the portal and vestibule of palazzo Mezzabarba (1728–30), columns of the porch of Teatro Fraschini (1771–73), shafts and piers of the Aula Magna of the university (1850), shafts of Salone del Mercato (1880–82) and architectural elements of the façade of palazzo Saporiti (Moraglia, 1828) in Vigevano.

The use in Piedmont was significant in Turin: shafts and piers (7.7 m high) of the entrance of Mole Antonelliana (A. Antonelli, 1862–89), together with the shafts of the third order (3.8 m high); the "new" façade of palazzo Carignano (G. Bollati, 1867); the monument of Vittorio Emanuele (sculptor P. Costa, 1899); and the colonnaded porch on the side of the Duomo (Novara), together with the columns of the roof, the fluted shafts of the porch (A. Antonelli, 1863–69). To make these fluted shafts (14.7 m high, 1.7 m diameter), the architect arranged twelve specially cut pieces to be disposed in three rows of four pieces each together with a thirteenth circular piece on the top; the cut among the pieces was hidden in the bottom of the flute.

The use of granite in Rome shows many significant works: the narthex of the quadripartite porch of San Paolo *extra muros* (shafts about 10 m high – L. Poletti, 1833–69); some shafts on the façade of the building of the former Istituto Poligrafico dello Stato (G. Burba, 1913). The most interesting occurrence concerns two obelisks (10.28 m long, more than 22 metric tons in weight) of villa Torlonia (via Nomentana, Rome, 1842).

The use of Baveno granite in the 20th century was continuous in Milan and abroad for column shafts, blocks, voussoirs, slabs of veneer, etc. The base of the Corriere della Sera building (L. Beltrami, 1903/04), the monumental entrance with columns of Università Cattolica (G. Muzio, 1929); the monumental entrance with piers and arches of Palazzo dell'Arte (G. Muzio, 1932); the portal of Palazzo Fidia (A. Andreani, 1930); the cladding of the tower of the former Palazzo dei Fasci milanesi (P. Portaluppi, 1935–40); the Palazzo Vittoria (D. Frisia, 1935). Together with several apartment buildings, such as via Senato 11 (M. Zanuso, 1947–50); corso Genova 4 (P. Bottoni, 1950), piazza Velasca 4 (Asnago and Vender, 1950–52), via Bossi 2 (P. Lingeri, 1956–57). A particular case is the via Ulrico Hoepli, a short street near the Galleria Vittorio Emanuele and the theatre alla Scala, where the whole buildings, completed in the sixties, show a cladding made of slabs of Baveno granite (i.e. the building by G. Muzio, via Hoepli 10, 1957). The building by Figini e Pollini (via Hoepli 5, 1955/57) shows the granite cladding in alternation with *béton brut*.

Another important field was the funerary architecture between the late 19th century and the first decades of the 20th century: i.e. "Edicola Pinardi" of Cimitero Monumentale (Riparto V, 92 – U. Stacchini, 1903/05) where the granite is worked according to unusual sinuous shapes.

Finally, it is worth to note the imitation of granite columns using brick masonry coated with plaster opportunely painted with pink ground and black spots: i.e. column shafts and piers on the nave of San Fedele (late 16th century).

DECAY MORPHOLOGIES: scaling or detachment of three-dimensional fragments, mainly in the lower parts of column shafts.

Addendum

A particular kind of granite showing the same features and origin of Baveno pink granite and resembling Aswan red granite also comes from the Mottarone area: it is called "Granito rosso imperiale" (red colour, pointed by black spots). Small quarry sites are present on the north-western side of Mottarone mountain (north of Omegna – Agrano).

The use in architecture flourished in the 20th century: in Milan it is worth it to note the base and the staircase of Arengario (G. Muzio, 1940).

Granito di Montorfano

COMMERCIAL NAME: the name comes from an isolated mountain (Italian "orphan mountain") of lower val d'Ossola, between Lake Maggiore and Lake Mergozzo. The ancient term "miaròlo bianco" comes from the term "miarœu" (granite) of the Milanese dialect.

REFERENCES

Scamozzi: part 2, book 7, chap 8, Pietre vive (p. 199, line 6)
Rondelet: book 1, sec. 2, art. 8, "Montorfano" (p. 90)
Amati: Granito (Note – Marmi moderni, book 2, chap. 7, p. 49)
Cantù: Geografia fisica; Granito bianco di Montorfano presso Baveno
Jervis: Provincia di Novara, Circondario di Pallanza; §251 Mergozzo (p. 88)
Salmojraghi: §568 Provincia di Novara; Graniti; Verbano e Cusio (p. 371)
Marmi italiani: Piemonte; Granito bianco del Mont'Orfano e di Mergozzo (p. 21)

PETROGRAPHIC CLASSIFICATION: igneous plutonic rock (granite)

MACROSCOPIC FEATURES: mainly white colour with black spots; medium grain size

MINERALOGICAL COMPOSITION: quartz, K-feldspar, plagioclase (oligoclase), biotite and iron sulphides (pyrrhotite)

MICROSCOPIC FEATURES: granular texture with anhedral crystals of quartz showing non-uniform extinction, K-feldspar with micro-perthitic texture, zoned plagioclase crystals with polysynthetic twinning, crystals of biotite with tabular habit

GEOLOGICAL SETTING: the pluton of Montorfano is a Hercynian pluton embedded into metamorphic rocks (gneisses and micaschists) of the "Serie dei Laghi". Together with the pluton of Mottarone (Baveno granite) it pertains to "Graniti dei Laghi". Outcrops on Montorfano, a mount with almost circular shape (axes 2.4 km and 1.7 km) between Lake Mergozzo (north) and the Toce river course (south) close to the western shore of Lake Maggiore.

GEOLOGICAL REFERENCES: CGI, Foglio 30 Varallo (1927) and Foglio 31 Varese (1932)

QUARRY: mainly located on the southern side of the Montorfano. The works started in the first decades of the 16th century, and the quarries are still working. The stones were transported on boats through Lake Maggiore, river Ticino and the waterway to Milan (Naviglio Grande) and Pavia (Naviglio Pavese).

STONEWORK CATEGORIES: block, slab, monolithic shaft, lintel, door jamb

SURFACE FINISHING: the traditional bush hammering for outside use was replaced, in modern times, by a perfect polishing

USE IN ARCHITECTURE: the medieval use was only local, as in Gravellona Toce (i.e. San Maurizio, 12th century; the ruined Castello del Motto, 12th century)

An extensive use started in the early 16th century in the same way of Baveno – Mottarone pink granite, but to a lesser extent. Some Milanese examples are the huge pilasters with base and capital together with the entablature of Cappella Trivulzio (Bramantino, since 1512); some shafts of the portico of Lazzaretto (early 16th century); the shafts of the cloister (early 16th century) of the former monastery of San Pietro in Gessate, now included in the "Liceo Leonardo da Vinci". The use of Montorfano granite was ample in the 17th century also: shafts and obelisks of the façade of Sant'Angelo (early 17th century); shafts of porch and loggia of the cloisters (middle 17th century) of the former monastery of San Vittore, now Museo della Scienza; shafts of porch and loggia of the main court (G. Mangone and F. Ricchino, 1624–35) of the former Ospedale Maggiore, now Università degli Studi.

After a period of decrease (late 17th century–mid-18th century), the use of granite to make column shafts was once again ample during the redevelopment of the late 18th century when neo-classical style was rising. Some Milanese examples are shafts of the former Caserma dei Vèliti (G. Rossi, 1811); shafts of the Ospedale Fatebenefratelli (P. Gilardoni, 1836, demolished in 1936) and of the Ospedale Fatebenesorelle (G. Alvisetti, 1836).

The coupling of white and pink granite to enhance the chromatic difference was a particular matter: the landings of the staircase of palazzo Reale (G. Piermarini, 1778) show squared slabs of Montorfano granite in alternating with squared slabs of Baveno granite; in this staircase the steps were made using the pink granite of Baveno. A coupling of granite is still perceptible in the steps of the double staircase of Palazzo di Brera (G. Piermarini, 1774), despite the changes made over the years.

White granite was also used in Piedmont during the 19th century: i.e. the colonnade of Palazzo Orelli in Novara (L. Orelli, 1817–40) and the bases of the columns (Baveno granite) in the porch of the Duomo in Novara (A. Antonelli, 1863–69).

The most significant use of white Montorfano granite out of Lombardy was the reconstruction of San Paolo *extra muros* (Rome) after the fire of 1823. All the original shafts (Pavonazzetto and other marbles) of the nave and aisles were slowly replaced by shafts of Montorfano (P. Belli and L. Poletti, since 1825). Original documents report strong competition with the granite suppliers of Elba Island. Two bigger shafts (almost 13 m high) were set out (1829) to support the arch of Galla Placidia, between the naves and the presbytery. Finally, the Montorfano granite was employed in the quadripartite porch in front of this basilica (F. Vespignani, 1892).

In the 20th century the stone was still employed with structural purposes: some shafts of the naves of Madonna di Lourdes (Milan, A. Campanini, 1902) together with other granite shafts. The use for façade cladding started at the same time: Casa Cirla (G. Sommaruga, 1908); Palazzo Fidia (A. Andreani, 1924–30); Palazzo RAI (G. Ponti, 1929).

DECAY MORPHOLOGIES: apart from the scaling as other silicate rocks, white granite shows brownish stains in relation with iron sulphides associated with biotite laminae; this peculiar phenomenon strongly influences the surface aspect. On a vertical surface, the stains show an elongated shape (axis about 3–4 cm) with the top corresponding to the sulphide crystals; on a horizontal surface, the stains show an almost circular shape (diameter 1–2 cm) around the sulphide crystals. The oxidation process of iron sulphide involves the oxidation of ferrous iron to ferric iron through the oxygen of the atmosphere and of meteoric waters, ferric iron precipitates as rust, first as amorphous iron hydroxide followed by the slow crystallization of goethite, ochreous brown in colour.

Granito di Mergozzo (Ausolo)

COMMERCIAL NAME: the name comes from a village in the lower val d'Ossola, west of Lake Maggiore. Another name refers to the toponym included in the municipality of Mergozzo.

REFERENCES

Marmi italiani: Piemonte; Granito verde e verde rosso macchiato del Montorfano (p. 21)

PETROGRAPHIC CLASSIFICATION: igneous plutonic rock (granite)

MACROSCOPIC FEATURE: greenish colour with some white spots

MINERALOGICAL COMPOSITION: albite, chlorite, quartz, sericite

MICROSCOPIC FEATURES: granular texture of plagioclase (albite), chlorite and quartz together with apatite and titanite

GEOLOGICAL SETTING: this "green granite" pertains to the Hercynian pluton of Montorfano white granite ("Graniti dei Laghi"), and it is the result of a transformation of a pre-existing intrusive rock. The outcrop occupies a narrow zone on the northern part of the pluton; the contact with the white granite is gradual through a thick zone featuring a pale pink granite.

GEOLOGICAL REFERENCES: CGI, Foglio 30 Varallo (1927)

QUARRY: a small site was opened in the northern side of Montorfano, facing Lake Mergozzo; the quarries of Granito di Montorfano are located on the opposite side of Montorfano.

STONEWORK CATEGORIES: block, slab

SURFACE FINISHING: bush hammering to polishing

USE IN ARCHITECTURE: the use started in the 1930s, mainly for public or corporate buildings in Milan: i.e. the base of the Uffici Comunali (via Larga), the base of former

ENEL headquarters (via Ceresio), the stairways of Stazione Centrale. In the last decades of the 20th century, the stone was also used to make slabs for cladding (i.e. apartment buildings at via Prina 1, via Annunciata 2, via Cappuccini 11). Finally, this granite was used for her own sculptures by the Danish artist Eva Sørensen (1940–2019), exhibited in the Danish Pavilion at "XL Biennale Arti Visive" (Venice, 1982).
DECAY MORPHOLOGIES: scaling

METAMORPHIC ROCKS

Beola

COMMERCIAL NAME: the term refers to the Latin name of the quarry site (Béura)
REFERENCES
Rondelet: book 1, sec. 2, art. 15, §258 (p. 196)
Cantù: Geografia fisica, Pietra di Bévola
Jervis: Provincia di Novara, Circondario di Domodossola; §244 Beura (p. 86), §245 Cardezza, §246 Villadossola, §247 Pallanzeno (p. 87)
Salmojraghi: §568 Provincia di Novara; Gneiss; Val d'Ossola. Beole (p. 371)
Marmi italiani: Piemonte; Gneiss di Beura, Villadossola, Vogogna (p. 34)
PETROGRAPHIC CLASSIFICATION: regional metamorphic rock (gneiss) of the Pennidic units; fine to medium grain size (0.05–1 mm or 0.2–3.0 mm)
MACROSCOPIC FEATURES: mainly greyish colour featuring a thin alignment of dark strips
MINERALOGICAL COMPOSITION: quartz, K-feldspar, muscovite
MICROSCOPIC FEATURES: equant composite aggregates of quartz with undulose extinction; equant feldspar crystals; alignments (foliation) of muscovite and biotite tabular crystals
GEOLOGICAL SETTING: "Falda del Monte Rosa", an upper Pennidic unit made of gneisses coming from granite metamorphism
GEOLOGICAL REFERENCES: CGI, Foglio 15 Domodossola (1938)
QUARRY: the main site is located near the village of Bèura (Domodossola, upper Val d'Ossola, north-eastern Piedmont). Different varieties are traded according to the colour and the texture: Bianca or Ghiandonata (quarry Pianasca), Grigia (Trontano), Favalle (Crevoladossola).
STONEWORK CATEGORIES: slab (step for staircase, platform for balcony)
SURFACE FINISHING: quarry split face
USE IN ARCHITECTURE: the use of wide and thin slab, with high resistance to wear, spreads from the quarry area around Domodòssola to Milan and elsewhere. Quadrangular slabs were present on the roofs; rectangular slabs, placed vertically and fixed by a clamp, covered the lower part of the masonries to masque the capillary rise and to avoid plaster decay (i.e. the base of the former "Pensionato Marianum" of the Università Cattolica, G. Muzio, 1938). Beola slabs were also widely employed to make staircases in hundreds of apartment buildings, thanks to its durability and non-slippery surface: stairs and landings were made of slabs a few centimetres thick. The same kind of wide and thick slab was also largely employed in balconies (i.e. the balcony of the court of the Orologio at palazzo Litta); brackets made of granite or of iron, together with iron railings, were employed to support these slabs.
DECAY MORPHOLOGIES: scaling, exfoliation

Grigio del Boden

COMMERCIAL NAME: the name comes from a sanctuary (called “Madonna del Boden”) a few kilometres south-west of Orvanasso in the lower val d’Ossola. The term “marmo bastardo” (bastard marble) was employed by Scamozzi (“L’idea dell’architettura universale”, Venice, 1615).

REFERENCES

Scamozzi: part 2, book 7, chap. 5, Marmi e misti (p. 188, line 57)

Marmi italiani: Piemonte; Marmi di Ornavasso (Marmo granitico del Boden) (p. 30)

PETROGRAPHIC CLASSIFICATION: regional metamorphic rock (silicate marble)

MACROSCOPIC FEATURES: light grey colour, weak alignment featuring darker silicate veins associated with pyrite crystals; medium to fine grain size

MINERALOGICAL COMPOSITION: calcite, quartz, phlogopite, tremolite-actinolite, pyrite and opaque

MICROSCOPIC FEATURES: calcite crystals with polysynthetic twinning in alternating with quartz (single or monocrystalline and composite or polycrystalline) with undulose extinction; slightly pleochroic tabular crystals of phlogopite; polygonal crystals of opaque minerals

GEOLOGICAL SETTING: marble lenses in the “Dioritico – kinzigitica” formation (medium- to high-grade metamorphic rocks) pertaining to Ivrea-Verbano zone (Hercynian age). The outcrop occupies a narrow area between Biella (south-west) to Locarno (Switzerland, north-east), the area is wider between Val Sesia and Val d’Ossola.

GEOLOGICAL REFERENCES: CGI, Foglio 30 Varallo (1927)

QUARRY: some sites along the road from Ornavasso to the sanctuary of Madonna del Boden (a quarry face is still visible immediately below the sanctuary) on the western slopes of lower val d’Ossola

STONEWORK CATEGORIES: block, slab, small shaft, base and capital, carving

SURFACE FINISHING: polishing

USE IN ARCHITECTURE: the stone is present at the Santuario della Madonna del Boden, close to the quarry area, as steps and door frames. The main use was to make bases for columns, exploiting the great resistance of stone pieces cut along the planes of foliation and set with these planes perpendicular to the load. The bases were always coupled with granite shafts (both Montorfano or Baveno) and Viggiù capitals in porches and loggias of Milan’s buildings since the early 16th century. Another use was as a slab for cladding: i.e. the façade of the church of Monastero Maggiore (F. Piròvano, 1574–81); the façade of Santa Maria dei Miracoli (G. Alessi and M. Bassi, 1556–80). Moulded elements were also used for portals: i.e. Cappella Trivulzio (Bramantino, 1512); Palazzo Arcivescovile (piazza Duomo 16, Pellegrini, late 16th century). Monolithic elements were employed as a pier for the quadrilateral porch of Santa Maria dei Miracoli (C. Cesariano, early 16th century); for the ionic cloister of the Monastery of Sant’Ambrogio (C. Solari, circa 1513); as column shaft in the so-called Falconiera, a narrow loggia of Castello Visconteo (Vigevano – late 15th century). This kind of use soon pointed out a phenomenon of cracking according to the main foliation of the stone elements (in the case of column or pier shafts, the cracks go from the top to the bottom); metallic rings were set out to prevent a complete collapse of these architectural elements (i.e. the ionic cloister of the Monastery of Sant’Ambrogio).

DECAY MORPHOLOGIES: disaggregation, sulphate skin formation; rust stains in the form of strips along the veins containing pyrite crystals; cracking along foliation

Marmo di Candoglia

COMMERCIAL NAME: the name comes from a village in the middle val d'Ossola
REFERENCES
Scamozzi: part 2, book 7, chap. 5, Marmi e misti (p. 188, line 44)
Rondelet: book 1, sec. 2, art. 11, §141 (p. 138)
Amati: Ornavasso e Candoglia (Note – Marmi moderni, book 2, chap. 7, p. 50)
Cantù: Geografia fisica; Marmo di Gandoglia
Jervis: Provincia di Novara, Circondario di Pallanza; §251 Mergozzo (p. 88)
Salmojraghi: §568 Provincia di Novara; Calcari saccaroidi (p. 369)
Marmi italiani: Piemonte; Marmo di Candoglia (p. 29)
PETROGRAPHIC CLASSIFICATION: metamorphic limestone (marble)
MACROSCOPIC FEATURES: white or pink colour, grey irregular veins; medium grain size
MINERALOGICAL COMPOSITION: calcite (main component); quartz, muscovite, pyrite, hematite, phlogopite (accessories)
MICROSCOPIC FEATURES: equi-granular calcite crystals with irregular boundaries, rhombohedral cleavage and polysynthetic twinning; silicate veins with pyrite crystals and fine-grained calcite crystals; rounded quartz crystals
GEOLOGICAL SETTING: marble lenses included into "Dioritico – kinzigitica" formation, medium- to high-grade metamorphic rocks pertaining to the Ivrea-Verbano zone (Hercynian age). Small outcrops are distributed from Piedmont to Canton Ticino (southern Switzerland) or from Valle Strona (south-west) to val Leventina (north-east).
GEOLOGICAL REFERENCES: CGI, Foglio 30 Varallo (1927)
QUARRY: some sites are located near Candoglia (municipality of Mergozzo, southern val d'Ossola) at different elevation on the slopes of Corni di Nibbio (left bank of river Toce). The most important quarry is called "Cava madre" (about 580 m a.s.l.): more than 80 m long, 25 m wide and about 40 m high; the cavity was provided (C. Ferrari da Passano, 1969) by a series of reinforced concrete arches to avoid failures of the rock body; marble blocks are moved from inside using a crane connected to the concrete arches. Marble reached Milan (about 85 km south-east) by boat using the river Toce, Lake Maggiore until Sesto Calende, the river Ticino (for about 23 km south of Sesto Calende) and the Naviglio, a waterway dug out since the late 12th century between Tornavento and Milan – Porta Ticinese; the length of Naviglio is about 50 km. The water transport was disused since the end of World War II and replaced by truck transport.
STONEWORK CATEGORIES: block, slab, small shaft, carving, statue
SURFACE FINISHING: polishing
USE IN ARCHITECTURE: the first use is reported in Roman period: some stelae at Milan (Museo del Castello Sforzesco) and Novara (Museo archeologico). The equestrian monument of the Duke of Milan Barnabò Visconti (Milano, Castello Sforzesco – mid-14th century) witnesses the first medieval use of this marble in the Milanese area; a few years later the marble was exclusively devoted to the construction of the cathedral (Duomo) according to the edict of Gian Galeazzo Visconti Duke of Milan (1387). The quarries worked for centuries to supply the countless statues, reliefs, decorative elements, etc. forming the Duomo, and they are still working to supply pieces of marble in order to maintain the aspect of the whole construction.
This kind of use exclusively reserved to one building only left a few materials for other edifices; the most important of these surprising employs is the church of the Certosa of

Pavia, built in 1396 for the will of Gian Galeazzo Visconti: the marble was utilized for blocks, decorative elements, statues and floor tiles. Other accidental uses are the moulded cornice of the base of the façade of San Petronio (Bologna – early 15th century); the lions of the southern porch of Santa Maria Maggiore (Bergamo – early 15th century); the squared tiles on the geometric cladding of Cappella Colleoni (Bergamo, G.A. Amadeo, 1472–76). In the 20th century the use for cladding of Casa Wassermann (Milan, P. Portaluppi, 1934–36) is noticeable.

The use for sculptures ran parallel to the use for buildings: different statues made of Candoglia are present not only at Duomo but also on the façade on the church of Certosa (16th century, Pavia). The marble was employed for several statues of Cimitero Monumentale (late 19th–early 20th centuries) and it was an important material for the sculptor Adolfo Wildt: i.e. the group "Il Santo, il Giovane, la Saggezza" (a fountain in the garden of Villa Reale, via Palestro, 1912) shows three statues originally carved in two blocks of marble (about 2.68 x 2.50 x 1.20 m).

DECAY MORPHOLOGIES: surface erosion, calcium sulphate skin formation

Marmo di Ornavasso (Rosa Valtoce)

COMMERCIAL NAME: the name comes from a village of lower val d'Ossola, on the western slope; the name "Rosa Valtoce" was employed in the 20th century

REFERENCES:

Amati: Ornavasso e Candoglia (Note – Marmi moderni, book 2, chap. 7, p. 50)

Cantù: Geografia fisica; Marmo di Ornavasco

Jervis: Provincia di Novara, Circondario di Pallanza; §249 Ornavasso (p. 87). This marble is wrongly reported about the capitals and the bases of the columns of the main courtyard of Ospedale Maggiore (Milano); actually, a direct observation shows capitals made of Viggiù limestone and bases made of Grigio Boden.

Salmojraghi: §568 Provincia di Novara; Calcari saccaroidi (p. 369)

Marmi italiani: Piemonte; Marmi di Ornavasso (Rosa Valtoce) (p. 30)

PETROGRAPHIC CLASSIFICATION: metamorphosed limestone (marble)

MACROSCOPIC FEATURES: white colour (grain size 1.0 mm) or pink to orange colour (grain size 2.0 mm); grey thick veins. In many cases, white, pink, orange in alternating with grey strips gives to the cut surfaces a zoned aspect.

MINERALOGICAL COMPOSITION: calcite; quartz, muscovite, pyrite

MICROSCOPIC FEATURES: heteroblastic calcite crystals featuring irregular boundaries, rhombohedral cleavage and polysynthetic twinning; silicate veins with pyrite crystals; rounded quartz crystals

GEOLOGICAL SETTING: marble lenses included in "Dioritico – kinzigitica" formation, medium- to high-grade metamorphic rocks pertaining to the Ivrea-Verbano zone (Hercynian age). The outcrop spread from Biella (south-west) to Locarno (Switzerland, north-east).

GEOLOGICAL REFERENCES: CGI, Foglio 30 Varallo (1927)

QUARRY: different galleries were dug on the mountain, on the right bank of the river Toce, opposite the quarries of Candoglia marble. A quarry (Moschini), with a long tunnel to reach the marble lens, was opened at a curve of the road from Ornavasso to the Sanctuary of Madonna del Boden, just above the Sanctuary of Madonna della Guardia. The blocks were carried, as at Candoglia quarries, for some 10 m, to the Toce riverbank, then transported by boat through Lake Maggiore, the Ticino River and the waterways to Milan and Pavia.

STONEWORK CATEGORIES: block, carving, statue
SURFACE FINISHING: polishing
USE IN ARCHITECTURE: this marble was employed at the same time and for the same purpose as Candoglia marble, both in the cathedrals of Milano and Pavia. A good example of the different aspect of this marble is the floor of the Sanctuary of Madonna del Boden: the marble slabs vary from pink to grey with a dozen arrangements of dark veins. The quality is often scarce (variegated colour, very thick veins, coarse grain size) compared with the quality of Candoglia marble. The marble was employed, in Milan during the 20th century, for cladding, both outside and inside, in different public buildings (i.e. San Camillo de' Lellis – S.M. Chiappetta, 1902; Cassa Assicurazioni Sociali – M. Piacentini, 1929–31; "la Rinascente" department store – F. Reggiori, 1950) and several apartment buildings (i.e. Casa Feltrinelli, A. and L. Belgioioso, 1935; casa Wassermann, P. Portaluppi, 1936). Another kind of use was the funerary architecture at Cimitero Monumentale: i.e. Edicola Pirelli (Riparto IX, 541 – L. Beltrami, 1919/21) with two shafts on the portal, together with bases and capitals made of pietra d'Oira.
DECAY MORPHOLOGIES: erosion, sulphate skin formation, rust stain linked the dark veins with iron sulphide minerals. The stain phenomenon is strictly related to the cut of each stone element: if the cut is perpendicular to the veins, almost circular brown stains are visible around the iron sulphide minerals; if the cut is parallel to the veins, the coalescence of stains produces a brown nuance on the whole surface of the stone element. A pregnant example of this marble decay is visible at the parish church (San Nicola) of Ornavasso (late 16th century) where the rust affects the veins of each block of the façade and the white colour is faded.

Marmo di Crévola

COMMERCIAL NAME: the name comes from a village in the upper Val d'Ossola (Crevoladòssola)
REFERENCES
Scamozzi: part 2, book 7, chap. 5, Marmi e misti (p. 188, line 52)
Milizia-Antolini: part 1, book 4, chap. 12, art. 3, Ordine e gradazione dei marmi moderni, Bianco – Marmo bianco di Crevola
Amati: Crevola (Note – Marmi moderni, book 2, chap. 7, p. 50)
Cantù: Geografia fisica; Marmo di Crèvola
Jervis: Provincia di Novara, Circondario di Domodossola. §229 Crevola d'Ossola (p. 83)
Salmojraghi: §568 Provincia di Novara; Calcari saccaroidi (p. 369)
Marmi italiani: Piemonte; Marmo di Crevola d'Ossola (p. 30)
PETROGRAPHIC CLASSIFICATION: regional metamorphic rock (dolomite marble)
MACROSCOPIC FEATURES: whitish to greyish ground furrowed by dark, dense veins with sinuous trend; medium to fine grain size
MINERALOGICAL COMPOSITION: dolomite, phlogopite
MICROSCOPIC FEATURES: euhedral crystals of dolomite (0.4–0.8 mm) with polysynthetic twinning and rhombohedral cleavage and some intergranular fissures; euhedral crystals of dolomite (0.2–0.6 mm) with polysynthetic twinning and rhombohedral cleavage together with quartz as sub-rounded crystals (0.2–0.6 mm) or as polycrystalline aggregates (0.6–1.2 mm); euhedral crystals of dolomite (0.2–0.3 mm) with a clear alignment together with thin tabular crystals of phlogopite (0.2–0.3 mm); in some cases phlogopite crystals are disposed to form irregular veins [Palissandro].

GEOLOGICAL SETTING: marble lenses included in the "Falda di Monte Leone", a lower Pennidic unit (pre-Triassic crystalline basement) featuring orthogneisses

GEOLOGICAL REFERENCES: CGI, Foglio 15 Domodossola (1938)

QUARRY: north of Crevoladossola (upper Val d'Ossola), at the mouth of val Divedro. The quarry is still working and now are sold some varieties called "Palissandro classico", Oniciato, Bronzetto, Bronzo, Bronzo brown, Platino, Tigrato

STONEWORK CATEGORIES: block, slab, shaft

SURFACE FINISHING: polishing

USE IN ARCHITECTURE: the use is well documented around the quarries (i.e. some shafts of the porches in Domodossola). A significant Milan employ is represented by eight monolithic shafts (more than 10 m high) of the Arco della Pace (L. Cagnola, 1807–38); the capitals were made of white Apuanian marble, sculpted by C. Cattori (Canton Ticino, †1826). The larger employ of this marble is visible into Duomo of Pavia (since the 15th century), with eight huge piers under the dome, four piers in the vestibule and six piers in the apses (two per apse); each pier was made using many superimposed blocks with mouldings. Crèvola marble was also employed for masonries or decorative elements.

The marble had different uses in Milan during the 20th century, mainly for cladding: the mezzanine of a building (via Cosimo del Fante 16 – G. Brusa, 1923); the capitals of the porch of the Planetario Hoepli (Giardini Pubblici – P. Portaluppi, 1929–30); the lower floor of Assicurazioni RAS (via delle Asole – P. Portaluppi, 1935–38); the lower part of the Malugani building (piazza Repubblica – G. Muzio, 1936); the upper floor of casa Portaluppi (via Morozzo della Rocca – P. Portaluppi, 1935–39).

DECAY MORPHOLOGIES: erosion till crumbling; sulphate skin formation

Pietra d'Oira

COMMERCIAL NAME: the name comes from a village on the western shore of Lake Orta

REFERENCES

Jervis: Provincia di Novara, Circondario di Novara; §254 Nonio (p. 92)

Salmojraghi: §568 Provincia di Novara; Ofioliti; Marmo d'Oira (p. 370)

Marmi italiani: Piemonte; Verde Oira (p. 25)

PETROGRAPHIC CLASSIFICATION: metamorphic rock (meta-peridotite)

MACROSCOPIC FEATURES: dark green colour with some lighter irregular veins; medium grain size

MINERALOGICAL COMPOSITION: lizardite, clinochlore, olivine, pyroxene

MICROSCOPIC FEATURES: large olivine crystals crossed by irregular veinlets serpentine (antigorite)

GEOLOGICAL SETTING: lenses of metamorphosed mafic and ultramafic rocks (meta-basalts, meta-gabbroes, meta-ultramafites) together with amphibolites, coming from volcano-clastic sediments along the contact between "Scisti dei Laghi" and "Strona-Ceneri zone" (paragneiss and fine-grained gneiss). These two sub-units form the geologic unity called "Serie dei Laghi", outcropping west of Lake Maggiore, from Lake Orta (south-west) to Canton Ticino (Switzerland, north-east).

GEOLOGICAL REFERENCES: CGI, Foglio 30 Varallo (1927)

QUARRY: a small site is located on the western shore of Lake Orta (Cusio), near the village of Oira (municipality of Nònio, province of Novara)

STONEWORK CATEGORIES: block, carving

SURFACE FINISHING: bush hammering to polishing

USE IN ARCHITECTURE: the ancient local use is witnessed by the sculpted ambo in the basilica of San Giulio in an island (Isola San Giulio) of Lake Orta (municipality of Orta San Giulio). This ambo, located near the fourth pier of the left nave, shows four columns and the railing carved with the symbols of the four evangelists and allegorical figures (about 1110–1120). Examples of use until the 19th century are almost lacking: the statue of a lion, set on the top of the column (1626, made of superimposed drums of Ceppo) in front of the façade of San Babila in Milan. Starting from the late 19th century, the most important use seems to be the replacement of badly decayed dark-coloured stone materials. These replacements involved the most representative buildings of the Lombard architecture during the major restoration works carried out in the first decades of the 20th century. There are well-documented cases: cornices and mouldings of the façade of the church of Certosa (Pavia), made of black Saltrio limestone, were replaced by moulded elements of "pietra d'Oira"; ashlar of black limestone (probably Saltrio limestone) of the façade of Duomo (Monza) were replaced by block of "pietra d'Oira". Other use in Milan buildings is probably referred to conservation works: square panels of the façade of San Raffaele; the lintel of the main portal together with some elements of the columns of the façade of Santa Maria delle Grazie. The stone was also employed to make capitals of re-assembled shafts in the western court (Cortile dei Canonici) of the basilica of Sant'Ambrogio after the destruction of World War II (1943). Another field of use was the funerary architecture and decoration until the first decade of the 20th century (Milan, Cimitero Monumentale): i.e. the Edicola Izar (Riparto IV, 86 – sculptor F. Bialetti) composed by an arch of pietra d'Oira standing on two bases of Sienite della Balma carrying bronze statues. Moreover, Cimitero Monumentale (C. Maciachini, 1863–66) shows a very particular use of the stone, already reported by ancient authors: a few superimposed hollow cylinders, about 50 cm long, are inserted into one another, using an apposite joint on the tip to favour the connection, and set as a gutter along the façades.

DECAY MORPHOLOGIES: artefacts of this stone exposed to the atmosphere are subject of a strong chromatic alteration: the dark green surfaces fade to a very light green. This is caused by a selective chemical leaching of magnesium ions from the microstructure of the serpentine in the first 0.5 mm of the surface; a surface re-crystallization of silicon oxide also takes place. This phenomenon does not remarkably affect the resistance of the stone material, apart from the formation of micro-cracks and the increasing detachment of thin scales.

Serizzo dell'Ossola

COMMERCIAL NAME: the name comes from the term "Sarìzz" of the Milanese dialect, meaning "schist"

REFERENCES

Scamozzi: part 2, book 7, chap. 8, Pietre vive (p. 199, line 9)

Jervis: Provincia di Novara, Circondario di Domodossola; §231 Varzo, §232 Preglia, §234 Caddo, §236 Schieranco (p. 85), §242 Domodossola, §243 Trontano (p. 86); Circondario di Pallanza; §248 Rumianca (p. 87)

Salmojraghi: §568 Provincia di Novara, Gneiss granitico; Valle d'Antigorio (p. 371)

Marmi italiani: Piemonte; Gneiss del Sempione o d'Antigorio (p. 33)

PETROGRAPHIC CLASSIFICATION: regional metamorphic rock (gneiss), fine to medium grain size

MACROSCOPIC FEATURES: grey and white in alternating thin wavy bands, uneven foliation

MINERALOGICAL COMPOSITION: K-feldspar, quartz, muscovite

MICROSCOPIC FEATURES: equant crystals of potash feldspar, anhedral crystals of quartz (polycrystalline, undulose extinction), some intergrowth of vermicular quartz and feldspar; streaks of muscovite surrounding quartz and feldspar

GEOLOGICAL SETTING: "Falda Antigorio", a lower Pennidic unit, featuring "augen" gneisses with feldspar porphyroblasts. Outcrops are present in the upper val d'Ossola and in valle Antigorio, along the Italian–Swiss border.

QUARRY: many sites are spread in the outcrop area, from Vogogna, Piedimulera, Villadossola, to Preglia, Rivasco and Foppiano, along the Strada Statale 659 di Valle Antigorio e Val Formazza. Different varieties are available: Serizzo Antigorio Passo or Serizzo Antigorio Oira (quarry Al Passo); Serizzo Formazza (quarry Bort); Bianco Walser (quarry Bort); Pietra Sempione (quarry Balmoreglio).

STONEWORK CATEGORIES: block, slab

SURFACE FINISHING: quarry face, split face, smoothing

USE IN ARCHITECTURE: the use for paving roads, reported by Scamozzi in the 17th century, is now completely disappeared. From the Middle Ages until the late 15th century, the stone was largely employed to make column shafts: i.e. the quadrilateral porch of the court of palazzo Carmagnola (now "Chiostro Nina Vinchi" of the Piccolo Teatro) contains shafts of Serizzo coupled with bases and capitals made of Viggiù. A very significant change occurred in the early 16th century: Serizzo shafts were replaced by granite ones. A significant example is the façade of "Ca' Granda" the former Ospedale Maggiore (now Università degli Studi): the 15th-century part shows the whole colonnade made of Serizzo, but the adjacent part (17th century) shows the colonnade made of Montorfano and Baveno granites. Another example is Palazzo Recalcati and its two courtyards: one (early 16th century) shows shafts of Serizzo, another one (late 17th century) shows shafts of Baveno granite. Another well-documented employ is the block in the base of the Duomo (together with Ceppo, limestone, granite, etc.). About the Duomo, it is worth it to note the use of chips of Serizzo to fill the core of the huge piers of nave, aisles and transept.

Different varieties of Serizzo, during the 20th century, were valorized through particular cuts and surface finishing and largely employed for cladding in dozens of apartment buildings.

Outside of Lombardy, the use was important in Piedmont together with other metamorphic rocks, specifically used in this region (Diorite di Malannaggio). In the 1930s, Serizzo was employed for the shafts of the new street (via Roma) linking the railway station to the Royal Palace.

DECAY MORPHOLOGIES: scaling along the foliation

Marmo di Valle Strona (Grigio Perla)

COMMERCIAL NAME: the name comes from an alpine valley, running eastwards to the northern tip of Lake Orta (Omegna, eastern Piedmont)

REFERENCES

Scamozzi: part 2, book 7, chap. 5, Marmi e misti (p. 188, line 50)

Amati: Luzzogno (Note – Marmi moderni, book 2, chap. 7, p. 50)

Jervis: Provincia di Novara, Circondario di Pallanza; §256 Sambughetto (p. 93)

Salmojraghi: §568 Provincia di Novara; Calcari saccaroidi; V. Strona (p. 369)

Marmi italiani: Piemonte; Marmi di Valle Strona (p. 30)
PETROGRAPHIC CLASSIFICATION: metamorphic limestone (marble)
MACROSCOPIC FEATURES: grey colour, sometimes with rectilinear grey veins; coarse grain size
MINERALOGICAL COMPOSITION: calcite (main component); quartz (accessory)
MICROSCOPIC FEATURES: large heteroblastic calcite crystals featuring irregular boundaries with polysynthetic twinning and rhombohedral cleavage; rounded quartz as single crystals or polycrystalline aggregates
GEOLOGICAL SETTING: marble lenses included in the "Dioritico – kinzigitica" formation, medium- to high-grade metamorphic rocks pertaining to the Ivrea-Verbano zone (Hercynian age). Outcrops are distributed from Valle Strona (south-west) to Locarco (north-east) as seen about the marbles of Candoglia and Ornavasso.
GEOLOGICAL REFERENCES: CGI, Foglio 30 Varallo (1927)
QUARRY SITES: the main site was located near the village of Sambughetto, in the middle of Valle Strona.
STONEWORK CATEGORIES: block, slab, carving
SURFACE FINISHING: polishing
USE IN ARCHITECTURE: the use flourished in Milan in the 20th century, mainly as slab for cladding. The most significant employ was the cladding of the façades of Palazzo di Giustizia (M. Piacentini, 1932–40). Very far from the quarry site (Naples), slabs of this marble form the cladding of the curvilinear façade of "Palazzo delle Poste" (G. Vaccaro, 1933–36) and of the main room Banco di Napoli (via Toledo 178, M. Piacentini, 1939).
DECAY MORPHOLOGIES: disaggregation, sulphate skin formation

Part II

Other Italian regions

Chapter 11

Stones from Apulia

Compact limestones are quarried in two sites of Apulia showing the same features of stones quarried in Lombardy, so their use was introduced in Milan only in the late 20th century. The stone trade excluded the soft limestone (called "tufo") used in the whole territory of Apulia but showing the worst mechanical characters.

SEDIMENTARY ROCKS

Pietra di Apricena

COMMERCIAL NAME: the name comes from a village of northern Apulia (province of Foggia)
REFERENCES
Jervis: Provincia di Capitanata, Circondario di San Severo; §906 Apricena (p. 225)
Salmojraghi: §612 Provincia di Foggia; Calcari; Monte Gargano (p. 424)
Marmi italiani: Puglia; Calcare bianco (p. 136)
PETROGRAPHIC CLASSIFICATION: chemical sedimentary rock (limestone)
MACROSCOPIC FEATURES: light brown colour, sometimes with very thin red stylolites; fine grain size
MINERALOGICAL COMPOSITION: calcite
MICROSCOPIC FEATURES: microcrystalline calcite, stylolithes with iron oxides, rare bioclasts
GEOLOGICAL SETTING: "Calcari di Sannicandro" formation (lower Cretaceous – upper Jurassic) pertaining to Carbonatic unit of the Apulian Platform (Cretaceous) and featuring crypto-crystalline limestones, whitish to pinkish in colour, massive or well bedded, rarely with whitish dolomites and oolitic limestones
GEOLOGICAL REFERENCE: CGI, Foglio 155 San Severo (1969)
QUARRY: several sites are located about 8 km north-west of Apricena (province of Foggia) along the road to Poggio Imperiale. Three different levels are quarried starting from the base: Fiorito, Biancone, Serpeggiante. Different varieties are available: Biancone, Bronzetto, Filettato, Filetto rosso, Fiorito, Moganato, Ondagata, Serpeggiante, Silvabella. The presence of laminae (mainly undulated) and reddish stylolites allows to show, according to different directions of cut, two different patterns: one with almost parallel lamination (edge bedding, cut perpendicular to the bedding), another one with almost concentric lamination (face bedding, cut parallel to the bedding).
STONEWORK CATEGORIES: block, slab

DOI: 10.1201/9781003361008-13

SURFACE FINISHING: polishing
USE IN ARCHITECTURE: the recent use in Milan is mainly for flooring
DECAY MORPHOLOGIES: erosion, scaling

Pietra di Trani

COMMERCIAL NAME: the name comes from a town on the Adriatic Sea coast (province of Bari)
REFERENCES
Jervis: Provincia di Terra di Bari, Circondario di Barletta; §933 Trani (p. 229), §937 Bisceglie, §938 Molfetta (p. 230); Circondario di Bari delle Puglie; §939 Giovinazzo (p. 230)
Salmojraghi: §613 Provincia di Bari delle Puglie; Calcari; Compatti, liassici o cretacei (p. 426)
Marmi italiani: Puglia; Calcare bianco (p. 136)
PETROGRAPHIC CLASSIFICATION: chemical sedimentary rock (limestone)
MACROSCOPIC FEATURES: light brown in colour, sometimes with thin reddish veins; fine grain size
MINERALOGICAL COMPOSITION: calcite
MICROSCOPIC FEATURES: microcrystalline calcite, stylolithes with iron oxides, rare bioclasts and intraclasts (Serpeggiante); microcrystalline and sparry calcite together with several bioclasts and intraclasts (Cocciolato, Perlato)
GEOLOGICAL SETTING: "Calcare di Bari" formation (Turonian-Barremian, Cretaceous) pertaining to Carbonatic unit of the Apulian Platform (Cretaceous) and featuring fine-grained detritic limestones, white to brown in colour, massive or well bedded and sometimes thin bedded; dolomitic limestones and dolomites; massive limestones. The outcrops occupy almost all of the territory of Apulia.
GEOLOGICAL REFERENCE: CGI, Foglio 176 Barletta (1970)
QUARRY: several sites are located at Contrada La Fondola and Contrada Gesù e Maria, about 3 km west of Trani. Different varieties are available: Avorio, Biancone, Bronzetto, Cocciolato, Cremamore, Fiorito, Ondagata, Perlato, Rosvedo, Serpeggiante, Tigrato. A different cut with regard to the bedding (face bedding or edge bedding) shows a different arrangement of the typical laminae of the stone.
STONEWORK CATEGORIES: block, slab
SURFACE FINISHING: polishing
USE IN ARCHITECTURE: these stones were used in Milan mainly for flooring in the last decades of the 20th century and later on
DECAY MORPHOLOGIES: scaling

Chapter 12

Stones of Friuli Venezia-Giulia

This north-eastern region of Italy was added to the Italian territory after World War I (1919), and it can be separated into two parts: Friuli together with Carnia (the majority of the present-day territory); Venezia Giulia together with Karst and Istria (the south-eastern part). Karst and Istria were mainly assigned to the former Republic of Yugoslavia since 1948, and now they pertain to the Republic of Slovenia and the Republic of Croatia.

STONES FROM CARNIA

The north-western part of the region Friuli – Venezia Giulia (close to the borders of Austria and Slovenija) is called "Carnia". The most ancient stones of the whole Italian territory outcrop in this specific area and some of these were quarried as building material and were used in Italy and abroad in the 20th century.

Sedimentary rocks

Grigio Carnico

COMMERCIAL NAME: the name comes from the colour
REFERENCES
Jervis: Provincia di Udine, Distretto di Tolmezzo; §576 Paluzza (p. 163)
Salmojraghi: §549 Provincia di Udine; Calcari del Trias (p. 338)
Marmi italiani: Venezia Giulia; Marmi di Timau (p. 60)
PETROGRAPHIC CLASSIFICATION: semi-crystalline limestone
MACROSCOPIC FEATURES: colour varies from pink to red, from grey to black (different colours refer to different varieties), fine grain size
MINERALOGICAL COMPOSITION: calcite
MICROSCOPIC FEATURES: microcrystalline calcite with irregular cavities; veins (some millimetres wide) filled by calcite spar (crystal size about 1 mm)
GEOLOGICAL SETTING: "Calcari massicci di Monte Coglians" formation, featuring bioherm and biostrome massive light-grey limestones (metric beds), algal limestones, Amphipora limestones, rare crynoid calcarenites and dolomitic limestones (Devonian)
Outcrops are located in the northern part of Friuli (Carnia), a few kilometres north of Tolmezzo.
GEOLOGICAL REFERENCE: CGI, Foglio 4c Monte Cavallino (1971); CARG, Foglio 031 Ampezzo (2009)

DOI: 10.1201/9781003361008-14

QUARRY: a few sites are opened near Malga di Pramosio, on the southern slopes of Creta di Timau, a mountain north of Timau di Paluzza, close to the Italy–Austria border.
STONEWORK CATEGORIES: slab
SURFACE FINISHING: polishing
USE IN ARCHITECTURE: the use was spread in Milan mainly in the 1930s as slab for flooring or cladding in apartment buildings
DECAY MORPHOLOGIES: erosion, chromatic alteration

Marmi di Verzegnis

COMMERCIAL NAME: the name comes from a village in the central Friuli
REFERENCES
Salmojraghi: §549 Provincia di Udine; Calcari; Diversi (p. 339)
Marmi italiani: Venezia Giulia; Marmi di Verzegnis (p. 60)
PETROGRAPHIC CLASSIFICATION: chemical sedimentary rock (marly limestone)
MACROSCOPIC FEATURES: brown to red colour, fine to medium grain size
MINERALOGICAL COMPOSITION: calcite, haematite as pigment
MICROSCOPIC FEATURES: uniform microcrystalline calcite with ossicles of crinoid sterns and microfossil tests; thin cracks with calcite spar (Rosso Porfirico)
GEOLOGICAL SETTING: "Rosso Ammonitico Veneto" formation (lower-middle Jurassic) featuring red nodular limestones with Ammonoida, lenses and beds of red flint; dark red marly limestones; well-bedded grey limestones, fossiliferous calcarenites (Foraminifera, Crinoidea, Brachiopoda, Belemnites, algae). Outcrop in the mountain area of Verzegnis (Tolmezzo), on the right bank of river Tagliamento.
GEOLOGICAL REFERENCE: CGI, Foglio 4c Monte Cavallino (1971); CARG, Foglio 031 Ampezzo (2009)
QUARRY: the main site is located on Monte Lovinzola, a few kilometres south-west of Tolmezzo, at an elevation of about 1700 m. The stone material was carried to the national road at Sella Chianzutan (elevation 954 m) using a cableway. Different varieties were sold, according to different colours and textures (Noce Radica, Porfirico fiorito, Rosso Porfirico, Rosso Radica).
STONEWORK CATEGORIES: slab
SURFACE FINISHING: polishing
USE IN ARCHITECTURE: the use was spread in Milan mainly in the 1930s as slabs for flooring in apartment buildings and villas (Milan, Villa Necchi, P. Portaluppi, 1935)
DECAY MORPHOLOGIES: erosion, chromatic alteration

Metamorphic rocks

Fior di pesco carnico

COMMERCIAL NAME: the name comes from the resemblance to an ancient coloured marble from Greece featuring a peach colour (Heretria, Evia Island); "Carnico" is referred to the geographical area (Carnia)
REFERENCES
Jervis: Provincia di Udine, Distretto di Tolmezzo; §68 Forni Avoltri (p. 161)
Salmojraghi: §549 Provincia di Udine; Calcari; Marmi policromi (p. 338)
Marmi italiani: Venezia Giulia; Marmi del Bellunese (p. 59)

PETROGRAPHIC CLASSIFICATION: low-grade metamorphic crystalline limestone (marble)
MACROSCOPIC FEATURES: light grey ground sometimes with pinkish or purplish spots (fine grain size) and veins (coarse grain size)
MINERALOGICAL COMPOSITION: calcite, opaque minerals
MICROSCOPIC FEATURES: calcite crystals of different shape and size (0.2–1.0 mm) showing rhombohedral cleavage and polysynthetic twinning
GEOLOGICAL SETTING: banded reef-limestones (Devonian) of the "Palaeozoico carnico" (Palaeozoic of Carnia). Outcrops are present around the Italy–Austria border, between monte Peralba (west) and Monte Coglians (east).
GEOLOGICAL REFERENCE: CGI, Foglio 4c Monte Cavallino (1971); CARG, Foglio 031 Ampezzo (2009)
QUARRY: the main site is located near Pierabec, north of Forni Avoltri (Canale di Gorto, province of Udine), very close to the Italy–Austria border. Two different varieties are sold: Fior di Pesco (purple-red spots) and Grigio Carnico (grey with white calcite veins)
STONEWORK CATEGORIES: slab
SURFACE FINISHING: polishing
USE IN ARCHITECTURE: the use in Milan was spread in the 20th century, mainly for cladding: atrium of Palazzo della Provincia, (G. Muzio, 1942); staircase of "la Rinascente" shopping centre (F. Reggiori, 1950)
DECAY MORPHOLOGIES: surface erosion, sulphate skin formation

STONES FROM FRIULI

The eastern part of the region shows a series of pre-Alpine valleys going to the border between Italia and Slovenija and featuring Tertiary sedimentary rocks.

Sedimentary rocks

Pietra Piasentina

COMMERCIAL NAME: the name comes from the regional term "piaser" ("to like" in English), meaning "a stone who likes". According to another hypothesis, the name comes from the resemblance to a cheese called "Piacentino".
REFERENCES
Salmojraghi: §549 Provincia di Udine; Pietre Piacentine (p. 341)
Marmi italiani: Venezia Giulia; Pietra Piacentina (p. 60)
PETROGRAPHIC CLASSIFICATION: clastic sedimentary rock (clastic limestone, breccia)
MACROSCOPIC FEATURES: uniform grey colour with thin white veins and some brownish spots; fine grain size
MINERALOGICAL COMPOSITION: calcite
MICROSCOPIC FEATURES: intraclasts and fragments of fossil tests (size less than 1.0 mm) with angular edges; cement of calcite spar
GEOLOGICAL SETTING: "Flisch del Grivò" (Lower Eocene, Ypresian) pertaining to "Turbiditic Succession of Paleogene" and featuring marls and sandstones in alternating with thick beds of coarse calcareous breccias, fine breccias, arenaceous limestones, marly

limestones. Outcrops occupy a wide area of the Prealps (Prealpi Giulie) around Cividale del Friuli.

GEOLOGICAL REFERENCE: CGI, Foglio 25 Udine (1929); CARG, Foglio 66 Udine (2008)

QUARRY: different sites are spread in the pre-Alpine area from Faedis, to Torreano, to San Pietro al Natisone present-day quarries. The main site, owned by "Consorzio Produttori Pietra Piasentina", is located on the left side of the valley of the Chiarò river, near Casali Laurini, a few kilometres north of Torreano (Cividale del Friuli, province of Udine).

STONEWORK CATEGORIES: slab

SURFACE FINISHING: saw finishing, polishing

USE IN ARCHITECTURE: local use has been documented since the Roman times; several noble palaces (Antonini, now Banca d'Italia; della Frattina, etc.) and churches (Santa Maria di Castello) of Renaissance and Baroque at Udine, show columns, windows, claddings and decorations made of this stone. At the same time, some palaces at Cividale del Friuli (Provveditori Veneti, now Museo Archeologico; de Nordis, etc.) and churches (Duomo, San Francesco) show a rusticated cladding. The use in Milan is very recent (last decade of the 20th century) mainly as slab for paving.

DECAY MORPHOLOGIES: scaling

STONES FROM KARST

The Karst ("Carso" in Italian and "Kras" in Slovenijan languages) is a calcareous plateau named for the phenomena of limestone dissolution causing a wide complex of caves and caverns, such as the world famous Postumia-Postojna and San Canziano-Skocjan. This plateau pertained to the Repubblica di Venezia for centuries; later on it was part of Austro-Hungarian Empire since 1797 and part of Regno d'Italia since 1920. The territory was divided (1948) between Italy and Yugoslavia (Slovenija after 1992). Stones from Karst have been quarried since the Roman period.

Sedimentary rocks

Pietra di Aurisina

COMMERCIAL NAME: the name comes from a village (Nabrežina in Slovenijan language), north of Trieste

REFERENCES

Salmojraghi: §632 Litorale Illirico; Calcari del Trias (p. 448)

Marmi italiani: Venezia Giulia; Marmi di Aurisina (p. 62)

PETROGRAPHIC CLASSIFICATION: biochemical sedimentary rock (limestone, biosparite-wackestone)

MACROSCOPIC FEATURES: greyish colour with fine dark spots and irregular white spots, evidence of fossil shells (complete, fragmented, shredded) made of Rudistae (reef-forming bivalves) and other bivalves. The size and sorting of the fragments feature the different varieties of the stone: fine grain size and well sorting (Roman stone and Aurisina chiara), medium grain size and less well sorting (Granitello), coarse grain size and bad sorting (Fiorito).

MINERALOGICAL COMPOSITION: calcite

MICROSCOPIC FEATURES: skeletal components (bioclasts, consisting of prismatic calcite crystals), together with micritic sediment and calcite spar cement

GEOLOGICAL SETTING: "Calcari di Aurisina" formation (upper Cretaceous) featuring massive bedded fossiliferous limestone, sometimes marly or bituminous and with calcareous-dolomite at the base. The total thickness varies from 1800 m (Karst of Gorizia) to 4000 m (Karts of Trieste), and the outcrops are spread in a long strip, parallel to the Adriatic Sea coast, from Monfalcone to Padriciano (a few kilometres north of Trieste). The sedimentary environment is neritic with shallow marine waters with the construction of a large geologic body (bioherma); the demolition of the bioherma, operated by wave-motion, produced the fossil fragments, then sedimented to form the rock. The bioherma, was formed by thousands of shells of Rudistae (Bivalvia, family Hippuritidae) typical of the Cretaceous. The shell of Rudistae is made of a valve shaped like a horn (frustum of cone), about 20–40 cm long; the second valve is smaller and has the function of operculum. The big valve is generally clearly visible, though deformed, on the cut surface of stone slab.

GEOLOGICAL REFERENCE: CGI, Foglio 40 Gorizia (1951)

QUARRY: huge sites with tall vertical faces are located a few hundred metres north of the village of Aurisina along the railway and about 15 km north-west of Trieste. The quarry works started in the Roman period and have been well developed since the second half of the 19th century. The different texture (size of skeletal components) leads to different varieties on sale: "Aurisina fiorito" (fossil shells), "Aurisina granitello" (fossil fragments), "Aurisina chiara" (small fragments), "Roman stone" (very small fragments).

STONEWORK CATEGORIES: block, slab, carving

SURFACE FINISHING: bush hammering, polishing

USE IN ARCHITECTURE: early uses are documented in Roman architecture of *Tergeste* (Trieste), *Aquileia* (Aquileia), *Altinum* (Altino) and other nearby towns and villages. Despite the long distance (about 400 km), this stone was also used in *Mediolanum* (Milan): funerary stelae, formerly set out in Archi di Porta Nuova (via Manzoni) and now located at Civico Museo Archeologico; squared blocks in the masonry of the tower of Carrobio, remain of the Roman urban walls. A very important example of the Byzantine period is the big monolithe placed as the dome of the Mausoleum of Teodorico in Ravenna (11 m in diameter, 3.2 m high, about 300 metric tons in weight). The medieval reuse involved different towns in northern Italy: some buildings in Venice; crypt of San Giovanni in Conca in Milan (shafts and capitals, 13th century); Duomo in Piacenza (caryatides of side portals, 13th century); Cathedral and Ghirlandina in Modena (blocks in the masonries, 11th–14th centuries).

The stone was largely employed in architecture of northern Italy (Trieste, rose windows of San Giusto) and other countries of the whole world (Austria, Great Britain, Egypt, Singapore, New Zealand, etc.), in the late 19th century and in the first decades of the 20th century. In particular, it is worth it to note the sarcophagus of San Marco (found in 1811) in the crypt of St Mark's Basilica (Venice); the façade of the second building of Banca Commerciale (Milan, L. Beltrami, 1913–23); the multiform façades of the Stazione Centrale (Milan, U. Stacchini, 1912–31).

Other significant examples are Tempio Voltiano (Como, A. Sant'Elia, 1928) and Monumento ai Caduti (Como, L. Sant'Elia, 1933) built using this sole stone as memory of the Karst, scene of many bloody battles during the World War I.

A more recent Milanese use (1990) concerns the flooring and the cladding (Granitello variety) of the stations of the underground line 3.
DECAY MORPHOLOGIES: disaggregation, sulphate skin formation

Gabria Tomadio Lipos

COMMERCIAL NAME: the name comes from the village of Tomadio-Tomaj near San Daniele del Carso-Štanjel, north of Trieste; since 1945 it lies in the territory of the Republic of Slovenija.
REFERENCES
Marmi italiani: Venezia Giulia. Gabria-Tomadio-Lipos (p. 61)
PETROGRAPHIC CLASSIFICATION: biochemical sedimentary rock (limestone, biosparite-wackestone)
MACROSCOPIC FEATURES: black ground with irregular white spots and evidence of fossil remains (Rudistae); fine to coarse grain size
MINERALOGICAL COMPOSITION: calcite
MICROSCOPIC FEATURES: skeletal components (bioclasts, made of drusy calcite mosaic) together with micritic sediment and a cement made of calcite spar
GEOLOGICAL SETTING: "Lipica" formation (upper Cretaceous) featuring thick-bedded and massive-bedded limestones with rare Rudist biostromes. Outcrops are located in the central Karst from Duttogliano-Dutovlije (west) to Storie-Štorje (east).
GEOLOGICAL REFERENCE: CGI, Foglio 40 Gorizia (1951)
QUARRY: the main site is located near the village of Tomadio-Tomaj in the Slovenian territory, about 7 km north of Monrupino
STONEWORK CATEGORIES: block, slab, carving
SURFACE FINISHING: polishing
USE IN ARCHITECTURE: the local use was important over the centuries, but in Northern Italy it was increased only in the first half of the 20th century. Outstanding examples are the floor of Villa Necchi (Milan, P. Portaluppi, 1932–35) and the vestibule of Cassa Assicurazioni (Milan, M. Piacentini, 1930).
DECAY MORPHOLOGIES: disaggregation, sulphate skin formation

Nero del Carso

COMMERCIAL NAME: the name comes from the colour and the geographical location of the quarry
RFERENCES
Marmi italiani: Venezia Giulia; Neri del Carso (p. 61)
PETROGRAPHIC CLASSIFICATION: chemical sedimentary rock (limestone, micrite, mudstone)
MACROSCOPIC FEATURES: black to dark grey colour sometimes with dark chert nodules or with white shades; very fine grain size
MINERALOGICAL COMPOSITION: calcite
MICROSCOPIC FEATURES: uniform microcrystalline calcite
GEOLOGICAL SETTING: "Calcari di Monte San Michele *Auct*" (Cenomanian, upper Cretaceous) pertaining to the geological series of the western Friuli featuring compact grey to blackish compact thin-bedded limestones and grey dolomite limestones. Outcrops are

spread north-east of Trieste (from Ronchi dei Legionari to Monrupino) across the Italia–Slovenija border.

GEOLOGICAL REFERENCE: CGI, Foglio 40 Gorizia (1951)

QUARRY: different sites were opened in the outcrop territory

STONEWORK CATEGORIES: slab

SURFACE FINISHING: honed

USE IN ARCHITECTURE: the use in Northern Italy was increased only in the first half of the 20th century. The stone was employed in the floor of Villa Necchi (Milan, P. Portaluppi, 1932–35).

DECAY MORPHOLOGIES: chromatic alteration, erosion, exfoliation

Repen

COMMERCIAL NAME: the name comes from a village of Karst north of Trieste (Monrupino in Italian language), near the Italy–Slovenija border

REFERENCES

Marmi italiani: Venezia Giulia; Breccia di Monrupino o Monte Rupino o Reppen (p. 61)

PETROGRAPHIC CLASSIFICATION: biochemical sedimentary rock (fossiliferous limestone, calcarenite)

MACROSCOPIC FEATURES:

- Light grey ground with black rod-shaped fossil fragments; fragments arranged along sub-parallel beds ("Repen Classico Chiaro", biomicrite, scarce skeletal components)
- Dark grey ground with black rod-shaped fossil fragments; fragments arranged along sub-parallel beds ("Repen Classico Zolla", biosparite, abundant skeletal components)
- A breccia made of dark organic debris up to 2–3 cm ("Fior di Mare", biolithite)

MINERALOGICAL COMPOSITION: calcite

MICROSCOPIC FEATURES: skeletal components (bioclasts, consisting of drusy calcite mosaic or of foliated internal structure), together with micritic sediment and calcite spar cement

GEOLOGICAL SETTING: "Calcare di Monrupino" formation (upper Cretaceous, Turonian) featuring massive-bedded fossiliferous limestones. Outcrops are spread in a long strip, parallel to the Adriatic Sea coast, of the Karst plateau along the Italia–Slovenija border (province of Trieste).

GEOLOGICAL REFERENCE: CGI, Foglio 40 Gorizia (1951)

QUARRY: the main site is located near Monrupino-Repentabor, a few kilometres north of Trieste, close to the Italia–Slovenija border

STONEWORK CATEGORIES: block, slab

SURFACE FINISHING: polishing

USE IN ARCHITECTURE: the use is less documented than the contiguous limestone of Aurisina. The stone was employed in the architecture of northern Italy and Milan in the late 19th century and in the first decades of the 20th century: the stone was mainly used for floor tiles and for cladding of apartment buildings (i.e. the vestibule of Cassa Assicurazioni – Milan, M. Piacentini, 1930). A recent use is the cladding of Deutsche Bank (Valle Associati, 1997–2005) and of piazza "Tre Torri" (Citylife, 2017).

DECAY MORPHOLOGIES: disaggregation, sulphate skin formation

Stalattite del Carso

COMMERCIAL NAME: the name comes from the petrographic classification
REFERENCES
Marmi italiani: Venezia Giulia; Alabastro calcareo o Stalattite del Carso (p. 63)
PETROGRAPHIC CLASSIFICATION: non-clastic deposit of calcium carbonate from supersaturated water in limestone caves (speleothems)
MACROSCOPIC FEATURES: yellow to reddish colour with evident zoning; medium to coarse grain size
MINERALOGICAL COMPOSITION: calcite
MICROSCOPIC FEATURES: rod-shaped calcite crystals arranged in concentric rows
GEOLOGICAL SETTING: Quaternary deposits, outcropping in the caves of the Karst plateau
GEOLOGICAL REFERENCE: CGI, Foglio 40 Gorizia (1951)
QUARRY: very small sites are spread in different caves along the border between Italy and Slovenija, north of Trieste
STONEWORK CATEGORIES: slab
SURFACE FINISHING: polishing
USE IN ARCHITECTURE: the use in Milan was strictly connected with the 1930s, mainly for flooring in apartment buildings
DECAY MORPHOLOGIES: erosion if exposed to weathering

STONES FROM ISTRIA

Istria (Istra in the Croatian language) is a calcareous peninsula of the northern part of the Adriatic Sea; it pertained for centuries to the Repubblica di Venezia, then it was part of the Austro-Hungarian Empire (1797), part of Regno d'Italia (1920) and part of Yugoslavia (1947, Slovenija and Croatia after 1992). Stones from Istria contributed specifically to the construction of Venice.

Sedimentary rocks

Giallo d'Istria

COMMERCIAL NAME: the name comes from the specific colour
REFERENCES:
Marmi italiani: Venezia Giulia; Giallo d'Istria (p. 63)
PETROGRAPHIC CLASSIFICATION: chemical sedimentary rock (limestone Nerinea)
MACROSCOPIC FEATURES: yellowish colour, sometimes with reddish spots, very fine grain size
MINERALOGICAL COMPOSITION: calcite, iron hydroxides
MICROSCOPIC FEATURES: uniform microcrystalline calcite
GEOLOGICAL SETTING: compact or coarse-grained limestones with some fossiliferous (Nerinea) beds (upper Cretaceous, Cenomanian). Outcrops are located in a narrow area running along the southern bank of torrente Dragogna to punta Salvore on the Adriatic Sea.
GEOLOGICAL REFERENCE: CGI, Foglio 53 Trieste (1953)

QUARRY: some sites are located around Madonna del Carso, a few kilometres north-east of Umago/Umag; other sites were probably located on the central coast of Istria between Poreč-Parenzo and Rovinj-Rovigno

STONEWORK CATEGORIES: slab

SURFACE FINISHING: polishing

USE IN ARCHITECTURE: the use was strictly connected with the 1930s, according to the increase in the use of stone in architecture and the exploitation of stone materials coming from peripheral areas (the Istria peninsula was an Italian territory from 1919 to 1945) and made available thanks to the network of paved roads. A significant example are the slabs for cladding in the apartment building (corso Matteotti 5,9. Milan, E. Lancia, 1933–36).

DECAY MORPHOLOGIES: erosion, scaling, sulphate skin formation

Pietra d'Istria

COMMERCIAL NAME: the name comes from the peninsula in the northern tip of the Adriatic Sea

REFERENCES

Vasari (On technique, chap. 1): "(. . .) in Istria (. . .) una pietra bianca livida".

Scamozzi: part 2, book 7, chap. 8, Pietre vive (p. 198, line 22)

Rondelet: book 1, sec. 2, art. 15, §307, §308, §309 (p. 202)

Salmojraghi: §632 Litorale Illirico. Pietra d'Istria (p. 448)

Marmi italiani: Venezia Giulia; Orsera (p. 62)

PETROGRAPHIC CLASSIFICATION: chemical sedimentary rock (limestone, micrite, mudstone)

MACROSCOPIC FEATURES: ivory to whitish colour, conchoidal fracture; very fine grain size

MINERALOGICAL COMPOSITION: calcite

MICROSCOPIC FEATURES: uniform microcrystalline calcite with few round cavities

GEOLOGICAL SETTING: a Mesozoic sedimentary platform occupies the whole Istria peninsula, spreading from the upper Jurassic to the Cretaceous and Eocene; in particular, the south-western coast of Istria, between Vrsar/Orsera and Rovinj/Rovigno, involves a massive-bedded reef limestone (upper Jurassic, Titonian)

GEOLOGICAL REFERENCE: geological map of the Republic of Croatia at the scale 1:300.000 by Ivan Hećimović

QUARRY: some sites are mainly located east of Vrsar (Montižana), around the Limski Kanal/Canal di Leme (a narrow gulf about 10 km long) and north of Rovinj. A maritime transport from Rovinj to the yards of Venice and other Italian towns on the Adriatic Sea (from Trieste to Ancona) was put into practice for centuries.

STONEWORK CATEGORIES: block, slab, carving

SURFACE FINISHING: smoothing

USE IN ARCHITECTURE: the good qualities of this limestone have been appreciated by architects since the Roman times; an outstanding example is the amphitheatre of Pula/Pola. The coast of Istria shows this stone in Romanesque, Gothic and Renaissance buildings. The same long-term use is visible in Venice; before linked to brick masonries and employed only in some important parts of the buildings (portals, windows, roses, cornerstones), after used as block for masonries and also in the decoration (i.e. mouldings, reliefs, etc. of Palazzo Ducale and Ca' d'Oro); the use of this limestone was also combined with ancient coloured marbles (Santa Maria dei Miracoli). From the 16th century to the early 19th century, dozens of palaces were built by religious orders (Scuola di San Marco, Scuola di San Rocco), by noble families (Grimani, Vendramin Calergi, Rezzonico, Pesaro) or by the Republic (Procuratie nuove, Libreria vecchia, Zecca, Ponte di Rialto). At the same time, dozens of churches were built using this white limestone (San Giorgio Maggiore, Redentore, San Francesco della Vigna, Santa Maria della Salute). Other very important constructions made of pietra d'Istria are connected to the protection of the whole town from the attack of seawaters, waves and tides as embankments, piers, wharfs, breakwaters, etc. An outstanding example is the so-called Murazzi, a dam many kilometres long, built in the 18th century along the coast of the Lido and Pellestrina banks.

Finally, the use of this limestone was also noticeable in Trieste (buildings of piazza Verdi, 19th century), Ravenna (Renaissance to Baroque buildings) and Ancona (different buildings from the 15th to 18th centuries).

DECAY MORPHOLOGIES: erosion, scaling, sulphate skin formation

Chapter 13

Stones from Latium

Despite the long distance, some stones from Latium were traded to Milan. There are stones with very peculiar features, very different from the stones of Lombardy.

IGNEOUS ROCKS

Basaltina

COMMERCIAL NAME: the name comes from "Basalto", the Italian term to indicate the basalt (volcanic rock)

REFERENCES

Jervis: Provincia di Roma, Circondario di Viterbo; §1181 Bolsena (p. 336), §1186 Viterbo (basalto – p. 338)

Salmojraghi: §599 Provincia di Roma; Lave; Vulcani (spenti) (p. 412)

Marmi italiani: Lazio; Lave (p. 123)

PETROGRAPHIC CLASSIFICATION: volcanic igneous rock (leucitite lava)

MACROSCOPIC FEATURES: uniform grey colour with some darker points, medium grain size

MINERALOGICAL COMPOSITION: plagioclase (labradorite), leucite

MICROSCOPIC FEATURES: porphyritic texture with phenochrysts of plagioclase and leucite

GEOLOGICAL SETTING: "Scorie e lave leucititico-tefritiche di Monterado" pertaining to the Serie Superiore Settentrionale of the Apparato Vulsino (Pleistocene), the former volcano of Monti Vulsini (northern part of province of Viterbo). Lava flows from a wide plateau on the eastern side of Monti Volsini, towards the valley of river Tevere. Between Bagnoregio and Ponzano, trachyte flows contain leucite ("Trachite leucitica").

GEOLOGICAL REFERENCE: CGI, Foglio 137 Viterbo (1971)

QUARRY: the main site is located between Bagnoregio (ancient Bagnorea, province of Viterbo) and Ponzano, about 8 km east of the shores of Lake Bolsena

STONEWORK CATEGORIES: slab for cladding, tile for paving

SURFACE FINISHING: polishing (cladding), rough (paving)

USE IN ARCHITECTURE: the use to make elements (frustum of pyramid) for paving is spread in the towns of Latium and Umbria; the use for flooring, in Milan, is very recent (Museo delle Culture – Mudec, D. Chipperfield, 2000–15)

DECAY MORPHOLOGIES: scaling

DOI: 10.1201/9781003361008-15

Peperino (Lavagrigia – Lavarossa)

COMMERCIAL NAME: the present-day name comes from the geological origin and from the colour. The traditional name "Peperino" (English translation: peppercorn) comes from the aspect due to black volcanic glass diffused as spots in the rock.
REFERENCES
Vasari (On Technique, chap. 1): "(. . .) la pietra detta Piperno".
Rondelet: book 1, sec. 2, art. 15, §315 (p. 204)
Jervis: Provincia di Roma, Circondario di Viterbo; §1186 Viterbo (trachite – p. 338)
Salmojraghi: §599 Provincia di Roma; Peperini associati ai tufi vulcanici (p. 411)
Marmi italiani: Lazio; Peperino (p. 123)
PETROGRAPHIC CLASSIFICATION: trachyte from pyroclastic flow
MACROSCOPIC FEATURES: grey to pinkish colour with dark oblong spots (called "fiamme"), porphyric texture without evident phenocrysts; medium grain size
MINERALOGICAL COMPOSITION: plagioclase (andesine-labradorite), low sodium sanidine, biotite, hypersthene, augite
MICROSCOPIC FEATURES: clasts of volcanic glass, lithic clasts (sedimentary rocks), angular crystals of plagioclase and pyroxene; irregular cavities
GEOLOGICAL SETTING: "Ignimbrite cimina *Auctt*" formation (lower Pleistocene) of the "Sintema Faggeta" (volcanic complex of Monti Cimini) featuring lenses or bands of ignimbrite with coarse bedding or agglomeration with a various degree of compactness in the volcanic sediments (tuffs) of Monti Cimini referred to the northern part of the "Provincia Vulcanica Tosco-Laziale"
GEOLOGICAL REFERENCE: CGI, Foglio 137 Viterbo (1971); CARG, Foglio 345 Viterbo (2017), Foglio 387 Albano Laziale (2009)
QUARRY: the main site, with several excavations (pit), is now located at "Pallone" (2 km south of Vitorchiano and about 8 km north-east of Viterbo), on the north side of Monti Cimini
STONEWORK CATEGORIES: block, thick slab
SURFACE FINISHING: sawn surface
USE IN ARCHITECTURE: use during the Roman times was very huge, mainly as block for the foundation of religious buildings (i.e. Tempio di Antonino e Faustina in the Roman Forum). The local use is witnessed in the medieval town of Viterbo, both for building and for decoration (i.e. Palazzo dei Papi and its loggia with seven arches, second half of the 13th century, restored in 1887–1919). After World War II the use mainly involved slab for cladding with sawn surface, as visible in some Milanese buildings.
DECAY MORPHOLOGIES: scaling
NOTE: Other similar rocks of Latium supplied building stones named "Peperino", such as phreatomagmatic grey, lithoidal deposits pertaining to "Peperino di Albano" formation (Volcanic complex of Colli Albani – Upper Pleistocene) and featuring leucite, pyroxene and biotite together with sedimentary and lava xenoliths. Important quarry sites were located near Marino (massive deposits; via dei Laghi, south-east of Rome) and near the ancient town of Gàbii (stratified deposits; via Prenestina, east of Rome) on the volcanic area of "Colli Albani".

SEDIMENTARY ROCKS

Travertino

COMMERCIAL NAME: the name comes from the Latin appellation *Lapis Tiburtinus*, meaning "from Tibur" or Tivoli (east of Rome). The importance of the stone material quarried around Tivoli gave the general name "travertine" to the deposits of the same origin spread in the whole territory of central Italy and abroad.

REFERENCES

Vasari (On technique, chap. 1): "(. . .) un'altra sorte di pietra chiamata travertino".

Scamozzi: part 2, book 7, chap. 8, Pietre vive (p. 197, line 6)

Rondelet: book 1, sec. 2, art. 15, §314 (p. 203)

Jervis: Provincia di Roma, Circondario di Roma; §246 Tivoli (p. 354)

Salmojraghi: §599 Provincia di Roma; Travertini (p. 410)

Marmi italiani: Lazio; Travertino (p. 120)

PETROGRAPHIC CLASSIFICATION: non-clastic deposit of calcium carbonate, precipitated from hot spring waters

MACROSCOPIC FEATURES: creamy white to dull yellowish-brown colour; compact bands (some cm thick) in alternating with very porous bands with deep irregular cavities, sometimes with spongy appearance

MINERALOGICAL COMPOSITION: fibrous calcite coming from chemical encrustation of microphyta and macrophyta (i.e. algal felts or leaves)

MICROSCOPIC FEATURES: coarse calcite crystals of irregular shape arranged in convoluted forms around irregular shaped cavities

GEOLOGICAL SETTING: "Travertino dei Bagni di Tivoli" formation (upper Pleistocene – Holocene) pertaining to "Supersintema Tiburtino" or deposits of continental environment after the regional uplift of the Thyrrenian edge. The formation includes deposits from cold waters, in an environment of waterfall or rapids, featuring high porosity with inserts of plant materials (leaves), low to strong cementation and irregular bedding.

GEOLOGICAL REFERENCES: CGI, Foglio 150 Roma (1967); CARG, Foglio 375 Tivoli (2012)

QUARRY: the most important site, where the Romans took the stone of their architectures, is located near Bagni di Tivoli (*Aquae Albule*), a few kilometres west of Tivoli. The modern quarries, owned by "Società del Travertino Romano", are located between Bagni di Tivoli (west) and Ponte Lucano (east) and they supply different varieties: "Valle Pillella" (Classico, Bianco, Alabastrino, Striato chiaro). "Barco" (Bianco, Noce, Noisette, Paglierino), "Le Fosse" (Paglierino, Bianco, Noisette). The quarry works involve a wide area, and the landscape features deep excavations (about 50 m high) on both sides of "Via Tiburtina".

STONEWORK CATEGORIES: block, drum, slab, carving

SURFACE FINISHING: sawn face (both face or edge bedding); modern use requires polishing with an epoxy resin to fill up the cavities

USE IN ARCHITECTURE: use has been very large since the Roman times. A great part of buildings in ancient Rome (i.e. the Colosseum, Theatre of Marcellus, Circus of Domitian)

was built using this stone. Travertine blocks form, in many cases, the masonry then covered by a cladding made of slabs of coloured marbles.

A wide reuse of Roman artefacts took place in the Middle Ages; during the Renaissance and later, the activity of the quarries had been constantly increased (basilica of San Pietro, Rome) since the Baroque period (piazza Navona). The modern use is spread, far from the quarry areas, in the whole Italian territory, starting from the second decade of the 20th century. The use, in the 1930s, was boosted by the policy of the Fascist government in order to promote a rebirth of the Roman Empire (buildings of the Public Administration and of the Fascist Party). One of the first examples of the employ of this stone in Milan, is the façade of the so-called Ca' Bruta (via Moscova, G. Muzio, 1922), then dozens of public buildings (i.e. Intendenza di Finanza, via Manin, E. Marelli, 1932), schools, apartment buildings and monuments. In some cases, a cladding with travertine slabs was imitated using a mortar with irregular alignments of cavities on the surface.

DECAY MORPHOLOGIES: erosion, sulphate skin formation, dust and soot deposits in the cavities

ADDENDUM

Other limestones from Latium

Two other limestones are significant because of their role in architecture: Cottanello in the past centuries, Coreno at present.

Cottanello

A brown limestone with broad white veins and narrow dark veins, was quarried in northern Latium, close to the border of Umbria. This limestone is called "Cottanello" from the name of the village, a few kilometres south of the quarry area (province of Rieti); it pertains to Scaglia Rossa formation (well-bedded marly-limestones, upper Cretaceous – middle Eocene), and it was widely employed in Rome, mainly to make column shafts for the Baroque churches. The use in Lombardy and in Northern Italy was irrelevant.

References

Milizia: part 1, book 4, chap. 12.3, Marmi moderni; §24 Cottanello
Amati: Cottanello (Note – Marmi moderni, book 2, chap. 7, p. 49)
Marmi Italiani: Lazio; Marmo di Cottanello (p. 124)

Coreno

A creamy limestone with whitish skeletal remains, is quarried in Ciociaria (Monti Aurunci, across the border of the provinces of Frosinone and Latina, southern Latium). This limestone is called "Perlato Royal Coreno" for the dispersion of fossil fragments on the surface, and it pertains to a formation (Miocene, Elvetian) originated from a marine transgression and featuring well-bedded detrital-shell limestones (pectinids, oysters, echinoids, corals, bryozoans, foraminifera, lithothamnia), white to brown in colour. The quarry area involves Coreno Ausonio and Ausonia (Frosinone); Santi Cosma e Damiano and Castelforte (Latina); different varieties are quarried according to different beds of the formation. The use in Milan started in the last decades of the 20th century, mainly as slab for flooring.

NOTE

A similar stone material is called "Perlato di Sicilia". This is a dark, spotted, creamy limestone pertaining to "Formazione Pellegrino" (upper Cretaceous) of the "Succession Monte Monaco – Monte Sparagio". This formation features biolithites with grey massive biocalcarenites containing Rudistae, corals, gastropods, crynoids, algae and benthos foraminifera (CARG, Foglio 593 – Castellammare del Golfo). The outcrops are present in the north-western part of Sicily (province of Trapani). Quarry sites are located north of Custonaci (west side of outcrop) and south of Monte Sparagio (south-eastern side of the outcrop). Polished slabs of this stone were used since the last decades of the 20th century, for flooring in apartment and public buildings (Milan, Stazione Centrale, floor of Galleria dei Viaggiatori, set during the renovation of 2005–08).

Chapter 14

Stones from Liguria

A few stones coming from the whole regional territory were used in Lombardy, each featuring some peculiar characters.

SEDIMENTARY ROCKS

Pietra di Finale

COMMERCIAL NAME: the name comes from a village of the western side of the Gulf of Genoa

REFERENCES

Jervis: Provincia di Genova, Circondario di Albenga; §1057 Final Marina (p. 251)

Salmojraghi: §574 Provincia di Genova; Calcare; Cristallino (p. 380)

Marmi italiani: Liguria; Pietra di Finale (p. 67)

PETROGRAPHIC CLASSIFICATION: clastic sedimentary rock (limestone; biosparite – packstone) of "Bacino del Finalese"

MACROSCPIC FEATURES: white to pinkish to reddish colour, coarse grain size

MINERALOGICAL COMPOSITION: calcite

MICROSCOPIC FEATURES: microcrystalline calcite with irregular skeletal fragments

GEOLOGICAL SETTING: "Calcare di Finale Ligure" formation (Pietra di Finale *Auctorum*) (middle-lower Miocene) belonging to post-orogenetic deposits (Bacino del Finalese). Cavernous bioclastic limestones with Anthozoa (corals) and Codiaceae (algae), white, pinkish or reddish colours, including beds or lenses of sandstones. The upper part of the formation is massive bedded, the lower part is well bedded. Fossils are represented by echinoidea, bivalvia and brachiopoda. Sometimes marls and polygenic conglomerates are present. Outcrops in the western coast of Liguria, near Finale Ligure and Orco-Feglino, between Savona and Albenga (province of Savona)

GEOLOGICAL REFERENCE: CGI, Foglio 92 Albenga (1971); CARG, Foglio 228 Cairo Montenotte (2017)

QUARRY: different sites are located around Finale Ligure: Val Ponci, Valle di Cornei, Cava Vecchia, Rocca di Perti, Cava Pianmarino, Cava Arma del Sambuco, Chiesa di Verezzi, Cava dell'Aquila, Cava Crocetta; Cava del Colle

STONEWORK CATEGORIES: slab

SURFACE FINISHING: chiselling

USE IN ARCHITECTURE: the stone was mainly employed in Milan for cladding, since the first half of the 20th century. Significant examples are the façade of the main building of

DOI: 10.1201/9781003361008-16

Università Luigi Bocconi (G. Pagano, 1941) or the façades of the Palazzo del Toro (E. Lancia, 1939)
DECAY MORPHOLOGIES: disaggregation, sulphate skin formation

Portoro

COMMERCIAL NAME: the name refers to the golden coloured veins as a distinguishing feature
REFERENCES
Scamozzi: part 2, book 7, chap. 8, Pietre vive (p. 197, line 44)
Milizia: part 1, book 4, chap. 12.3, Marmi moderni; §31 Nero e giallo, detto Portoro
Rondelet: book 1, sec. 2, art. 11, §135 (p. 137), §226 (p. 144)
Amati: Nero-Giallo, Portoro (Note – Marmi moderni, book 2, chap. 7, p. 50)
Jervis: Provincia di Genova, Circondario di Spezia; §1089 Porto Venere, § 1090 Spezia (p. 258)
Salmojraghi: §574 Provincia di Genova; Calcare; Portoro (p. 380)
Marmi italiani: Liguria; Portoro (p. 65); Bianco e nero de La Spezia (p. 67)
PETROGRAPHIC CLASSIFICATION: chemical sedimentary rock with nodular texture
MACROSCOPIC FEATURES: deep black ground with golden to yellow parallel and interlacing veins; grey ground with whitish interlacing veins
MINERALOGICAL COMPOSITION: calcite originating from aragonite muds with a partial substitution by dolomite crystals
MICROSCOPIC FEATURES: calcite microspar together with dolomite, crossed by veins of coarse-grained calcite spar and dolomitic veins
GEOLOGICAL SETTING: "Portoro" formation (upper Rhetian – lower Hettangian) of the northern Apennine (Falda Toscana – Tuscan nappe). The formation, without fossils, features grey to black calcilutites often with nodular or stylolite structures with white and yellow dolomite spots (about 1 m thick) in alternating with coarsely crystalline whitish to yellowish dolomite (called "Tarso", about 1 m thick). Outcrops are located in eastern Liguria (province of La Spezia) near the Tuscan border from Pignone (north-west) to Biassa, Portovenere, islands of Palmaria and Tino (south-east).
GEOLOGICAL REFERENCES: CGI, Foglio 95 Spezia (1928); CARG, Foglio 248 La Spezia (2005)
QUARRY: different sites are present on the promontory of La Spezia: from monte Castellana (close to forte del Muzzerone, north of Porto Venere) to island of Palmaria (southern tip called "Capo del'Isola") with a pier for stone loading on boats; to island of Tino (northern part). The different size of the yellow veins produces two varieties: "Portoro Macchia Larga" and "Portoro Macchia Fina".
STONEWORK CATEGORIES: slab
SURFACE FINISHING: polishing
USE IN ARCHITECTURE: mainly thin slab for the decoration of Baroque altars
DECAY MORPHOLOGIES: chromatic alteration if exposed to weathering

ADDENDUM

Another sedimentary rock, called "Arenaria di Lèvanto", was employed (Milan) for road paving in the late 19th century. This is a grey, fine-grained sandstone pertaining to "Formazione del Macigno", a turbiditic formation (middle to upper Oligocene) featuring beds of sandstone (a few centimetres to 5 m thick) in alternating with thin beds of grey siltites. Different quarry sites (Riccò del Golfo, Biassa, Monterosso) were worked in the Eastern Ligurian Riviera (province of La Spezia).

METAMORPHIC ROCKS – OPHICALCITES

Rosso di Lévanto/Verde di Levanto

COMMERCIAL NAME: the names come from the most important town in the quarry area (Riviera di Levante, eastern Liguria, province of La Spezia)

REFERENCES

Scamozzi: part 2, book 7, chap. 5, Marmi e misti (p. 188, line 26)

Jervis: Provincia di Genova, Circondario di Spezia; §1087 Bonassola, §1088 Lévanto (p. 257)

Salmojraghi: §574 Provincia di Genova; Oficalci; Rosso di Levanto (p. 380)

Marmi italiani: Liguria; Rosso di Lévanto (p. 69)

PETROGRAPHIC CLASSIFICATION: regional metamorphic rock (ophicalcite)

MACROSCOPIC FEATURES: dark red ground with very irregular white veins, a variety shows a dark green colour instead of red; coarse grain size

MINERALOGICAL COMPOSITION: serpentine, calcite (white veins)

MICROSCOPIC FEATURES: serpentine aggregates surrounded and crossed by irregular veins of calcite spar

GEOLOGICAL SETTING: "Breccia di Levanto" – "Breccia di Framura" formations (middle Jurassic) pertaining to the "Complesso vulcano-sedimentario del Borzonasco", (tectonic unit of Bracco/Val Graveglia; Supergruppo del Vara; Dominio Ligure interno – middle Jurassic – Paleocene). The lower part (middle Jurassic) includes ophiolite succession (serpentinitic rocks, mainly from peridotites of the mantle, featuring swarms of fractures, both parallel or crossing, enhanced by veins of calcite spar, often containing hematite) together with mafic and ultramafic rocks; the upper part (middle Jurassic – Paleocene) includes breccias with ophiolitic clasts, pelagic deposits and flysch.

GEOLOGICAL REFERENCES: CGI, Foglio 95 Spezia (1928); CARG Foglio 232 Sestri Levante (2014)

QUARRY: different sites spread along the coast of Eastern Liguria (province of La Spezia); the main site is located above Bonassola along the SS332 (100 m after the fork of the road to San Giorgio) and along the SP42 (road to Framura, 100 m after the fork of SS332), about 1 km north-west of the previous one, where the same outcrop was excavated on both sides of the mountain crest. The cuts made using the diamond wire are clearly visible on the quarry faces, but the bottom is now completely flooded by underground waters.

STONEWORK CATEGORIES: block, slab, carving

SURFACE FINISHING: polishing

USE IN ARCHITECTURE: the first use seems to be the frames of the sculpted panels (scenes of the life of the Virgin Mary) positioned in the upper part of the retro-choir, encircling the main altar of Milan's Duomo (arch. Pellegrini, late 16th century)

The stone went in use in the 1930s, mainly as slab for cladding and for flooring (Palazzo di Giustizia, Piacentini and Rapisardi, 1931–40; Casa Wassermann, Portaluppi 1935–36); moulded elements were also employed as jambs and lintels in portals and doors. After

World War II, the stone was still employed for flooring (i.e. "Salone affresco" at the ground floor of Museo Poldi Pezzoli, F. Reggiori 1950).

DECAY MORPHOLOGIES: scaling, disaggregation of calcite, roughening, chromatic alteration

Verde Polcévera

COMMERCIAL NAME: the name comes from a valley of central Liguria (north of Genoa)

REFERENCES

Scamozzi: part 2, book 7, chap. 5, Marmi e misti (p. 188, line 23)

Jervis: Provincia di Genova, Circondario di Genova; §1074 Campomorone (p. 253)

Salmojraghi: §574 Provincia di Genova; Oficalci; Verde Polcevera (p. 380)

Marmi italiani: Liguria; Verde Polcevera (p. 68)

PETROGRAPHIC CLASSIFICATION: regional metamorphic rock (ophicalcite)

MACROSCOPIC FEATURES: green ground with very irregular white veins, a variety shows a dark red colour instead of green; coarse grain size

MINERALOGICAL COMPOSITION: serpentine, calcite (white veins)

MICROSCOPIC FEATURES: brecciated texture, serpentine aggregates, serpentinized olivine crystals together with irregular veins of calcite spar

GEOLOGICAL SETTING: "Metaoficalciti di Pietralavezzara" formation of the "Unità Tettono-metamorfica Figogna" (Dogger? – Malm?) featuring tectonic breccias of the oceanic ground with serpentinite clasts and carbonate cement. Outcrops are scarce and concentrated around Pietralavezzara. Tectono-metamorphic units (Figogna, Cravasco-Voltaggio, Palmaro-Caffarella, Voltri) contain ophiolite successions, including parts of the gabbro-peridotite basement together with volcano-sedimentary covers. In particular "Unità Figogna" includes (starting from the bottom): serpentinites, metagabbroes, metabasalts, breccias, metaophicalcites, ialoclastites (Jurassic) and metalimestones, limestones, argillaceous schists (Cretaceous).

QUARRY: an old site was located a few hundred metres south of Pietralavezzara on the road of "Passo della Bocchetta" (Strada Provinciale 5, from Genova-Campomorone to Gavi)

GEOLOGICAL REFERENCE: CGI, Foglio 82 Genova (1971); CARG, Foglio 213 Genova (2008)

STONEWORK CATEGORIES: slab, carving

SURFACE FINISHING: polishing

USE IN ARCHITECTURE: a wide use is witnessed by Baroque altars of the churches of Genoa, but the exact distinction among ophicalcites is almost impossible. The ophicalcite of Polcevera, as the ophicalcite of Levanto, went in use in Milan during the 1930s. Slabs for cladding and tiles for flooring or moulded elements were also employed as jambs and lintel in portals and doors.

DECAY MORPHOLOGIES: scaling, disaggregation of calcite, roughening, chromatic alteration

METAMORPHIC ROCKS – SLATE

Verde Roja (Vert de Levant)

COMMERCIAL NAME: the name comes from a valley of western Liguria, a territory partially included in the French Republic since 1945

REFERENCES
Salmojraghi: §571 Provincia di Cuneo; Scisti; Cristallini o filladici (p. 378)
Marmi italiani: Piemonte; Verde Roja o Pietra di Roja (p. 35)
PETROGRAPHIC CLASSIFICATION: regional metamorphic rock (slate, clayey schist)
MACROSCOPIC FEATURES: uniform green colour associated with darker silicate veins; easily divisible into slabs; fine grain size
MINERALOGICAL COMPOSITION: chlorite, muscovite, quartz
MICROSCOPIC FEATURES: cryptocrystalline texture with some alignments, rare angular crystals of quartz
GEOLOGICAL SETTING: "Scisti gneissici" formation (Permian) of the "Dominio Delfinese-Provenzale". Outcrops are present in the valley of river Roja, close to the border between Italy and France.
GEOLOGICAL REFERENCE: CGI, Foglio 91 Boves (1934)
QUARRY: located in the upper valley of the river Roja near Tenda (Colle di Tenda); a former Italian territory assigned to France (département Alpes Maritimes, région Provence-Alpes-Côte d'Azur) after World War II
STONEWORK CATEGORIES: block, slab
SURFACE FINISHING: smoothing
USE IN ARCHITECTURE: slabs for cladding or flooring were employed locally and in the Ligurian territory; the use in Milan was concentrated in the 1930s: Casa Fiocchi, cladding (via Cernaia 6, M. Fiocchi, 1924–25); Stazione Centrale, flooring together with other coloured stones (U. Stacchini, 1912–31); Cassa Assicurazioni, flooring (M. Piacentini, 1930). In other cases, thin slabs or moulded elements were employed as window frames (Casa Ponti, via Randaccio 9, G. Ponti and E. Lancia, 1924–25; Casa dei Sindacati Fascisti, corso Porta Vittoria 43, A. Bordoni, L.M. Caneva, A. Carminati, 1930–32).
DECAY MORPHOLOGIES: scaling

ADDENDUM

Another kind of slate, rarely used in Milan, but diffused in Genoa and in the whole coast of Liguria, is called "Ardesia" (from the French term "Ardoise"). This black stone may be split in thin slabs and employed for the roofs and for the cladding of walls exposed to the sea winds. Vasari (On technique, chap. 1) reported, "These slates are a product of the sea coast of Genoa, in a place called Lavagna; they are excavated in pieces ten *braccia* long and are made use of by artists for their oil paintings". Jervis, Salmoiraghi and Marmi italiani pointed out these features and the use as blackboard for schools. The outcrop is concentrated in Val Fontanabuona, north of Chiàvari (province of Genoa) and the quarries are mainly located on the left side of the valley (Orero, Cicagna, Mocònesi). The rock pertains to the "Ardesie di Monte Verzi" formation (upper Cretaceous, Campanian) featuring marls and marly limestones together with intercalations of medium to thin beds of dark pelites (CARG, Foglio 214 Bedonia, 2005). On the microscopic point of view, Ardesia shows a very fine grain size together with a strong foliation, including calcite crystals and thin alignments of clay minerals. It is worth it to note the employ of Ardesia to produce world famous billiard table beds.

Chapter 15

Stones from Piedmont

Few stones of this region were used in Lombardy, despite their large availability.

IGNEOUS ROCKS

Sienite di Balma

COMMERCIAL NAME: the name comes from a village north of Biella (north-western Piedmont)
REFERENCES
Jervis: Provincia di Novara, Circondario di Biella; §196 Campiglia Cervo, §197 San Paolo Cervo (p. 78), §198 Quittengo (p. 79)
Salmojraghi: §568 Provincia di Novara; Sienite; Granito della Balma o di Biella (p. 371)
Marmi italiani: Piemonte; Sienite di Balma (p. 23)
PETROGRAPHIC CLASSIFICATION: igneous plutonic rock (syenite)
MACROSCOPIC FEATURES: more or less dark purple colour with black spots; medium to fine grain size
MINERALOGICAL COMPOSITION: K-feldspar, plagioclase (30–35% anortite), quartz, amphibole (hornblende), biotite, titanite
MICROSCOPIC FEATURES: granular textured cryptoperthitic K-feldspar, small crystals of plagioclase, wide prismatic crystals of hornblende, euhedral (angular rhombic sections) crystals of titanite
GEOLOGICAL SETTING: "Plutone di Biella", a late-alpine pluton (Oligocene) linked to a magmatic activity in the inner part of the Alpine range. The pluton is made of monzogranite ("porphiyric granites"), syenite, monzonite, aplite and dicks. The host rocks belong to the "Sesia-Lanzo" zone (metamorphic rocks, mainly micaschists). The outcrop of the pluton shows elliptical shape (about 35 square km) in the middle of the valley of river Cervo (province of Biella).
GEOLOGICAL REFERENCE: CGI, Foglio 43 Biella (1967)
QUARRY: the main site is located near the village of Balma on the right bank of river Cervo. Another quarry, located near San Paolo Cervo on the road of the sanctuary of San Giovanni, is still working.
STONEWORK CATEGORIES: block, slab
SURFACE FINISHING: polishing
USE IN ARCHITECTURE: starting from the quarry area (sanctuary of San Giovanni, 17th century; paving of the of streets of Biella) the use spread to the whole Italian territory in

DOI: 10.1201/9781003361008-17

the 20th century. The most relevant use in the first half of the 20th century is the so-called Chiesa Nuova of the Sanctuary of Oropa (north of Biella), based on a plan of I.A. Galletti (1887): the columns of the front (about 20 m high) together with the pediment, were made using several drums or moulded elements of Sienite; this front was slowly built in the first decades of the 20th century, and the same stone was employed later to build the walls of the church. The use in Milan was very large: slabs for cladding of apartment buildings, mainly after World War II. The more significant use was the Palazzo delle Colonne (G. Greppi and G. Muzio, 1933–40): the cladding of the façade was made using thick slabs with polished surfaces on the ground floor, very thick rusticated slabs with pointed surface on the other floors.
DECAY MORPHOLOGIES: scaling

Rosso Pantheon

COMMERCIAL NAME: the name refers to the famous temple in Rome, dedicated to all the gods. Some columns of this temple are made of red granite (from Aswan, southern Egypt), quite similar to this one.
REFERENCES
Jervis: Provincia di Torino, Circondario di Ivrea; §154 Valperga (p. 71)
Salmojraghi: §569 Provincia di Torino; Rocce massicce (p. 373)
Marmi italiani: Piemonte; Graniti rosso e roseo di Belmonte (detti "Pantheon") (p. 23)
PETROGRAPHIC CLASSIFICATION: igneous plutonic rock (granite)
MACROSCOPIC FEATURES: dark red colour scattered of light grey spots; coarse grain size
MINERALOGICAL COMPOSITION: quartz, K-feldspar, plagioclase, biotite
MICROSCOPIC FEATURES: granular texture of quartz and K-feldspar crystals, with sodium plagioclase and biotite; some thin cracks
GEOLOGICAL SETTING: "Graniti rossi e grigi of Serie del Canavese" (Permian-Triassic) including red and grey granites barren of micas. Outcrops are connected to the ridge going from Levone, Valperga and Cuorgnè (south-west) to Vidracco, Alice and Lessolo (north-east) close to the edge of the alluvial plain.
GEOLOGICAL REFERENCES: CGI, Foglio 42 Ivrea
QUARRY: the main site is located near San Colombano Belmonte, a few kilometres west of Valperga, in the Canavese region (northern part of Torino province)
STONEWORK CATEGORIES: block, slab
SURFACE FINISHING: polishing
USE IN ARCHITECTURE. the stone was in use in the 1930s, mainly for façade cladding of public and apartment buildings. Significant Milanese examples are slabs on the portals and the cladding of the first floor under the porch of Palazzo INA (Milan, piazza Diaz, P. Portaluppi, 1934–37); the balcony of Federazione Fascista – palazzo Castani (piazza San Sepolcro, P. Portaluppi, 1937); the balcony of Palazzo dell'Informazione (piazza Cavour, G. Muzio 1938/42)
DECAY MORPHOLOGIES: scaling

Diorite di Vico Canavese

COMMERCIAL NAME: the name refers to a village of the ancient region of Canavese (northern Piedmont, province of Torino)

REFERENCES
Salmojraghi: §569 Provincia di Torino; Rocce massicce (p. 373)
Marmi italiani: Piemonte; Diorite di Vico Canavese (p. 24)
PETROGRAPHIC CLASSIFICATION: igneous plutonic rock (diorite)
MACROSCOPIC FEATURES: homogeneous dark grey colour with black spots; medium grain size
MINERALOGICAL COMPOSITION: plagioclase, hornblende, biotite, quartz
MICROSCOPIC FEATURES: granular texture with twinned crystal of plagioclase, prismatic crystals of hornblende and biotite, anhedral crystals of quartz
GEOLOGICAL SETTING: "Pluton of Traversella" featuring diorite and granodiorite and pertaining to Tertiary alpine plutonism; the pluton is surrounded by hornfels and embedded in the micaschists of the Sesia-Lanzo zone. The outcrops, mixed with glacial deposits, are scattered in a small area between Traversella (west) and Brosso (east) or between monte Gregorio (north) and Vico Canavese (south).
GEOLOGICAL REFERENCES: CGI, Foglio 42 Ivrea
QUARRY: the main site is located in a small outcrop some 100 m west of Vico Canavese
STONEWORK CATEGORIES: block, slab
SURFACE FINISHING: polishing
USE IN ARCHITECTURE: the stone went in use in Milan during the 1930s for flooring and for cladding (palazzo Crespi, corso Matteotti 1, P. Portaluppi, 1930). A significant example is the pavement, renovated after World War II, of the big niche of Galleria Vittorio Emanuele on piazza della Scala: slabs of diorite di Vico are put together with slabs of Musso (white) and Verona (red).
DECAY MORPHOLOGIES: scaling

METAMORPHIC ROCKS

Bargiolina

COMMERCIAL NAME: the name comes from the village of Barge located at the border between Torino and Cuneo provinces (western Piedmont)
REFERENCES
Jervis: Provincia di Cuneo, Circondario di Saluzzo; §80 Barge (quarzite – p. 48)
Salmojraghi: §571 Provincia di Cuneo; Quarzoscisto; Bargiolina (p. 377)
Marmi italiani: Piemonte; Quarzite di Barge o Bargiolina (p. 35)
PETROGRAPHIC CLASSIFICATION: regional metamorphic rock (quartzite)
MACROSCOPIC FEATURES: homogeneous colour with different hues (yellow, white, grey, green) according to different layers; fine grain size
MINERALOGICAL COMPOSITION: quartz, sericite (minutely crystalline muscovite)
MICROSCOPIC FEATURES: interlocking mosaic of grains of quartz, interstitial sericite
GEOLOGICAL SETTING: "Quartzites and micaschists of Ensemble Sampeyre" (Permian – Triassic) pertaining to the Poly-metamorphic basement. The outcrop occupies the most elevated sector of Monte Bracco, circa 1200 m above sea level; Monte Bracco is located at the edge of the plain (south of Barge, province of Cuneo) as the easternmost relief of Alpi Cozie.
GEOLOGICAL REFERENCES: CGI, Foglio 67 Pinerolo (reprint 1951)

QUARRY: different sites along a narrow road starting from Rocchetta (municipality of Sanfront, across the bridge over the river Po), on the western slopes of Monte Bracco
STONEWORK CATEGORIES: thin slab
SURFACE FINISHING: rough (split face)
USE IN ARCHITECTURE: the local use was devoted to rural architecture featuring irregularly thick slabs in masonries. The use started in Milan during the 20th century, and it was mainly devoted to external paving (courtyards run across by cars) and for wall cladding, always using thin slabs (about 1 cm thick), shaped as irregular polygons. The stone was also used to pave the so-called passeggiate a mare (walkways along the seashore) in many towns of Liguria (i.e. Rapallo, Santa Margherita Ligure, etc.). Outside the Italian territory, the stone was employed in England, for flooring or for cladding, in the 20th century (Bear and Staff public house, Charing Cross, London or Radiant House Gas Company, Bold Street, Liverpool).
DECAY MORPHOLOGIES: scaling on surface

Pietra di Luserna

COMMERCIAL NAME: the name comes from a village between the provinces of Torino and Cuneo (western Piedmont)
REFERENCES
Jervis: Provincia di Cuneo, Circondario di Saluzzo; §80 Barge (gneiss – p. 48), §85 Bagnolo Piemonte (p. 51); Provincia di Torino, Circondario di Pinerolo; §86 Luserna San Giovanni (p. 52)
Salmojraghi: §569 Provincia di Torino; Gneiss (p. 374)
Marmi italiani: Piemonte; Gneiss di Luserna (p. 35)
PETROGRAPHIC CLASSIFICATION: regional metamorphic rock (gneiss)
MACROSCOPIC FEATURES: grey, yellowish, bluish or greenish colour with strong foliation; medium grain size
MINERALOGICAL COMPOSITION: quartz, K-feldspar, muscovite
MICROSCOPIC FEATURES: bands of quartz and K-feldspar crystals with granular habit, surrounded by streaks of muscovite
GEOLOGICAL SETTING: "Orthogneiss of Freydour type" (Pre-Triassic) or meta-intrusive rocks (granitic and dioritic composition) pertaining to the Poly-metamorphic basement. The outcrop involves the upper valley of Torrente Luserna from Luserna to Rorà (province of Turin) to Montoso – Bagnolo Piemonte (province of Cuneo).
GEOLOGICAL REFERENCES: CGI, Foglio 67 Pinerolo (reprint 1951)
QUARRY: the most important sites are located near Montoso, at the end of a road coming from Bagnolo Piemonte and close to the border between the provinces of Torino and Cuneo; the outcrop involves the north-eastern slopes of Pizzo d'Ostanetta
STONEWORK CATEGORIES: slabs of different sizes, small block, cube
SURFACE FINISHING: rough (split face)
USE IN ARCHITECTURE: the almost unique use was for paving thanks to the very high resistance to wear. The sidewalks of Torino are paved with large slabs of Luserna, and also the stairs of the ancient houses are made of this stone. In Milan, the stone was used for the parvis of San Francesco al Fopponino (G. Ponti, 1964). Since the 21st century, the use is well documented in Milan, mainly for paving new urban spaces (i.e. paving of piazza Tre Torri, Citylife, 2017).
DECAY MORPHOLOGIES: scaling on surface

Moncervetto

COMMERCIAL NAME: the name comes from a topographical site of Southern Piedmont (province of Cuneo).
REFERENCES
Scamozzi: part 2, book 7, chap. 5, Marmi e misti (p. 188, line 33)
Rondelet: book 1, sec. 2, art. 11, §140 (p. 140)
Jervis: Provincia di Cuneo, Circondario di Mondovì. §33 Monastero di Vasco (p. 41)
Salmojraghi: §571 Provincia di Cuneo; Calcari; Saccaroidi o ceroidi (p. 376)
Marmi italiani: Piemonte; Moncervetto (p. 32)
PETROGRAPHIC CLASSIFICATION: regional metamorphic rock (marble)
MACROSCOPIC FEATURES: grey to greenish colour with white veins of irregular thickness; fine grain size
MINERALOGICAL COMPOSITION: calcite
MICROSCOPIC FEATURES: alignments of elongated calcite crystals (0.5–1.0 mm) with undulose extinction and rare opaque minerals
GEOLOGICAL SETTING: lenses of crystalline limestone embedded into "Calcescisti" (calcschist – lower Jurassic) of the Ligurian Alps
GEOLOGICAL REFRENCES: CGI, Foglio 80 Cuneo (1935)
QUARRY: a site (owned by Carbocalce) is located south of Monastero di Vasco (between Mondovì and Frabosa, province of Cuneo), on the western slope of Val Corsaglia (Strada Provinciale 183 from Vasco to Corsaglia, a few hundred metres after the fork to Bassi)
STONEWORK CATEGORIES: slab
SURFACE FINISHING: polishing
USE IN ARCHITECTURE: the stone was employed in the Baroque architecture of Piedmont, mainly as slab for floor (i.e. Santuario of Vicoforte, A. Vittozzi and F. Gallo, since the late 16th century) a few kilometres south of Mondovì. The stone went in use in Milan during the 1930s, mainly for cladding of the façades of apartment buildings and for flooring (Banca Popolare, piazza Meda, G. Greppi 1931)
DECAY MORPHOLOGIES: disaggregation, sulphate skin formation

OPHICALCITES FROM PIEDMONT AND VALLE D'AOSTA

(Pietre Verdi – green stones)

Verde Alpi, Verde di Cesana

COMMERCIAL NAME: the generic name comes from the mountain chain; the peculiar name (Verde di Cesana) comes from a village of Valle di Susa (western Piedmont)
REFERENCES
Rondelet: book 1, sec. 2, art. 11, §267 (p. 147)
Jervis: Provincia di Torino, Circondario di Susa; §118 Bussoleno (p. 62)
Salmojraghi: §569 Provincia di Torino; Oficalce; Verde di Susa (p. 373)
Marmi italiani: Piemonte; Verde di Cesana (p. 28)
PETROGRAPHIC CLASSIFICATION: regional metamorphic rock (ophicalcite)
MACROSCOPIC FEATURES: dark green colour with several light green elements and a large grid of white calcitic veins; medium grain size
MINERALOGICAL COMPOSITION: serpentine, magnetite and calcite (white veins)

MICROSCOPIC FEATURES: fine-grained serpentine clasts crossed or surrounded by irregular veins of calcite spar

GEOLOGICAL SETTING: "Oficalciti e brecce basaltiche" formation (Jurassic) pertaining to tecto-stratigraphic "Unità del Lago Nero" ("Unità Ofiolitiche" of the Pre-Pliocenic basement), including ophicalcites, basaltic breccias and serpentine sandstones. Small outcrops near Cesana Torinese in the upper part of Valle di Susa, near the French border (Torino province).

GEOLOGICAL REFERENCES: CGI, Foglio 66 Cesana Torinese (1962); CARG, Foglio 171 Cesana Torinese (2012)

QUARRY: the main site was located west of Cesana Torinese, on the road to Clavière

STONEWORK CATEGORIES: slab

SURFACE FINISHING: polishing

USE IN ARCHITECTURE: the stone went in use in Milan during the 1930s, mainly as slab for cladding and for flooring (Casa Wassermann, viale Lombardia 17, P. Portaluppi, 1935–36); moulded elements were also employed as jambs and lintel in portals and doors.

DECAY MORPHOLOGIES: scaling, roughening, chromatic alteration

Verde Aver, Verde Champ-de-Praz; Verde Chatillon, Verde Issorie, Verde Patrizia (Valle d'Aosta)

COMMERCIAL NAME: similar stones are distinguished according to the name of Alpine villages

REFERENCES

Jervis: Provincia di Torino, Circondario di Aosta; §175 Châtillon, §177 Champ-de-Praz (p. 76)

Salmojraghi: §569 Provincia di Torino; Ofiolite (p. 373)

Marmi italiani: Piemonte; Verde Champ de Praz, Verdi di Chatillon (p. 26)

PETROGRAPHIC CLASSIFICATION: regional metamorphic rock (ophicalcite)

MACROSCOPIC FEATURES: green ground with irregularly arranged white veins; medium grain size

MINERALOGICAL COMPOSITION: serpentine, calcite (white veins), magnetite, clinoypiroxene (diopside), amphibole (tremolite) and Mg-chlorite

MICROSCOPIC FEATURES: aggregates of fibro-lamellar antigorite, irregularly crossed by veins of calcite spar

GEOLOGICAL SETTING: the stone belongs to different ophiolitic units ("Unità del Combin" and "Unità Zermatt-Saas") pertaining to "Zona piemontese" (calcschists and green stones). The "Unità del Combin" (Jurassic-Cretaceous) includes calcschists with prasinites, serpentinites and ophicalcites (serpentine breccias with carbonate matrix); the "Unità Zermatt-Saas" (Jurassic-Cretaceous) includes ophiolites and meta-sedimentary covers (ophicalcites or serpentine breccias with carbonate matrix). Outcrops are widely diffused on both sides of the eastern part of Valle d'Aosta. The "Unità del Combin" occupies the left side of the valley (Verrayes); the "Unità Zermatt-Saas" occupies the right side of the valley (Champdepraz).

GEOLOGICAL REFERENCES: CGI, Foglio 29 Monte Rosa

QUARRY: different sites are present between Verrayes (north-west), Chatillon (centre) and Champdepraz (south) in the eastern part of Valle d'Aosta

STONEWORK CATEGORIES: slab

SURFACE FINISHING: polishing

USE IN ARCHITECTURE: the stone went in use in the 1930s, mainly as slab for cladding and for flooring; moulded elements were also employed as jambs and lintel in portals and doors.

DECAY MORPHOLOGIES. scaling, disaggregation of calcite, roughening, chromatic alteration

ADDENDUM

A particular kind of ophicalcite outcrops in the area of Sizan – Challant Saint Victor (lower part of Val d'Ayas, a few kilometres north-east of Verrès). This stone is called "Malachite" and its colour resembles, in fact, the colour of the mineral of the same name (basic copper carbonate). Serpentinites and serpentinoschists connected to Peridotites, talc and chlorite schists (Triassic igneous rocks) are present in this area. The stone was quarried in the 1930s and later, and some examples are present in Milan; in particular, the architect P. Portaluppi used it for flooring (i.e. Casa Wassermann, 1934–36; Casa Portaluppi, 1935–39; Facoltà di Architettura – first building, 1953–61). Ancient authors ignore the stone; Pieri [Pieri, 1964] listed it under the name "Verde malachite", referring to "Verde Chatillon".

Chapter 16

Stones of Trentino-Alto Adige

Different stones come from this region of north-eastern Italy, and it is very important to point out some historical details influencing the stone trade: the region was ruled for centuries by prince-bishops as "Vescovato di Trento" (southern part) and "Vescovato di Bressanone" (northern part); the whole territory was annexed to the Austro-Hungarian Empire (1815), and it was joint to the Italian territory only after World War I (1918). Since 1948 the regional territory is split between the Italian-speaking province of Trento (south) and the German-speaking province of Bolzano/Bozen (north).

IGNEOUS ROCKS

Porfido del Trentino

COMMERCIAL NAME: the name refers to the petrographic nature of the stone
REFERENCES
Salmojraghi: §633 Trentino; Porfidi quarziferi permiani (p. 449)
Marmi italiani: Venezia Tridentina; Porfido (p. 64)
PETROGRAPHIC CLASSIFICTION: volcanic igneous rock (ignimbrite – rhyodacite)
MACROSCOPIC FEATURES: red to purple to pink to brownish colour; coarse grain size
MINERALOGICAL COMPOSITION: quartz, K-feldspar, plagioclase, sanidine, and biotite together with pyroxene, apatite and zircon.
MICROSCOPIC FEATURES: phenocrystals of quartz and K-feldspar within a matrix mainly made of glass (vitreous texture)
GEOLOGICAL SETTING: "Porfiriti e porfidi Auct" pertain to a thick series of Permian vulcanites, due to huge ignimbritic events; the series includes dacite, andesite and trachyandesite, rhyodacite and again dacite, andesite and trachyandesite ("Complesso inferiore"); dacite, rhyodacite and rhyolite ("Complesso superiore"). The outcrop spreads from Merano (north) to Trento (south) and Agordo (east).
GEOLOGICAL REFERENCE: CGI, Foglio 21 Trento (1969)
QUARRY: different districts are present in the region Trentino – Alto Adige; the most important ones are Val Cembra (Trentino) and Bolzano (Alto Adige/Süd Tirol). Sites of Val Cembra (a valley of the left bank of Adige River, north of Trento) are located near the villages of Albiano, Lases (Montegorsa) and Baselga di Pinè; Villa di Giovo; sites of Bolzano (confluence between Adige/Etsch and Isarco/Eisack) are located on the west bank of the Adige/Etsch river (San Michele Appiano/Eppan) and the east bank of the Isarco/Eisack

DOI: 10.1201/9781003361008-18

river (Cornedo/Karned, Laives/Leifers, Bronzolo/Branzoll). Other similar stones come from Valle di Fiemme (Predazzo, Forte Buso), east of Val Cembra.
STONEWORK CATEGORIES: small cube, slab
SURFACE FINISHING: quarry face, sometimes polishing
USE IN ARCHITECTURE: this stone, cubes with sides less than 10 cm long, was used to pave the roads because of its mechanical features
DECAY MORPHOLOGIES: scaling, cracking

Predazzo (Granito rosa di Predazzo)

COMMERCIAL NAME: the name comes from a village of Val di Fiemme (north-east of Trento)
REFERENCES
Marmi italiani: Venezia Tridentina. Graniti (p. 64)
PETROGRAPHIC CLASSIFICTION: hypoabissal igneous rock (sieno-granite; granite with alkaline feldspars)
MACROSCOPIC FEATURES: reddish to orange colour with dark rounded crystals, granular texture, medium to fine grain size
MINERALOGICAL COMPOSITION: K-feldspar, quartz, plagioclase, biotite; black tourmaline
MICROSCOPIC FEATURES: irregular crystals of quartz, zoned plagioclases, perthitic texture (lamellae of sodium-rich feldspar in a host of potassium-rich feldspar)
GEOLOGICAL SETTING: "Intrusive body of Predazzo" (Permian – Carboniferous), a magmatic body featuring granites and granosyenites (Granito rosa di Predazzo), potassic syenites, monzonites and pyrossenites
GEOLOGICAL REFERENCES: CGI, Foglio 22 Feltre (1971)
QUARRY: the main site is located near Predazzo (road to Moena) in the middle of Val di Fiemme (north-eastern part of the Trento Province)
STONEWORK CATEGORIES: block, slab
SURFACE FINISHING: polishing
USE IN ARCHITECTURE: the use was firstly local (basement of Dante's statue, C. Zocchi, 1896, Trento); later the stone was sent abroad and employed for cladding (i.e. Milan, Palazzo del Toro, E. Lancia and R. Merendi, 1937–40; Venice, Santa Lucia railway station, 1954)
DECAY MORPHOLOGIES: scaling

SEDIMENTARY ROCKS

Rosso di Trento

COMMERCIAL NAME: the name comes from the peculiar colour of the stone
REFERENCES
Rondelet: book 1, sec. 2, art. 11, §237 (p. 145)
Salmojraghi: §633 Trentino; Calcari compatti (p. 449)
Marmi italiani: Venezia Tridentina; Marmi colorati del Trentino (p. 51)
PETROGRAPHIC CLASSIFICATION: chemical sedimentary rock (nodular limestone, biomicrite)

MACROSCOPIC FEATURES: red to pink to white to pale green colour in a nodular texture with shells of Ammonoidea and Bivalvia; very fine-grain size
MINERALOGICAL COMPOSITION: calcite; haematite (pigment), illite (clay fraction)
MICROSCOPIC FEATURES: microcrystalline calcite matrix with micritic nodules together with bioclasts (molluscs and foraminifera); clay minerals and iron-oxides in stratigraphic discontinuities
GEOLOGICAL SETTING: “Rosso ammonitico” formation (upper Jurassic) of the sedimentary series Southern Alps (Venetia); wide outcrops are located in the valley of river Adige, mainly around the town of Trento
GEOLOGICAL REFERENCE: CGI, Foglio 21 Trento (1969); CARG, Foglio 060 Trento (2010)
QUARRY: some sites were located inside the town of Trento (close to Castello del Buon Consiglio) and in some boroughs as Sòlteri (north) and Pila (east); other quarries were located near Terlago (about 5 km north-west of Trento)
Different varieties were sold: Rosso Moro, Rosso Pila, Rosso Terlago (red colour), Ciresolo San Martino (pink), Bianco Pila (white), Verdello (pale green)
STONEWORK CATEGORIES: bock, slab, carving
SURFACE FINISHING: polishing
USE IN ARCHITECTURE: the use of these materials was very important in the region (buildings in Trento, Rovereto, etc.), but also in the neighbouring region of Veneto, a region easily accessible through waterways. The use was increased in the 20th century, thanks to faster transportation systems. The features of this stone are very similar to the features of the more diffused Verona red limestones, so it is difficult to distinguish each stone by visual observations only.
DECAY MORPHOLOGIES: erosion, chromatic alteration, sulphate skin formation

Giallo Mori

The mountainous area around Brentonico, Castione and Mori (close to Val Lagarina, the valley of river Adige in the southern part of the Trento province) about 8 km south-west of Rovereto, supplied coloured stones used by skilled craftsmen all around the Europe (17th and 18th centuries) for the decorations of altars, churches, palaces, etc.
COMMERCIAL NAME: the name comes from the peculiar colour of the stone and from a village of southern Trentino.
REFERENCES
Scamozzi: part 2, book 7, chap. 8, Pietre vive (p. 199, line 47)
Rondelet: book 1, sec. 2, art. 11, §183 (p. 141), §215 (p. 143)
Marmi italiani: Venezia Tridentina; Marmi colorati del Trentino (p. 51)
PETROGRAPHIC CLASSIFICATION: chemical sedimentary rock (oolite limestone)
MACROSCOPIC FEATURES: yellow colour with irregular whitish spots and a grid of reddish veins; very fine grain size
MINERALOGICAL COMPOSITION: calcite, iron hydroxides
MICROSCOPIC FEATURES: microcrystalline calcite with peloids, rhombohedral dolomite, interstitial iron oxides
GEOLOGICAL SETTING: a yellowish limestone (micrite, pellets and echinoids) with regular beds (15–35 cm thick) totally replaces, in the eastern side of the valley of river Sarca, the “Oolite di San Vigilio” formation (oolitic limestones, Toarcian – lower Aalenian,

Jurassic). Outcrops are located in the mountain area between the upper part of Lake Garda (Torbole, west) and the valley of river Adige (Ala, east).

GEOLOGICAL REFERENCE: CGI, Foglio 26 Schio (1968); CARG Foglio 080 Riva del Garda (2005)

QUARRY: the main site (San Rocco) was located near Castione, between Brentonico and Mori, about 7 km south-west of Rovereto (Province of Trento), but other small quarries were present in the surroundings

STONEWORK CATEGORIES: slab

SURFACE FINISHING: polishing

USE IN ARCHITECTURE: the use of this stone, as other regional stone materials, spread in the 1930s when the search of new materials was also encouraged by the Italian Fascist government with a policy enhancing the national products. The most important use in Milan are the cubic tesserae of the floor of Galleria Vittorio Emanuele (1967). The coloured stones of this area were also present in Venice (floor of Basilica San Marco, restored areas) and Trento (altare del Crocifisso, Duomo).

DECAY MORPHOLOGIES: erosion

Other varieties of stone of Val Lagarina are: Grigio di Castione, Giallo di Brentonico, Rosso di Francia del Monte Giovo, Rosso scuro di Prada. The stones pertain to geological formations from Middle Jurassic to Lower Cretaceous.

ADDENDUM

A sedimentary rock coming the province of Trento seems to be employed in Lombardy (Bergamo), but the identification is uncertain. In fact, a stone called "Pietra di Arco" was documented on the upper loggia of Cappella Colleoni (Bergamo) during the restoration works, made by the architect A. Cattò in 1875–76.

Geological literature reports "Pietra di Arco" as white oolitic limestone ("Calcare oolitico di Massone" formation, lower Jurassic) quarried at San Martino d'Arco and Massone, some kilometres eastward from Arco di Trento, on the northern tip of Lake Garda.

Scientific analyses carried out on samples coming from the upper part of Cappella Colleoni (restoration works, 1988) report the presence of a nummulitic limestone. A nummulitic limestone outcrops near Tòrbole ("Calcare di Tòrbole" formation, Middle Eocene), a few kilometres south-east of Arco di Trento.

METAMORPHIC ROCKS

Marmo di Lasa

COMMERCIAL NAME: the name comes from a village (Laas in German) of Val Venosta/Vinschgau (province of Bolzano/Bozen)

REFERENCES

Marmi italiani: Venezia Tridentina; Marmi di Lasa (p. 51)

PETROGRAPHIC CLASSIFICATION: regional metamorphic rock (marble)

MACROSCOPIC FEATURES: white colour with soft bands of different colours (grey to black due to graphite; green due to chlorite; pink due to zoisite); sometimes groups of little elongated black spots with shaded rims are present, and an appropriate cut may produce a particular graphic effect (called "Fantastico"); medium to fine grain size

MINERALOGICAL COMPOSITION: calcite, graphite, chlorite and zoisite

MICROSCOPIC FEATURES: anhedral calcite crystals (polysynthetic twinning and rhombohedral cleavage) with rectilinear boundaries (white ground, grain size 0.2–2.0 mm) or with sutured boundaries (grey bands, grain size 0.2–0.4 mm). Presence of veins of very fine calcite crystals (0.05–0.1 mm) accompanied by rounded quartz crystals (0.3 mm).

GEOLOGICAL SETTING: "Laas Unit" of "Ortles-Campo Nappe" (Austro-Alpine system, Pre-Permian metamorphic basement) including micaschists; paragneisses with muscovite-biotite and quartz-feldspar bands; amphibolites and white marbles with different grain sizes and sometimes folded with evidence of grey, green and yellow bands

GEOLOGICAL REFERENCE: CGI, Foglio 9 Monte Cevedale (1951); CARG, Foglio 025 Rabbi (2009)

QUARRY: different sites on the right slope of the narrow valley of Lasa-Laas (val Venosta-Vinschgau – Bolzano-Bozen province), the quarries were exploited underground, and they were connected to each other. The most important one (Weisswasser) was located at an elevation of 1600 m and the marble came down the hill using an incline railway. Different kinds of Lasa marble were sold: "Lasa bianco perla", "Lasa bianco nuvolato" (with graphite and tourmaline), "Lasa venato vena d'oro" (with titanite and zoisite) and "Lasa venato vena Verde" (with chlorite).

STONEWORK CATEGORIES: block, slab

SURFACE FINISHING: polishing

USE IN ARCHITECTURE: a local use was developed in the Middle Ages and later on; a strong exploitation grew since the middle of the 19th century for construction in Munich (in this time the quarries were located in the territory of the Austro-Hungarian Empire). The use in Milan is limited to the 20th century, mainly as polished slabs for cladding; in some cases, the slabs are disposed in "open book style" (Casa Rustici, P. Lingeri and G. Terragni 1935; Torre Rasini, E. Lancia and G. Ponti, 1934).

DECAY MORPHOLOGIES: surface erosion, sulphate skin formation

Chapter 17

Stones from Tuscany and Sardinia

An important kind of building stone comes from the islands of the Arcipelago Toscano, off the south-western coasts of Tuscany (Tyrrhenian Sea).

IGNEOUS ROCKS – GRANITES

Granito dell'Elba

COMMERCIAL NAME: the name refers to the main island of the Arcipelago Toscano
REFERENCES
Vasari (On technique, chap. 1): "Di questo granito bigio . . . nell'isola dell'Elba".
Milizia: part 1, book 4, chap. 12.3, Marmi modern; §28 Granito – isola d'Elba
Rondelet: book 1, sec. 2, art. 8, "île d'Elbe" (p. 89)
Jervis: Provincia di Livorno, Circondario di Porto Ferraio; §1143 Marciana Marina (p. 315)
Salmojraghi: §588 Provincia di Livorno; Granito (p. 397)
Marmi italiani: Toscana; Granito (p. 113)
PETROGRAPHIC CLASSIFICATION: intrusive igneous rock (granodiorite)
MACROSCOPIC FEATURES: whitish to greyish ground; large, rectangular-shaped white crystals are present in the external area of the pluton; medium to coarse grain size
MINERALOGICAL COMPOSITION: plagioclase (andesine), K-feldspar (rectangular crystals), quartz, biotite
MICROSCOPIC FEATURES: subhedral crystals of K-feldspar, twinned and discontinuously zoned crystals of plagioclase, anhedral crystals of quartz, tabular crystals of biotite
GEOLOGICAL SETTING: Tertiary plutonites of Tuscany (Miocene-Pliocene), subdivided into six different intrusive bodies; the outcrop of Elba concerns the western part of the island (Monte Capanne)
GEOLOGICAL REFERENCES: CGI, Foglio 126 Isola d'Elba (1969); CARG, Foglio 316 Isola d'Elba (?)
QUARRY: the main site was located on the southern slope of Monte Capanne, 100 m west of the village of San Piero in Campo (Campo nell'Elba, province of Leghorn, Tuscany). Ancient sites of granite quarrying are spread westward, on the mountain slope above Càvoli: unfinished columns and other architectural elements are visible together with rustic shelters made out under eroded boulders. The quarry works involve the boulders shaped by erosion and surrounded by trees and shrubs.
STONEWORK CATEGORIES: block, shaft
SURFACE FINISHING: polishing

DOI: 10.1201/9781003361008-19

USE IN ARCHITECTURE: a first use is witnessed during the Roman period (Roman Forum); reuse prevails in the Middle Ages, and it is witnessed by column shafts in Rome (San Gregorio Magno) and in some cathedrals, such as Pisa (Tuscany), Cefalù and Monreale (Sicily). The use in Milan is circumscribed to the 1930s for column shafts and slab for cladding.
DECAY MORPHOLOGIES: scaling

Granito del Giglio

COMMERCIAL NAME: the name refers to the southernmost island of the Arcipelago Toscano, off the west coast of Tuscany (Tyrrhenian Sea)
REFERENCES
Jervis: Provincia di Grosseto, Circondario di Grosseto; §1146 Giglio (p. 317)
Salmojraghi: §592 Provincia di Grosseto; Graniti (p. 402)
Marmi italiani: Toscana; Granito (p. 113)
PETROGRAPHIC CLASSIFICATION: intrusive igneous rock (granodiorite) of Tertiary plutonites of Tuscany
MACROSCOPIC FEATURES: greyish ground interspersed by white and grey crystals; medium to fine grain size
MINERALOGICAL COMPOSITION: plagioclase (andesine), K-feldspar, quartz, biotite, muscovite
MICROSCOPIC FEATURES: crystals of plagioclase and fine aggregates of sericite, subhedral crystals of K-feldspar, anhedral crystals of quartz, tabular crystals of biotite and shreds of muscovite
GEOLOGICAL SETTING: Tertiary plutonites of Tuscany (Miocene-Pliocene), subdivided into six different intrusive bodies; the outcrop of Giglio concerns the whole surface of the island apart from a western promontory (Promontorio del Franco)
GEOLOGICAL REFERENCES: CGI, Foglio 126 Isola d'Elba (1969)
QUARRY: the main sites were located on the oriental coast of the island around Giglio Porto (province of Grosseto): Cala delle Cannelle (south) and Cala dell'Arenella (north)
STONEWORK CATEGORIES: shaft, slab
SURFACE FINISHING: polishing
USE IN ARCHITECTURE: the use is witnessed in the Roman period for column shafts (Ostia, Forum) and in the Middle Ages (Rome, Santa Croce in Gerusalemme and San Lorenzo in Lucina). The use in Milan is circumscribed to the 1930s for column shafts and slab for cladding.
DECAY MORPHOLOGIES: scaling

SEDIMENTARY ROCKS – RED STONES

Different kinds of red colour stone outcrops in Tuscany (Apuanian Alps) and Umbria (Assisi), pertaining to "Rosso Ammonitico" formation (lower Jurassic) and to "Scaglia" formation (Eocene). The use of these stones in Milan is documented in the late 19th century and later. Other red stones outcrops in Tuscany (i.e. Rosso of Monte Amiata), but the use in Milan is not documented.

Rosso di Assisi

COMMERCIAL NAME: the name comes from the town of Saint Francis (central Umbria)
REFERENCES
Jervis: Provincia dell'Umbria, Circondario di Foligno; §1199 Assisi (p. 343)
Salmojraghi: §598 Provincia di Perugia; Calcari; Compatti; Mandorlati (p. 407)
Marmi italiani: Umbria; Mandorlati (p. 116)
PETROGRAPHIC CLASSIFICATION: chemical sedimentary rock (marly limestone)
MACROSCOPIC FEATURES: colour from red with white spots to yellowish to pink, fine grain size
MINERALOGICAL COMPOSITION: calcite
MICROSCOPIC FEATURES: microcrystalline calcite together with some thin veins of calcite spar
GEOLOGICAL SETTING: "Scaglia Bianca e Rossa" formation (Eocene-Cretaceous), pertaining to the marine formations of the Umbro-Marchigiana facies from the Triassic (Calcare massiccio) to the Miocene (Conglomerato di Schifanoia). The formation contains white or pink to red marly limestone: the white one features scale fracture, beds of low thickness, clay layers and flint nodules; the second one (Scaglia Rossa) features beds of moderate thickness (20–30 cm) and a colour varying from pink (bottom) to red with red chert (top).
GEOLOGICAL REFERENCE: CGI, Foglio 123 Assisi (1969)
QUARRY: some sites were located close to Assisi (Colle di San Rufino) or in the nearby Monte Subasio (on the east).
STONEWORK CATEGORIES: slab
SURFACE FINISHING: honed
USE IN ARCHITECTURE: one of the most important stones of Central Italy, largely employed for the historical buildings of Assisi (Duomo, Santa Chiara, etc.). In Milan it was introduced for cladding on Portici settentrionali and Portici meridionali, built in 1870–75 around piazza del Duomo as the completion of the Galleria Vittorio Emanuele. This use seems to be the sole use in Milan's architecture.
DECAY MORPHOLOGIES: erosion, chromatic alteration

Rosso di Castelpoggio

COMMERCIAL NAME: the name comes from a village of the Apuanian Alps (north-western Tuscany, province of Massa-Carrara)
REFERENCES
Salmojraghi: §585 Provincia di Massa e Carrara; Calcari (p. 392)
Marmi italiani: Toscana; Rosso e Viola di Castelpoggio (p. 95)
PETROGRAPHIC CLASSIFICATION: chemical sedimentary rock (limestone, micrite)
MACROSCOPIC FEATURES: red colour with some white parallel white veins; very fine grain size
MINERALOGICAL COMPOSITION: calcite
MICROSCOPIC FEATURES: microcrystalline calcite with some irregular veins made of calcite spar; presence of some undulose veins of clay minerals
GEOLOGICAL SETTING: "Rosso Ammonitico" formation (upper Pliensbachian – upper Sinemurian, lower Jurassic) pertaining to "Unità Tettonica della Falda Toscana" of

the "Dominio Toscano", including pink calcilutites and nodular marly limestone with Ammonoidea; thin beds (about 10 cm) with red marly interbedding. Outcrops are present in a narrow area, going north-west from Carrara to Gragnana and Castelpoggio.

GEOLOGICAL REFERENCES: CGI, Foglio 96 Massa (1971); CARG, Foglio 249 Massa Carrara (2019)

QUARRY: some small sites were located near the village of Castelpoggio, on the road to from Carrara and Gragnana to Campocecina

STONEWORK CATEGORIES: slab

SURFACE FINISHING: polishing

USE IN ARCHITECTURE: the use was witnessed by documents in the 1930s, as slab for flooring and for veneering; on the other hand it is very difficult to distinguish from other red marbles from the northern Tuscan territory (as Rosso Collemandina), also used in this period for the same purposes

DECAY MORPHOLOGIES: chromatic alteration on the surfaces exposed to weathering: the colour changes from red to pale pink

Rosso Collemandina

COMMERCIAL NAME: the name comes from a village of Garfagnana (north-western Tuscany, province of Lucca)

REFERENCES

Jervis: Provincia di Massa e Carrara, Circondario di Castelnuovo di Garfagnana; §1110 Villa Collemandina (p. 300)

Salmojraghi: §585 Provincia di Massa e Carrara; Calcari (p. 391)

PETROGRAPHIC CLASSIFICATION: chemical sedimentary rock (limestone)

MACROSCOPIC FEATURES: red colour including some white parallel veins; very fine grain size

MINERALOGICAL COMPOSITION: calcite

MICROSCOPIC FEATURES: microcrystalline calcite together with fossil fragments (Mollusca), microfossil tests and straight veins of calcite spar

GEOLOGICAL SETTING: "Rosso ammonitico" formation (lower Jurassic, lower Sinemurian – lower Pliensbachian) of "Unità tettonica della Falda Toscana" pertaining to "Dominio toscano", featuring red to grey nodular calcilutites sometimes rich of Ammonoida together with thin grey to red marl and argillite intercalations; the upper part of the formation includes well-bedded brown limestones. Very small outcrops on the left side of Garfagnana (valley of river Serchio), between Villa Collemandina and Corfino. These Jurassic outcrops are surrounded by outcrops of turbiditic sandstones ("Macigno" formation, Oligocene-Miocene).

GEOLOGICAL REFERENCES: CGI, Foglio 96 Massa (1971); CARG, Foglio 250 Castelnuovo Garfagnana

QUARRY: the main site is located a few kilometres north of Castelnuovo Garfagnana (province of Lucca) on the road to Passo delle Radici, in the municipal territory of Villa Collemandina (Sasso Rosso). The old quarry occupies the top of a hill, and it was connected to the main road by a paved way with small stone blocks; a new quarry is located below the old one.

STONEWORK CATEGORIES: slab

SURFACE FINISHING: polishing

USE IN ARCHITECTURE: the use was important in the 1930s, mainly as slab for flooring and for veneering
DECAY MORPHOLOGIES: chromatic alteration, from red to pale pink, of the surfaces exposed to weathering

OTHER SEDIMENTARY ROCKS

Pietra dorata

COMMERCIAL NAME: the name comes from the yellow colour of the stone
REFERENCES: there are no quotations about this stone in the marble treatises of the late 19th–mid-20th centuries
PETROGRAPHIC CLASSIFICATION: clastic sedimentary rock (sandstone)
MACROSCOPIC FEATURES: brown–yellowish ground with some dark brown corrugated veins of different thicknesses; medium to coarse grain size
MINERALOGICAL COMPOSITION: quartz, feldspar, quartz-calcareous cement
MICROSCOPIC FEATURES: angular crystals of quartz (0.2–0.5 mm) and K-feldspar, calcareous fragments, some cavities in the interstices; cement of calcite spar
GEOLOGICAL SETTING: "Arenarie a Scutella" formation (Miocene, Langhian) of the Neogenic sediments of the Central Apennines; featuring clays, conglomerates and calcareous sandstones with beds of irregular thickness, containing *Scutella striatula* (Echinoidea) and *Ostrea sp.* (Mollusca). The grey to yellowish-brown sandstone (quartz-calcareous cement) shows a strongly cemented conglomerate at the base, as the result of a marine transgression, featuring pebbles and cobbles coming from the flysch (Cretaceous-Paleogene); the same pebbles are sometimes present in the sandstones together with argillaceous beds.
GEOLOGICAL REFERENCE: CGI, Foglio 136 Tuscania (1970)
QUARRY: the quarry site, actively worked (Santafiora s.r.l.), is located at Poggio Sanopia, about 5 km south-east of Manciano (province of Grosseto, Tuscany) on the road to San Pietro and Farnese, near the border of Viterbo province (Latium)
STONEWORK CATEGORIES: slab
SURFACE FINISHING: honed
USE IN ARCHITECTURE: the stone was quarried in the late 20th century and use rapidly increased, mainly for cladding: i.e. Milano, viale Montesanto (Calzavara and Tintori, 1995); via Sacchi (2012); Terminal 1 of the International Airport of Malpensa (interior walls, E. Sottsass, 1998)
DECAY MORPHOLOGIES: erosion

Portasanta

COMMERCIAL NAME: the name refers to the ancient coloured marble, quarried in the island of Chios (Aegean Sea, Greece) and used to make jambs of the right door (called Porta Santa) of the Saint Peter's Basilica (Rome), this door only opens during the Holy Years
REFERENCES
Jervis: Provincia di Grosseto, Circondario di Grosseto; §1157 Gavorrano (p. 328)
Salmojraghi: §592 Provincia di Grosseto; Calcari; Compatti; Portasanta (p. 402)

Marmi italiani: Toscana; Portasanta (p. 108)
PETROGRAPHIC CLASSIFICATION: chemical rock (pseudo-breccia, limestone)
MACROSCOPIC FEATURES: red to purple with white, purple or pink irregular spots and very thin grey or purple veins; fine grain size
MINERALOGICAL COMPOSITION: calcite
MICROSCOPIC FEATURES: microcrystalline calcite, veins of different thicknesses made of calcite spar
GEOLOGICAL SETTING: "Calcare massiccio" formation (Hettangian, lower Lias) of the "Serie Toscana", including light grey or whitish massive limestone, sometimes oolitic, with Mollusc remains; the upper part of the formation features a pseudo-clastic texture, pink or reddish colour with purple or yellow shades (Portasanta) and fossils of Ammonoidea, Gasteropoda and Crinoidea. Outcrops are located in the southern part of the "Colline Metallifere" (north-west of Grosseto).
GEOLOGICAL REFERENCE: CGI, Foglio 127 Piombino (1969)
QUARRY: the main quarry site is called "La Crociona", and it is located a few hundred metres from the village of Caldana, on the road from Ravi and Gavorrano (province of Grosseto)
STONEWORK CATEGORIES: slab
SURFACE FINISHING: polishing
USE IN ARCHITECTURE: the use involves polished slabs for flooring in apartment buildings together with other coloured marbles (Milan, Casa Wassermann, P. Portaluppi, 1935–36). A significant use is displayed on the great staircases of Milan's Stazione Centrale (U. Stacchini 1912/31): the big shafts are made assembling moulded pieces of curvilinear shape and pieces of fluted shape.
DECAY MORPHOLOGIES: surface erosion, sulphate skin formation, chromatic alteration

Travertino of Tuscany

COMMERCIAL NAME: the name comes from the Latin appellation of this stone (*Lapis Tiburtinus*) meaning "from Tibur" (Tivoli, east of Rome). The importance of the stone material quarried around Tivoli named "travertine" the deposits of the same origin spread in the whole territory of central Italy.
REFERENCES
Scamozzi: part 2, book 7, chap. 8, Pietre vive (p. 197, line 8)
Rondelet: book 1, sec. 2, art. 15, §313 (p. 203)
Jervis: Provincia di Siena, Circondario di Siena; §1164 Rapolano (p. 331)
Salmojraghi: §591 Provincia di Siena; Travertino (p. 401)
Marmi italiani: Toscana; Travertino (p. 112)
PETROGRAPHIC CLASSIFICATION: non-clastic deposit of calcium carbonate, precipitated from hot spring waters
MACROSCOPIC FEATURES: creamy white to dark yellow colour; compact bands (some centimetres thick) in alternating with very porous bands with deep irregular cavities, sometimes with spongy appearance.
MINERALOGICAL COMPOSITION: calcite
MICROSCOPIC FEATURES: coarse elongated calcite crystals arranged in convoluted forms in alternating with irregular pores left by decaying of plant organisms (leaves, small branches, etc.)

GEOLOGICAL SETTING: "Travertini antichi, recenti ed attuali" sometimes with clay lenses (upper Pleistocene); a wide outcrop is located north and west of Rapolano (province of Sienna).
GEOLOGICAL REFERENCES: CGI, Foglio 121 Montepulciano (1969); CARG, Foglio 297 Asciano (?)
QUARRY: the site is located between Rapolano Terme (north) and Serre di Rapolano (south), about 30 km east of Sienna
STONEWORK CATEGORIES: block, ashlar, drum, slab, carving
SURFACE FINISHING: rough surface (both face bedding or edge bedding); modern use requires the polishing with cavities filled up with epoxy resin
USE IN ARCHITECTURE: it is very difficult to detect, on the basis of a macroscopic observation, the provenance of the Travertino set in a building. Original documents attest the use of Travertino from Rapolano in the Stazione Centrale of Milan (U. Stacchini, 1912–31): in this case the stone (slab) was used in some areas (i.e. cladding of the atrium, but it was replaced by a plaster of the same colour in the upper part of the walls).
NOTE: a light-brown banded travertine was also quarried in the southern part of the Tuscany (province of Grosseto), a few kilometres north of Montemerano, on the road to Terme di Saturnia.

ADDENDUM

Travertines from The Marches and Umbria

Travertines, coming from the same kind of Pleistocene deposits already described, outcrop in the territory of two regions of central Italy (namely The Marches and Umbria). These travertines present light differences (colour of the ground, colour of the band, concretion, cavity), and they were quarried for centuries from different sites and locally used for building purposes (i.e. the Roman, medieval and modern buildings of Ascoli Piceno or the Etruscan, Roman and medieval buildings of Perugia). The possible use of these stones far away from the quarry area is unknown.
REFERENCES (The Marches – Umbria)
Jervis: Provincia di Pesaro e Urbino, Circondario di Urbino; §807 Piòbbico (p. 209); Provincia dell'Umbria, Circondario di Perugia; §1197 Perugia (p. 343)
Salmojraghi: §594 Provincia di Pesaro e Urbino; Calcari; Compatti; Travertino (p. 403)
Marmi italiani: Marche; Travertino (p. 115)
QUARRY: different sites in each region
The Marches: the most important site is located in a region (Colle di San Marco) with deep valleys, escarpments and isolated hills, a few kilometres south of Ascoli Piceno. Another site is located between Piòbbico and Monte Nerone (province of Pesaro-Urbino, about 30 km south-west of Urbino) supplying both "Travertino femmina" (Quaternary travertine) and "Travertino" (actually a Mesozoic limestone).
Umbria: the most important site is located in a wide area, between Ellera and Sodi di Santa Sabina, at the base of the hill of Perugia, a few kilometres east of the town.

METAMORPHIC ROCKS – MARBLES

This term includes the world-renowned white marble from the Apuanian Alps and the coloured one from Sienna.

Marmi delle Alpi Apuane (white, grey, veined marbles)

COMMERCIAL NAME: The names come from the mountain range in the north-western part of Tuscany (province of Massa-Carrara). The usual name "marmo di Carrara" (Carrara marble, called *marmor Lunense* by the Romans) is partially erroneous, as it involves only the northern part of the mountain range (Carrara). The whole quarry area is larger, and it includes also the territories of the central part of the Apuanian Alps (Massa) and of the southern one (Pietrasanta, Seravezza and other sites of the province of Lucca).

REFERENCES

Strabo (book 5, chap. 2, sec. 5) reported the harbour of Luna surrounded by mountains with numerous and large quarries of marble, both white and marked with green. Strabo was also quoted by Scamozzi (part 2, book 7, chap. 5, p. 187, line 20) about the transport to Rome and the use for building.

Plinius (book 36, §14) reported the quarries of Luna with white marbles. Plinius was also quoted by Scamozzi (part 2, book 7, chap. 5, p. 187, line 19) about a comparison between Luna and Paros marbles.

Vasari (On technique, chap. 1): "Sono nelle montagne di Carrara, nella Carfagnana vicino ai monti di Luni, molte sorte di marmi".

Scamozzi: part 2, book 7, chap. 5. Marmi e misti (p. 187, line 13)

Milizia: part 1, book 4, chap. 12.3, Marmi moderni §20 Bardiglio; §35 Statuario

Rondelet: book 1, sec. 2, art. 11, §133, §134 (p. 137); §158, §159, §160, §161 (p. 139)

Amati: Statuario di Carrara; Bardiglio di Carrara (Note – Marmi moderni, book 2, chap. 7, pp. 49–50)

Jervis: Provincia di Massa e Carrara, Alpi Apuane; §1101 Carrara (p. 264), §1102 Massa (p. 283), §1103 Montignoso (p. 287); Circondario di Castelnuovo di Garfagnana; §1108 Vagli Sotto (p. 299); Provincia di Lucca, Valle della Versilia; §1104 Stazzema (p. 288), §1105 Seravezza (p. 292), §1106 Pietrasanta (p. 298)

Salmojraghi: §585 Provincia di Massa e Carrara; Calcari; Alpi Apuane (p. 390)

Marmi italiani: Toscana; Marmi della Regione Apuana; Bianco chiaro (p. 80), Bianco venato o Ordinario venato (p. 84), Bardiglio (p. 86), Statuario (p. 86), Statuario macchiato e venato (p. 87), Piastraccia (p. 87), Paonazzo (p. 87), Paonazzetto (p. 87), Calacata (p. 88), Bianco P (p. 88), Arabescato (p. 89), Fior di Pesco o Persichino (p. 89), Cipollini (p. 89), Brecce e mischi (p. 92), Bardiglio fiorito (p. 95)

PETROGRAPHIC CLASSIFICATION: regional metamorphic rocks (calcitic or dolomitic marble)

MACROSCOPIC FEATURES: mainly white colour, sometimes with irregular dark veins; very fine grain size (about 0.2 mm). The variations of thickening and of disposition of dark veins produce a great number of commercial varieties: statuario, bianco-ordinario, bianco-unito, bianco-venato, nuvolato, arabescato, etc. Bardiglio is the grey variety of marble sometimes with dark grey ground together with lighter parallel veins (Bardiglio Cappella) or with grey ground with darker veins (Bardiglio Imperiale).

MINERALOGICAL COMPOSITION: calcite (main component); dolomite, phyllosilicates, pyrite, graphite (accessory minerals)

MICROSCOPIC FEATURES: mainly polygonal homoblastic fabric formed by an equidimensional mosaic of calcite crystals showing straight to curved boundaries (average crystal size 0.2 mm); some varieties show heteroblastic texture with irregularly shaped calcite crystals; veined varieties show alternate "layers" of very fine-grained crystals (size 0.05)

GEOLOGICAL SETTING: "Marmi delle Apuane" formation (upper Pliensbachian – Hettangian; lower Jurassic) pertaining to Mesozoic and Tertiary covers of "Autoctono toscano *Auctt.*" The formation includes massive white and grey meta-limestones as result of a Tertiary metamorphism. Grey marbles are mainly diffused in the lower part (Bardiglio and Nuvolato) and in the upper one (Bardiglio with uniform grey colour). An example of the varieties of stones present in this succession is reported about the section of Orto di Donna (northern part of the Apuanian Alps): Scisti porfirici (meta-arenite, meta-arkose, chlorite-muscovite schist; middle Silurian); Verrucano (fine-grained quartzite, muscovite phyllite; Carnian – upper Ladinian); Formazione di Vinca (dolomite associated with quartz-muscovite phyllite; Norian-Carnian); Grezzoni (microcrystalline dolomite; Norian); marbles (meta-limestone; Retian-Hettangian – see below for details); Calcari rosati (marly meta-limestone; lower Lias); Calcare selcifero ad Ammoniti (meta-limestone including Ammonoidea; middle-upper Lias); Calcari foliati (meta-limestone with chert; upper Lias – Dogger); Diaspri e Scisti diasprini (meta-radiolarite and siliceous meta-limestone; Malm); Calcari selciferi a Entrochi (cherty meta-limestone; lower Cretaceous – upper Jurassic); Scisti sericitici varicolore (meta-pelite with muscovite and chlorite; Oligocene-middle Cretaceous); Pseudomacigno (micaceous meta-sandstone and meta-siltite; Chattian – upper Oligocene).
"Marmi delle Apuane" formation contains different marbles:

Marmo scistoso, light grey meta-limestone together with yellowish dolomite and greenish meta-pelite. Calcite crystals with twinning planes and size 0.03–0.05 mm (average) sometimes 0.2–0.3 mm; layers of phyllosilicates associated with dolomite.

Marmo dolomitico, grey meta-limestone together with light grey dolomite. Dolomite beds are about 1 m thick, rhombohedral dolomite crystals (0.2–0.4 mm in size); calcite beds are often alternated by whitish dolomite levels 1–10 cm thick, calcite crystals with twinning planes and size 0.1 mm (average), sometimes 0.3–0.8 mm.

Marmo venato, whitish meta-limestone with dense, dark grey vein. Calcite crystals with twinning planes are larger in the white spots (size 0.1–0.4 mm) and smaller into the veins (polygonal shape, size 0.1 mm).

Marmo venato forte, light grey with very thick black veins. Calcite crystals with twinning planes are larger in the white spots (size 0.1–0.4 mm) and smaller into the veins (polygonal shape, size 0.1 mm); opaque minerals concentrated in veins of microcrystalline calcite.

Marmo venato debole, whitsh meta-limestone with rare grey vein and white spot between the vein. Calcite crystals with average of 0.2–0.4 mm in size.

Brecce marmoree, monogenic breccia with white marble clasts (about 50 cm in size) in a dark grey ground. Very fine calcite crystals (<0.05 mm in size) in the matrix; larger calcite crystals with twinning planes in the clasts (0.2–0.4 mm in size).

Marmo bianco, white meta-limestone, homogeneous and lacking in veins. Calcite crystals very homogeneous in size (0.2–0.3 mm in size) with twinning planes.

Bardiglio, dark grey meta-limestone, homogeneous with some thin veins; the colour is due to microcrystalline pyrite and carbonaceous pigment. Calcite crystals with polygonal shape and 0.2–0.4 mm in size; opaque minerals are concentrated in the veins; sometimes quartz crystals are present (0.3–0.5 mm in size).

Marble outcrops are located in the whole Apuanian Alps, from Vinca (north) to Cardoso (south).

Some marble outcrops, north of Massa, refer to "Marmi del Monte Brugiana" formation (lower Ladinian – upper Anisian) pertaining to "Gruppo del monte Folgorito – monte Brugiana" of the "Unità Tettonica di Massa" (Autoctonous). There are white to greenish meta-limestones with Crinoids, mainly foliated in the lower part; the bottom of the formation includes dark grey meta-limestones and black meta-pelites.

GEOLOGICAL REFERENCES: CGI, Foglio 96 Massa (1971); CARG, Foglio 249 Massa Carrara (2019)

QUARRY: hundreds of sites are present in the whole mountain range, some already exploited in Roman times. A series of quarries is located close to the town of Carrara: Pescina-Boccanaglia, Torano (Piastra, Ravaccione); Miseglia (Fantiscritti), Colonnata (Gioia); a second series is located close to the town of Massa (Casette, Forno); a third one is located in the north-western territory of Lucca province (Seravezza, Stazzema, Arni).

The system to lift the marble blocks down from the quarries was, for centuries, the so-called lizzatura: marble blocks were let down along an incline way in order to reach the road for the carriages to the town of Carrara, at the foot of the mountain slopes; a railway line (Carrara – Miseglia – Torano – Fantiscritti) had been operating since 1876. The marbles went from Carrara to Avenza, on the coast of the Tyrrhenian Sea, less than 10 km from the quarry areas. An important remark is the use of the port of *Luna* (a few kilometres north, along the coast) during the Roman times, as the white marble was called "marble of Luna" ("marmor lunense"). The port of *Luna* was disused in the Middle Ages, so the marbles were loaded in the shore of Avenza using a very long pier on the sandy beach and small sailing boats. These boats went to the port of Livorno (Leghorn), then the marbles were loaded on other boats and followed different marine routes to reach the yards. For example, to reach the Padania plain (Mantua, Milan, Turin, etc.) the boats went around the Italian peninsula (Tyrrhenian and Adriatic Sea to Venice), then through the course of the river Po and their tributaries, the marbles arrived at the different towns.

STONEWORK CATEGORIES: block, slab, shaft, carving

SURFACE FINISHING: from pointed finish to the polishing

USE IN ARCHITECTURE: the marble was employed since the Roman period despite the great distance and the long journey needed to reach the towns in Italy and abroad. The use was pointed out to particular works when the very high quality of the marble was needed (sculpture, ornament, decoration). Some outstanding examples are monuments of Rome, such as Ara Pacis Augustae, the Imperial arches (Septimius Severus, Titus), the columns (Trajan and Marcus Aurelius); monuments in the Roman Italian territory (Trajan arch in Benevento), the temples (Vesta, Saturn, Castor and Pollux, Venus and Rome), the Basilicas (Julia, Aemilia, Maxentius), the theatres (Marcellus) and the baths (Caracalla, Diocletianus). White marble was employed in the Middle Ages in the churches of Central Italy (Pisa, Lucca, Pistoia, Florence, Sienna, Orvieto). White marble was employed in the Renaissance and Mannerist periods by sculptors such as Jacopo della Quercia, Verrocchio, Donatello, Michelangelo Buonarroti, Giambologna and for the architectural decorations in Florence (Santa Maria del Fiore), of Rome (San Marco), Venice (Palazzo Ducale) and in other towns in Northern (Cremona, Genoa) and Southern Italy (Messina). Hundreds of bas-reliefs with scenes from the Holy Bible were set on the church façades and inside (i.e. the panels around the choir of Milan's Duomo, early 17th century).

The use of white marble was always very large in the Baroque and Rococo periods for sculptures (Bernini, Borromini) and decorations (Basilica of San Pietro and Fontana di Trevi in Rome); the whole Italian territory was involved from north to south (Milan to Naples and Palermo). It is worth it to note a hundred Baroque altars where white marble was used as the base of the inlays made of coloured marbles and other materials; in the same way, the use of white marble in floors enhanced the patterns made of coloured materials. Brecciated marbles, mainly from Stazzema, were also diffused in Baroque altars both as column shaft and as decorative material. The neo-classicism magnified the whiteness of the Apuanian marble, and different masters of this artistic period (Canova, Thorvaldsen) created sculptures reaching a very high degree of refinement through the matchless qualities of this marble. White marble was employed to make sculptures evoking the Roman times (Milan, Arco della Pace, L. Cagnola, 1838). White marble was the protagonist of the Romantic sculpture with a separation from the cult of the Antiquity, as witnessed by Vela, Bartolini, Duprè, Butti, etc.

White marble was also used in the late 19th and early 20th centuries for the funerary monuments (Cemeteries of Milan, Brescia, Bergamo, Genoa, etc.). Finally, in the first decades of the 21st century, white marble is still employed for flooring and for cladding in public and private buildings.

The use of white marble in Lombardy followed the pattern described earlier: the sculptures of Cappella Colleoni (Bergamo), the sculptures and the reliefs of the church of the Certosa (Pavia), the Counter-Reformation altars, the Baroque and Rococo decorations of dozens of churches in the whole regional territory, the contemporary sculptures by Messina or Martini.

DECAY MORPHOLOGIES: disaggregation, cracking, sulphate skin formation

Cipollino apuano

COMMERCIAL NAME: the name refers to the arrangement of the undulating veins reminiscent of an onion ("cipolla" in Italian language). The original name refers to the Greek marble from Karistos (southern tip of the island of Euboea) with a very similar texture.

REFERENCES

Vasari (On technique, chap. 1): "I marmi cipollini sono un'altra specie, di grana e colore differente"

Marmi italiani: Toscana; Cipollini (p. 89)

PETROGRAPHIC CLASSIFICATION: regional metamorphic rock (crystalline limestone, marble)

MACROSCOPIC FEATURES: whitish ground with green undulating bands; very fine grain size. "Cipollino zebrino" shows perfectly straight black veins on a white to yellowish ground, like the coat of zebra.

MINERALOGICAL COMPOSITION: calcite, chlorite

MICROSCOPIC FEATURES: anhedral crystals of calcite together with thick irregular veins made of chlorite flakes and needles of varying sizes

GEOLOGICAL SETTING: "Scisti sericitici" formation pertaining to Mesozoic – Tertiary covers of "Autoctono Toscano". The formation is made of silica meta-pelites together with interbeds of grey meta-limestone sometimes with chert (lower Cretaceous, Aptian – upper Oligocene, Chattian). This formation includes also "Marmo Cipollino": light grey or greenish or pinkish meta-limestones together with interbeds of grey-greenish or purplish-red meta-pelites (upper Cretaceous, Turonian – Maastrichtian).

GEOLOGICAL REFERENCES: CGI, Foglio 96 Massa (1971); CARG, Foglio 249 Massa Carrara (2019)

QUARRY SITES: different sites of Apuanian Alps, such as the Carrara district or the Cardoso and Arni districts (Lucca province)

STONEWORK CATEGORIES: slab, shaft
SURFACE FINISHING: polishing
USE IN ARCHITECTURE: the use was spread in the 20th century, mainly as polished slabs for façade cladding in Italy and Milan (i.e. Palazzo Montecatini, G. Ponti, A. Fornaroli and E. Soncini, largo Donegani, 1936–38) and abroad (i.e. Vienna, Loos Haus, A. Loos, 1911). Shafts were also erected: an outstanding example in Milan is Stazione Centrale (U. Stacchini, 1912–31: polygonal shafts for lamps in Galleria delle Carrozze and in Padiglione Reale [platform 21]). Cipollino zebrino was employed in polished slabs for flooring (i.e. apartment buildings, Milan, piazza VI Febbraio 10, 1933; Palazzo delle Poste, Ferrara, 1930; Oratorio Santa Maria Assunta, Riomaggiore, 1941).
DECAY MORPHOLOGIES: erosion, disaggregation, cracking (along the veins), chromatic alteration, sulphate skin formation

Calacata

COMMERCIAL NAME: the name refers to a village of the Carrara district
REFERENCES
Marmi italiani: Toscana; Calacata (p. 88)
PETROGRAPHIC CLASSIFICATION: regional metamorphic rock (crystalline limestone, marble)
MACROSCOPIC FEATURES: white ground with gold-yellowish irregular veins; very fine grain size
MINERALOGICAL COMPOSITION: calcite
MICROSCOPIC FEATURES: polygonal calcite crystals with increased size along the veins
GEOLOGICAL SETTING: "Marmi delle Apuane" formation (upper Pliensbachian – Hettangian; lower Jurassic) pertaining to Mesozoic and Tertiary covers of "Autoctono toscano *Auctt.*"
GEOLOGICAL REFERENCES: CGI, Foglio 96 Massa (1971); CARG, Foglio 249 Massa Carrara (2019)
QUARRY: the main quarry site is located in the Carrara district (Calacata, Bacino di Boccanaglia, between Torano and Boccanaglia). Other quarry sites are spread throughout the whole Apuanian Alps (i.e. Vagli).
STONEWORK CATEGORIES: slab
SURFACE FINISHING: polishing
USE IN ARCHITECTURE: the use was spread in the 20th century for interior cladding (entrance of apartment buildings) and for flooring (Casa Wassermann, P. Portaluppi, Milan, 1934–36)
DECAY MORPHOLOGIES: erosion, disaggregation, sulphate skin formation

Breccia Medìcea (Breccia di Seravezza)

COMMERCIAL NAME: the name is referred to the Medici family, rulers of Florence for centuries
REFERENCES
Amati: Seravezza (Note – Marmi moderni, book 2, chap. 7, p. 50)
Jervis: Provincia di Lucca, Circondario di Lucca; §1104 Stazzema (p. 289)
Salmojraghi: §586 Provincia di Lucca; Brecce; Calcaree (p. 394)
Marmi italiani: Toscana; Brecce e mischi (p. 92)

PETROGRAPHIC CLASSIFICATION: regional metamorphic rock (marble breccia)
MACROSCOPIC FEATURES: white ground with purple spot with elliptical shape
MINERALOGICAL COMPOSITION: calcite, iron oxides, iron silicates, chlorite
MICROSCOPIC FEATURES: clastic texture with fragments of marble (crystalline calcite) in a microcrystalline calcite matrix
GEOLOGICAL SETTING: "Brecce di Seravezza e Scisti a cloritoide" formation (Hettangian – upper Norian) refers to the Mesozoic and Tertiary covers of "Autoctono *Auctt.*" The formation features polygenetic meta-breccias strongly heterogeneous with dolomite and limestone clasts varying from ochre to yellow to red to dark grey to black. The matrix is always rich in chloritoid with purple-red to dark green to yellow greenish to black colours. The outcrop is concentrated in a small area in the central part of Alpi Apuane (Massa) less than 10 km from Seravezza.
GEOLOGICAL REFERENCES: CGI, Foglio 96 Massa (1971); CARG, Foglio 249 Massa Carrara (2019)
QUARRY: some sites are located between Ponte Stazzemese (west) and Stazzema (east), others are present on Monte Corchia (road from Seravezza to Galleria del Cipollaio)
STONEWORK CATEGORIES: slab
SURFACE FINISHING: polishing
USE IN ARCHITECTURE: the use started in the Renaissance, under the support of the Medici family, for decorative purposes both in Italy (i.e. Florence, Cappella dei Principi, M. Nigetti, since 1604) and abroad (i.e. London, Westminster Cathedral, J.F. Bentley, 1895–1903). Noteworthy are the four shafts of the ciborium of Santa Croce (C. Arpesani, 1917).
DECAY MORPHOLOGIES: scaling, chromatic alteration

Marmi di Siena (Giallo di Siena)

Another important group of marbles comes from the area of Montagnola Senese, west of Sienna.
COMMERCIAL NAME: the name comes from the town of "Palio" horse race (central Tuscany)
REFERENCES
Scamozzi: part 2, book 7, chap. 5, Marmi e misti (p. 188, line 4)
Milizia: Marmi moderni; §27 Giallo di Siena (part 1, book 4, chap. 12.3)
Amati: Giallo di Siena (Note – Marmi moderni, book 2, chap. 7, p. 49)
Rondelet: book 1, sec. 2, art. 11, §135 (p. 137), §188 (p. 141), §196 (p. 142), §297 (p. 149)
Jervis: Provincia di Siena, Circondario di Siena; §1133 Casole d'Elsa (p. 311), §1159 Chiusdino, §1160 Sovicille (p. 329)
Salmojraghi: §591 Provincia di Siena; Calcari; Lucidabili, mesozoici; La Montagnola (p. 400)
Marmi italiani: Toscana; Giallo di Siena (p. 106)
PETROGRAPHIC CLASSIFICATION: regional metamorphic rock (crystalline limestone)
MACROSCOPIC FEATURES: from yellow colour with pale shades and thin black veins to grey colour with lighter shades; fine grain size
MINERALOGICAL COMPOSITION: calcite; quartz and phyllosilicates; pigments as limonite, hematite
MICROSCOPIC FEATURES: anhedral crystals of calcite with strong alignment, pyrite and iron hydroxides

GEOLOGICAL SETTING: "Marmi della Montagnola Senese" formation (lower Lias) pertaining to "Gruppo delle formazioni metamorfiche Mesozoico-Terziarie" of the "Unità tettonica di Monticiano – Roccastrada" of "Dominio Toscano". Grey and white massive marbles, changing in the upper part to yellow colour marbles; locally the upper part shows well-bedded yellow marbles, including some layers of purple chloritic schists.

GEOLOGICAL REFERENCE: CGI, Foglio 120 Siena (1969); CARG, Foglio 296 Siena (2009)

QUARRY: the sites are located in two sectors of the western side of the Montagnola senese (a hilly country a few kilometres west of Sienna): Marmoraja (Casole d'Elsa – North), Tegòia (Sovicille – South). The most important varieties are: Giallo Broccatello, Paonazzo, Giallo ocra, Giallo venato, Giallo avorio together with Bianco, Bianco arabescato and Bardiglio. In many cases, the size of the blocks is conditioned by tectonic and stratigraphic accidents.

STONEWORK CATEGORIES: slab, carving

SURFACE FINISHING: polishing

USE IN ARCHITECTURE: the use of this marble during the 17th and 18th centuries, in Lombardy and in other northern regions of Italy, was mainly for decorative purposes. Dozens of Baroque altars contain thin slabs of yellow marble as inlay, together with other coloured marbles and stones, in structures made of white marble or black limestone. The use increased, since the 1930s, for flooring of apartment buildings (i.e. the floor of the "Salone dell'Affresco" at the ground floor of Museo Poldi Pezzoli, F. Reggiori, 1950).

DECAY MORPHOLOGIES: surface erosion, sulphate skin formation

OTHER METAMORPHIC ROCKS

The Apuanian Alps supply another kind of metamorphic rock, apart from marbles: a meta-sandstone, used mainly in the 20th century.

Pietra del Cardoso

COMMERCIAL NAME: the name refers to a village of the Carrara district

REFERENCES

Jervis: Provincia di Lucca, Circondario di Lucca; §1104 Stazzema (Schisti argillosi – p. 290)

Salmojraghi: §586 Provincia di Lucca, Scisti; Alpi Apuane; Palaeozoici (p. 394)

Marmi italiani: Toscana; Pietra del Cardoso (p. 97)

PETROGRAPHIC CLASSIFICATION: regional metamorphic rock (meta-sandstone)

MACROSCOPIC FEATURES: silver grey to greenish colour with evident schistosity; very fine grain size

MINERALOGICAL COMPOSITION: quartz, sericite, chlorite

MICROSCOPIC FEATURES: minute shreds of sericite with ribbon texture in alternating lepidoblastic and granoblastic bands

GEOLOGICAL SETTING: "Pseudomacigno" formation (lower Miocene Aquitanian – upper Oligocene Chattian) referring to Mesozoic and Tertiary covers of "Autoctono Toscano *Auctt.*" The formation includes litho-feldspathic meta-sandstones, beds from decimetres to metres, and grey meta-pelites. Outcrops are spread in a wide area between Seravezza (west) and Castelnuovo Garfagnana (east), mainly in the northern side of the Apuanian Alps.

GEOLOGICAL REFERENCES: CGI, Foglio 96 Massa (1971); CARG, Foglio 249 Massa Carrara (2019)
QUARRY: some sites are located a few kilometres north of Cardoso, a village at the end of a small road from Seravezza, on the southern slope of Pania della Croce (province of Lucca)
STONEWORK CATEGORIES: slab
SURFACE FINISHING: rough face
USE IN ARCHITECTURE: the local use includes slabs for roofing and slabs for stairways. In Milan it is worth it to note the slabs used for cladding of the base of Palazzo Civita (piazza Duse 2, G. Zanini, 1932–34).
DECAY MORPHOLOGIES: scaling

STONES FROM THE REGIONS OF SARDINIA AND CORSICA

The islands of the Tyrrhenian Sea supply some building stones from the wide outcrops of granite and from some outcrops of limestones. Granites were traded to Milan and Lombardy after World War II; limestones were used for local purposes.

Igneous rocks – granites

The islands of the Tyrrhenian Sea are interested by a batholith, called "Batolite Sardo-Corso" (Hercinian Basement, upper Carboniferous – Permian). The outcrop involves the western part of Corsica (about 150 km in length and 65 km in width) and the eastern part of Sardinia (about 400 km in length and 100 km in width). The Sardinian outcrop spreads from the Arcipelago della Maddalena and Gallura to mount Limbara in the north, to Barbagia in the centre and from Ogliastra to Sarrabus in the south. This intrusive complex is subdivided in different intrusive units, including leucogranite (Gallura, province of Sassari; Sàrrabus, province of Sud Sardegna); monzogranite (Gocéano, province of Sassari, Gallura, Barbagia, province of Nuoro); monzogranitic granodiorite (Marghine, province of Nuoro, Gallura, Barbagia); tonalitic granodiorite (Ogliastra, province of Sud Sardegna, Gocéano); tonalite (Gocéano, Mandrolisai, province of Nuoro); gabbro (Gallura, Gocéano, Sàrrabus).

GEOLOGICAL REFERENCES: CGI, Foglio 167 Isola Rossa – La Maddalena (1974), Foglio 169 Isola Capreara (1974), Foglio 182 Olbia (1974), Foglio 194 Ozieri (1965), Foglio 195 Orosei (1967), Foglio 207 Nuoro (1976), Foglio 208 Dorgali (1971), Foglio 219 Lanusei (1960), Foglio 234 Cagliari (1951), Foglio 235 Villasimius (1951); CARG, Foglio 411 Santa Teresa di Gallura (2012), Foglio 412 La Maddalena (2016), Foglio 428 Arzachena, Foglio 443 Tempio Pausania; Sevizio Geologico Nazionale, Carta Geologica della Sardegna – Foglio Nord and Foglio Sud (1996)
The quarrying and trading of granites of Sardinia were developed in the Roman period, even if the use was mainly local. The Roman exploitation was pointed on the north-eastern coast (Santa Teresa – La Maddalena), featuring monolithic elements for column shafts (see next). The use of these granites was greatly increased in the late 19th century; the modern exploitation mainly involved other areas of Sardinia (Tempio Pausania), featuring slabs for cladding and flooring (i.e. apartment buildings of Milan and other towns of Lombardy and abroad).

The present-day quarry basins are located in different provinces of Sardinia.

Province of Sassari

Arzachena – Luogosanto, quarry sites of Bassacutena and Luogosanto. Monzogranite, heterogeneous grain size, small pink crystals of K-feldspar (size 1–2 cm). Commercial name: Granito rosa.

Tempio Pausania – Calangianus, quarry sites of Aggius, Luras, Sant'Antonio di Gallura, monte Lìmbara. Monzogranite, heterogeneous grain size, big pink crystals of K-feldspar (size 8–10 cm). Commercial name: Ghiandone.

Buddusò – Alà dei Sardi. Leuco-monzogranite, homogeneous grain size, white crystals of K-feldspar (size 1–3 cm). Commercial names: Granito grigio, Granito bianco.

Province of South Sardinia

Villasimius, quarry sites of Porto Giunco and Punta Molentis. Pink biotite granite with porphyritic structure; granite with two micas; white granite with coarse crystals of K-feldspar. Commercial name: Grigio.

NOTE

Orbicular diorite (called "Granito napoleonico" or "Corsite") was quarried from several dykes in Corse (Sainte Lucie de Tallano, Corse du Sud, France; about 20 km north-east of Sartène) and used for decorative purposes (i.e. vase, tabletop). This particular kind of igneous rock shows almost regular spheres of a few centimetres in diameter, featuring black hornblende and white plagioclase arranged in concentric circles around a nucleus.

Granito della Maddalena (Santa Teresa di Gallura)

COMMERCIAL NAME: the name comes from the group of granitic islands (Arcipelago della Maddalena), a few hundred metres from the north-eastern coast of Sardinia (Palau).

REFERENCES

Jervis: Provincia di Sassari, Circondario di Tempio; §1437 Santa Teresa Gallura, §1438 La Maddalena (p. 444)

Salmojraghi: §629 Provincia di Sassari; Graniti (p. 444)

Marmi italiani: Sardegna; Graniti (p. 153)

PETROGRAPHIC CLASSIFICATION: igneous plutonic rock (monzogranite)

MACROSCOPIC FEATURES: pale pink colour with some black spots; heterogeneous grain size

MINERALOGICAL COMPOSITION: quartz, K-feldspar, plagioclase, biotite

MICROSCOPIC FEATURES: euhedral crystals of K-feldspar (size 0.5–1.0 cm at La Maddalena or size 1.0–4.0 cm at Santo Stefano and Santa Teresa di Gallura), multiple twin plagioclase, biotite with tabular habit and small anhedral crystals of quartz

GEOLOGICAL SETTING: "Complesso intrusivo tardo-Palaeozoico" (upper Carboniferous) outcropping in the north-western area of Sardinia, including the islands of the Arcipelago della Maddalena. Different units of monzogranite outcrop in different sites: intrusive unit "La Maddalena", subunit "Guardia Vecchia", facies "Cala Francese" (La Maddalena); intrusive unit "Palau", facies "Palau" (Santo Stefano); intrusive unit "Longosardo", subunit "Monte La Colba", facies "Capo Testa" (Santa Teresa di Gallura).

GEOLOGICAL REFERENCES: CARG, Foglio 411 Santa Teresa di Gallura (2012), Foglio 412 La Maddalena (2016), Foglio 428 Arzachena

QUARRY: granites of La Maddalena are quarried in different islands of this archipelago (north-eastern coast of Sardinia); the most important sites are located on the southern coast of the island of Santo Stefano and on the western coast of La Maddalena (Cala Francese); other quarries are located on the nearby island of Caprera. These quarries have been worked since the late 19th century.

The quarries at Capo Testa (a small peninsula less than 5 km west from Santa Teresa di Gallura) were located, since the Roman times, at Punta Acuta, Baia Santa Reparata, Cala Spinosa. Some unfinished column shafts are still visible together with several rock surfaces marked by rows of little holes suitable to contain the wedges in order to split up rough blocks.

STONEWORK CATEGORIES: block, slab

SURFACE FINISHING: polishing

USE IN ARCHITECTURE: some column shafts of the monzogranite (Capo Testa) were identified in ancient Rome (i.e. porch of San Lorenzo in Lucina; left nave of Santa Maria in Aracoeli). Moreover, the granite of La Maddalena has been employed on the mainland since the late 19th century (i.e. Genoa, galleria G. Mazzini and the paving of via Roma; Rome, ponte Palatino, architect A. Vescovali, 1886). The stone has been known in Lombardy since the 1930s: the most important example is the Palazzo delle Colonne (architects G. Greppi and G. Muzio, 1934–42) with twenty polished column shafts (Santo Stefano).

DECAY MORPHOLOGIES: scaling

Chapter 18

Stones from Veneto

IGNEOUS ROCKS

Trachite euganea (Masegna)

COMMERCIAL NAME: the term "trachyte" (trachyte) refers to the petrographic classification together with the geographic area; the term "masegna" comes from the Venetian dialect referring to the stone blocks paving the whole streets and quays of Venice
REFERENCES
Scamozzi: part 2, book 7, chap. 8, Pietre vive (p. 200, line 37)
Rondelet: book 1, sec. 2, art. 15, §274 (p. 199)
Jervis: Provincia di Padova, Distretto di Monselice; §495 Battaglia (p. 148); Circondario di Padova; §503 Cervarese Santa Croce (p. 150)
Salmojraghi: §553 Provincia di Padova; Masegna; Euganei (p. 346)
Marmi italiani: Veneto; Trachite (p. 64)
PETROGRAPHIC CLASSIFICATION: volcanic igneous rock (trachyte)
MACROSCOPIC FEATURES: grey and uniform colour scattered with white round spots and very thin black laminae, sometimes the stone is crossed by thick variegated brown to yellow veins and spots (in this last case it is called "Calda variegata"); medium grain size
MINERALOGICAL COMPOSITION: sanidine, plagioclase, hornblende, biotite
MICROSCOPIC FEATURES: porphyric texture showing euhedral phenocrysts of sanidine (average 1 cm) together with sodium plagioclase (polysynthetic twinning); amphiboles and biotite; ground mass made of potash feldspar and plagioclase
GEOLOGICAL SETTING: "Monte Merlo member" of "Monte Venda formation" (lower Oligocene) of the "Provincia magmatica terziaria" of western Veneto, featuring trachytes and quartz-trachytes as sub-volcanic bodies and dykes; the thickness varies from 1 m to 100 m. Trachytes, mixed with per-silicic ryolites and latites, form a particular district (the sole district with acid nature) of the "Provincia magmatica terziaria". Limestones ("Biancone" of Jurassic-Cretaceous and "Scaglia rossa" of Cretaceous-Eocene) are mixed together with volcanic materials. The outcrop concerns the whole Euganean Hills, south of Padua.
GEOLOGICAL REFERENCES: CGI, Foglio 50 Padova (1947) and Foglio 64 Rovigo (?); CARG, Foglio 147 Padova Sud (2011)
QUARRY: different sites are present, mainly in the northern and in the eastern parts of the Euganean Hills, such as Monte Rosso, Montemerlo, Monte Lonzina, Monselice and Zovon di Vò. Each quarry formed a big slash on the woody slopes of the hills, and the

DOI: 10.1201/9781003361008-20

quarry works were strongly opposed in the name of the environment preservation, in the last decades of the 20th century; most of the quarries are now ceased and the quarry faces are partially hidden by vegetation. Different commercial varieties of stone were available according to the different quarries: Chiara, Grigio-scura, Roseo-fiorita, Zovonite.

STONEWORK CATEGORIES: block, slab, tiles

SURFACE FINISHING: bush hammering, sometimes polishing

USE IN ARCHITECTURE: the first uses go back to Roman period with blocks for masonries, but, thanks to the high resistance, the stone (polygonal elements) was mainly employed to pave the famous Roman roads crossing the whole of northern Italy. In Milan, a fragment of a Roman road is exposed in the "Stazione Missori" of the underground line "MM3". Blocks coming from Roman buildings were re-used as block or as column shaft in the Romanesque architecture of Middle Ages: i.e. blocks for the piers of the atrium of Sant'Ambrogio; shafts, bases or capitals in the crypt of San Giovanni in Conca, columns in the nave of San Celso. The use started again in the central decades of the 20th century (from the 1930s to 60s) mainly as slab for cladding of apartment buildings: (Torre Snia, piazza San Babila – A. Rimini, 1935/37) or as tile floor (Entrance building of Ospedale Maggiore, Niguarda – G. Marcovigi and G. Arata, 1935–39) or as portal (the main gate of the theatre Piccola Scala, M. Zavelani Rossi, 1949–55, demolished in 2002). Trachyte of the Euganean Hills was often used in buildings planned by the Studio BBPR: i.e. the staircase of Torre Velasca (1951–58) or the façades of different buildings (via Hoepli 7, 1958–69; via Nerino 15, 1968). The medieval reuse is also witnessed in Pavia: blocks at the base of Torre Civica and in other towers, cornerstones of San Teodoro.

Trachyte was widely employed in the Euganean Hills to make masonries (squared block), building elements (staircase, balconies, etc.), paving (small irregular blocks). Venice streets (called "salizzada") were sometimes paved with rectangular blocks of trachyte; since the middle of the 19th century, the trachyte paving was expanded to the whole urban area. Two important sights of Padua were made using squared blocks of trachyte: the Basilica of Sant'Antonio (buttresses on the sides, early 14th century); the pedestal of the equestrian statue of Erasmo da Narni – il Gattamelata (bronze by Donatello, 1453).

DECAY MORPHOLOGIES: scaling, crumbling

SEDIMENTARY ROCKS

Marmo di Chiampo

COMMERCIAL NAME: the name comes from a village located in the corresponding valley north of Vicenza

REFERENCES

Scamozzi: part 2, book 7, chap. 8, Pietre vive (p. 199, line 54)

Rondelet: book 1, sec. 2, art. 15, §269 (p. 198)

Jervis: Provincia di Vicenza, Distretto di Vicenza; §487 (p. 145)

Salmojraghi: §555 Provincia di Vicenza; Calcari; Compatti (p. 349)

Marmi italiani: Veneto; Marmi di Chiampo (p. 56)

PETROGRAPHIC CLASSIFICATION: biochemical sedimentary rock (limestone, biomicrite – packstone)

MACROSCOPIC FEATURES: white to yellow to pink ground colour scattered with small light spots, often interested by wavy black veins; clearly visible skeletal grains of foraminifera or tests of nummulites
MINERALOGICAL COMPOSITION: calcite
MICROSCOPIC FEATURES: a matrix mainly made of microcrystalline calcite; packed tests of nummulites featuring radial fibrous structure together with other bioclasts; syntaxial overgrowth cement on fossil grains
GEOLOGICAL SETTING: "Calcari nummulitici" formation (Eocene) of the Sedimentary series of the Southern Alps (Western Veneto), including nummulite limestones, calcarenites, reef limestones, sandstones and marls. Limestone beds are in alternating with basalt beds pertaining to Tertiary volcanites (basal flows, hyaloclastites, tuffs etc.; Oligocene – Paleocene). The outcrop occupies a square portion of the province of Vicenza from Schio (north) to San Bonifacio (south-west) to Vicenza (south-east); this territory is bounded by two parallel faults running north–south.
GEOLOGICAL REFERENCES: CGI, Foglio 49 Verona (1968)
QUARRY: some sites are located a few kilometres above Chiampo, on the right bank of river Chiampo (road to Vestenanova) or on the left bank (road to Mussolino). Many different varieties, based on the colour and the texture, were sold: Perla, Mandorlato, Paglierino, Rosa, Ondàgata, Porfirico.
STONEWORK CATEGORIES: block, slab, tile
SURFACE FINISHING: polishing
USE IN ARCHITECTURE: the use in Milan was large but exclusively linked to the 20th century; mainly for cladding and flooring: i.e. the cladding of the façade of Banca d'Italia (piazza Edison, G. Broggi, 1914), the cladding of the façade of Palazzo dell'Informazione (piazza Cavour – G. Muzio, 1938/42); the floor of the entrance building of Ospedale Maggiore (Niguarda – G. Marcovigi and G. Arata, 1935/39)
DECAY MORPHOLOGIES: erosion, sulphate skin formation

Lumachella di San Vitale

COMMERCIAL NAME: the name comes from the aspect of the stone with shell fragments linked to a village of Monti Lessini (province of Verona)
REFERENCES
Milizia: part 1, book 4, chap. 12.3, Marmi moderni §30 Lumachello
Rondelet: book 1, sec. 2, art. 11, §263 (p. 147)
Amati: Lumachello (Note – Marmi moderni, book 2, chap. 7, p. 50)
Jervis: Provincia di Verona, Val Pantena; §478 Grezzana (p. 143); Valle di Squaranto; §480 Roverè di Velo (p. 144)
Salmojraghi: §556 Provincia di Verona; Calcari; Compatti; San Vitale; Corallino con baccelli allungati (p. 351)
Marmi italiani: Veneto; Marmo di San Vitale (p. 54)
PETROGRAPHIC CLASSIFICATION: chemical sedimentary rock (shelly limestone) with skeletal components (biomicrite, mudstone)
MACROSCOPIC FEATURES: a matrix of different colours (grey, yellow, brown, pink) together with clear elongated fragments as pods or seedcases (shell of Lithiotis, Bivalvia) together with some irregular veins
MIMERALOGICAL COMPOSITION: calcite

MICROSCOPIC FEATURES: microcrystalline calcite together with recrystallized fragments of fossil shells (fibrous calcite); the cement is made of calcite spar

GEOLOGICAL SETTING: "Gruppo dei Calcari grigi" (former "Calcari grigi di Noriglio", lower and middle Lias, Jurassic) formation of the Sedimentary series of the Southern Alps; a series of formations featuring grey oolitic limestones, yellowish compact limestones, marls; calcareous beds with "Lithiotis problematica" and marly limestones. The beds with Lithiotis outcrop in a large area of Monti Lessini (province of Verona).

GEOLOGICAL REFERENCES: CGI, Foglio 49 Verona (1968)

QUARRY: small quarries, almost hidden by the growth of vegetation, were located in the upper part of "Vajo di Squaranto" near Roverè Veronese and San Vitale (part of the municipality of Roverè). Similar stones were also quarried in the southern part of "Val Pantena" (municipality of Grezzana).

STONEWORK CATEGORIES: moulded pieces, slab

SURFACE FINISHING: polishing

USE IN ARCHITECTURE: the amazing effect of the light spots in a darker ground was always used to make decorations in Baroque altars and in fireplaces (i.e. a fireplace at Sala delle Aquile, Palazzo Te, Mantua; first decades of the 16th century). Outstanding uses were performed in Cremona: polished "tondi" on the monolithic cistern (late 16th century, Rosso di Verona) in the baptistery; polished slabs in some Baroque altars of the Duomo (right transept).

DECAY MORPHOLOGIES: erosion, chromatic alteration if exposed to weathering

NOTE

According the ancient authors' quotations, another kind of "Lumachella" called "Lumachella di Verona" or "Astracane di Verona" or "Lumachella Occhio di Pernice" was quarried and used as decorative stone. The macroscopic features are: brownish yellow ground together with small dark-grey shells (Terebratula, Brachiopoda). The geological setting (Calcari Grigi, Lias) and the quarry sites (Grezzana, Val Pantena, province of Verona) are the same reported about the "Lumachella di San Vitale"

Pietra di Prun (pietra Perseghina)

COMMERCIAL NAME: the name comes from a village of Monti Lessini, north of Verona. The term "Pietra perseghina", means "stone coloured like a peach"

RFERENCES

Scamozzi: part 2, book 7, chap. 8, Pietre vive (p. 200, line 16)

Jervis: Provincia di Verona, Distretto di Verona; §467 Fumane, §468 San Pietro Incariano, §470 Prun (p. 139)

Salmojraghi: §556 Provincia di Verona; Calcari; Compatti (p. 351)

Marmi italiani: Veneto; Lastre marmoree naturali (p. 55)

PETROGRAPHIC CLASSIFICATION: chemical sedimentary rock (limestone)

MACROSCOPIC FEATURES: pale rose to white to red colour with curvilinear cracks

MINERALOGICAL COMPOSITION: calcite

MICROSCOPIC FEATURES: microcrystalline calcite whit some tests of pelagic foraminifera

GEOLOGICAL SETTING: "Scaglia" formation (lower Eocene – upper Cretaceous) of the Sedimentary series of the Southern Alps (western Veneto) including calcilutites and marly limestones with nodules of chert in the lower beds. The outcrop spread from Lake Garda to the valley of river Alpone of Monti Lessini (province of Verona), bordered by a fault running north–south.

GEOLOGICAL REFERENCES: CGI, Foglio 48 Peschiera del Garda (1969)

QUARRY: different sites in the northern (higher) part of Monti Lessini (north of Verona) were some valleys run parallel. The most important ones are located near the village of Prun.

STONEWORK CATEGORIES: slab

SURFACE FINISHING: quarry face, sawed face, polishing

USE IN ARCHITECTURE: the use in Lombardy started in the 20th century and it was pointed out on polished slabs for flooring or for cladding (i.e. apartment buildings made by architects Asnago and Vender in the 1930s)

DECAY MORPHOLOGIES: detachment of chips along the curvilinear cracks

Rosso di Verona

COMMERCIAL NAME: the name comes from the town of Romeo and Juliet (western Veneto)

REFERENCES

Vasari (On technique, chap. 1): "quella ch'è nei monti di Verona è molto più tenera".

Scamozzi: part 2, book 7, chap. 8, Pietre vive (p. 199, line 33)

Milizia-Antolini: art 1, book 4, chap. 12, art. 3, Ordine e gradazione dei marmi moderni, Rosso – Rosso di Verona, detto Mandorlato

Rondelet: book 1, sec. 2, art. 11, §142 (p. 138), §189 (p. 141), §234 (p. 145), § 248 (p. 140); art. 15, §267 (p. 198)

Jervis: Provincia di Verona, Distretto di Verona; §459 Malcesine, §460 Torri del Benàco (p. 133), §461 Garda, §462 Caprino Veronese, §463 Dolcè (p. 134), §464 Rivoli Veronese (p. 136), §466 Sant'Ambrogio di Valpolicella (p. 137)

Salmojraghi: §556 Provincia di Verona; Calcari; Compatti; Broccatello rosso, mandorlato (p. 351)

Marmi italiani: Veneto; Bianchi di Verona e Rossi di Verona (p. 52), Giallo Torri (p. 53), Breccia Pernice (p. 54)

PETROGRAPHIC CLASSIFICATION: chemical sedimentary rock (nodular limestone; biomicrite-wackestone)

MACROSCOPIC FEATURE: colour varies from red to white to yellow; nodular texture with calcareous nodules (shells of Ammonoidea – Cephalopoda and Bivalvia) surrounded by thin iron-oxides mineralizations; fine grain size

MINERALOGICAL COMPOSITION: calcite; haematite (pigment), illite (clay fraction)

MICROSCOPIC FEATURES: microcrystalline calcite matrix together with bioclasts (mainly broken shells and casts of bivalvia, cephalopoda, skeletal grains of foraminifera) aligned according to the bedding; iron oxides and clay minerals are present into stratigraphic discontinuities

GEOLOGICAL SETTING: "Rosso ammonitico" formation (lower Jurassic) of the Sedimentary series of the Southern Alps (western Veneto), including nodular limestones and cherty limestones. Outcrops are spread in a large area from Lake Garda (monte Baldo, west) to Altipiano di Asiago (east), from monti Lessini (south) to val Lagarina (Trento, north).

GEOLOGICAL REFERENCES: CGI, Foglio 48 Peschiera del Garda (1969)

QUARRY: many sites are located in Valpolicella, north-west of Verona (Sant'Ambrogio di Valpolicella, Domegliara). The most important varieties, distinguished on the basis of colour, are Nembro rosato (at the base), Broccatello, Biancone, Bronzetto and Verdello. A yellow variety (Giallo Torri) comes from the eastern shore of Lake Garda (Torri del Benaco). A brecciated variety (Breccia Pernice, pinkish ground and red spots) and another one (Rosa del Garda, pink ground and white veins) come from monte Pastello (between Valpolicella and Val Lagarina).

STONEWORK CATEGORIES: block, shaft, slab, floor tile, carving

SURFACE FINISHING: polishing

USE IN ARCHITECTURE: the use has been well documented since the Roman period both as block and slab, thanks to the perfect polishing. Lombard architecture enhanced the deep red colour in alternation with black and white elements. Significant examples in Milan are some column shaft, re-used from Roman buildings, in the crypt of San Giovanni in Conca (11th century); two sculpted panels of the apostles, from the church of Santa Maria Maggiore (9th century; now on the northern wall inside the Duomo); floor tiles in the vestibule (G. Pestagalli, 1810) and four shafts of the Scurolo of San Carlo Borromeo (F. Ricchino, 1610) of the Cathedral; and two shafts in the entrance of Banca d'Italia (piazza Edison, G. Broggi, 1914). It is also possible to point out in Bergamo pilaster of the façade of cappella Colleoni (G.A. Amadeo, 1472–76) showing a richly carved decoration. The stone was largely diffused (squared blocks, mouldings, reliefs) in the town of Pianura Padana, lacking in local stone materials but linked by waterways to the Verona area, such as Mantua, Cremona, Piacenza and Pavia. The most significant examples in Mantua are palazzo del Capitano and some other ducal buildings (since the early 14th century); basilica of Sant'Andrea (since the late 15th century); façade of the Duomo (N. Baschiera, 1756–61); portal of the Biblioteca (1763). The most significant examples in Cremona are: the portal of Duomo (late 13th century, but renovated in the 16th century); the baptistery (second half of the 13th century with some reconstruction in the middle 16th century) and the huge monolithic tank for the holy water in the centre (L. Trotti, 1527); the municipal building (first half of the 14th century); the palazzo Cittanova (mid-14th century, restored in 1910–27); palazzo Raimondi (1496). The municipal building at Piacenza, called "il Gotico" (1290) employed Verona limestone for the huge porch with pointed arches. Finally, the most significant examples in Pavia are some blocks on the bottom of Torre Civica (fell on 17 March 1989) reused from Roman buildings; the floor of San Michele made of cubic pieces (about 4 cm per side).

Three baptistery fonts or basins are worth noting, featuring a wide octagonal monolithic shape, sometimes with sculpted panels: Baptistery of Parma (diameter circa 2.7 m; 1.3 m high, 13th century) featuring red and white colours; Duomo of Verona (San Giovanni in Fonte, diameter circa 2.92 m, 13th century); Lodi (diameter circa 2.75 m, 16th century, today moved to the nearby piazza Broletto and used as a fountain). On the contrary, the Baptistery of Cremona shows a basin with eight panels of Rosso supporting a monolithic font, like a prism with an octagonal plan (1520, by G. Sacca and L. Trotti). During the 19th century, the stone was employed as ashlar (white variety) to build the Cimitero Monumentale (C. Maciachini, 1863–66) or as floor tile (white and red varieties) in the Galleria Vittorio Emanuele and in the neighbouring Portici Settentrionali and Meridionali (G. Mengoni, 1865–75).

In the late 19th–early 20th centuries, the stone was employed in some bases of civil buildings made of artificial stone. Moreover, polished slabs were diffused in floors of residential buildings, in many cases the red or white varieties were set in alternating.

DECAY MORPHOLOGIES: chromatic alteration (the stone surface change from deep red to light pink), differential roughening thanks to the presence of clay levels and fossil shells, scaling, sulphate skin formation

NOTE

Another outcrop area of "Rosso Ammonitico" is present around Trento (see Stones of Trentino-Alto Adige).

Rosso di Asiago

REFERENCES

Jervis: Provincia di Vicenza, Distretto di Asiago; §505 Roana (p. 150), §506 Rotzo, §508 Asiago (p. 151)

Salmojraghi: §555 Provincia di Vicenza; Calcari; Compatti (p. 348)

Marmi italiani: Veneto; Marmi di Asiago (p. 55)

Petrographic features, mineralogical composition, geological setting and decay are similar to Verona limestone. Outcrops are spread around the Asiago plateau [Geological References: CGI, Foglio 37 Bassano del Grappa (1948); CARG, Foglio 082 Asiago (2007)]. Some quarry sites are located on the north-western side of this plateau (Rotzo, Roana, Asiago). Different varieties are sold: Rosso Magnaboschi, Bianco, Rosa; Perlino (lower Cretaceous).

The aspect is very similar to the Rosso Ammonitico of Verona, so it is very difficult to clearly distinguish the single use of Asiago limestone in Lombard architecture. Surely, the use was implemented in the 20th century, thanks to the increased transport capacities. The Perlino variety was often used by the architects M. Asnago & C. Vender for cladding of apartment buildings (i.e. via Euripide 1).

Pietra di Vicenza (soft limestone of monti Berici)

COMMERCIAL NAME: this common term, linked to the town of Palladio, includes different kinds of soft limestones of the western part of Veneto. Nevertheless, is better to indicate the lithotypes quarried in this area with different names, mainly referring to a single site: "pietra di Nanto", "pietra di San Gottardo", "pietra di Costozza".

REFERENCES

Scamozzi: part 2, book 7, chap. 11, Pietre trattabili (p. 208, line 11)

Rondelet: book 1, sec. 2, art. 15, §276, §277 (p. 199), §294 (p. 100)

Jervis: Provincia di Vicenza, Distretto di Vicenza; §500 Longare, §501 Nanto (p. 149), §502 Arcugnano (p. 150)

Salmojraghi: §555 Provincia di Vicenza; Calcari; Colli Berici; Eocenici (p. 349)

Marmi italiani: Veneto; Pietra di Vicenza, detta anche pietra tenera o pietra morta (p. 58)

PETROGRAPHIC CLASSIFICATION: clastic sedimentary rock (biosparite – grainstone)

MACROSCOPIC FEATURES: homogeneous surface colour featuring complete fossil tests and fragments (until 1–2 cm in size). Each lithotype shows some differences in colour and in grain size: straw-yellow colour (10YR 8/2), medium to coarse grain size ("pietra di Nanto"); ivory-white (2.5Y 9/2) to yellowish colour (5Y 9/2), heterogeneous grain size ("pietra di San Gottardo", "pietra di Costozza").

MINERALOGICAL COMPOSITION: calcite

MICROSCOPIC FEATURES: four different varieties are described

- Packed nummulites, discocyclinids (up to 2 mm), echinoids; mosaic cement of calcite and syntaxial overgrowth; irregular cavities (Nanto quarries)
- Micrite walled miliolidae (0.5 mm), nullipores or coralline algae (0.4/0.6 up to 1.5 mm), echinoids; mosaic cement of calcite and syntaxial overgrowth; abundant microcrystalline matrix; irregular to rounded cavities (Zovencedo quarries)
- Nullipores (0.4–2 mm), nummulites (1 mm), echinoids; mosaic cement of calcite and syntaxial overgrowth; scarce microcrystalline matrix; irregular cavities (Zovencedo quarries)
- Nullipores (1.0–1.5 mm), micrite walled miliolidae (0.5–1.0 mm), nummulites (0.5 mm), echinoids; mosaic cement of calcite and syntaxial overgrowth, scarce microcrystalline matrix; interstitial to rounded cavities (Longare-Costozza quarries)

GEOLOGICAL SETTING: soft limestones of Veneto pertain to Tertiary formations (Sedimentary series of the Southern Alps) outcropping in a wide area between Monti Berici (province of Vicenza) and the eastern shore of Lake Garda, through the southern part of Monti Lessini (province of Verona). Both in the Berici area and in Lessini area each term of the stratigraphic succession often shows a great number of *facies* in the same geological stage with different features for different geographical sites.

- "Calcari nummulitici" formation (middle and upper Eocene) featuring microcrystalline limestones with nummulites and echinoids together with marly limestones ("pietra di Nanto")
- "Calcareniti di Castelgomberto" formation (Oligocene), featuring blioclastic and nulliporic massive-bedded limestones together with marly and calcite-marly intercalations; limestones with miliolidae, nummulites and echinoids ("pietra di San Gottardo" and "pietra di Costozza")

GEOLOGICAL REFERENCES: CGI, Foglio 50 Padova (1947)

QUARRY: Eocene lithotypes (pietra di Nanto) were exploited in the south-eastern sector of Monti Berici area; Oligocene lithotypes (pietra di San Gottardo" and "pietra di Costozza) were exploited in the north-eastern sector of the Monti Berici area (Longare-Costozza, Lumignano, Zovencedo). The current exploitation concerns the "pietra di Nanto".

STONEWORK CATEGORIES: block, slab, small shaft, carving

SURFACE FINISHING: honed

USE IN ARCHITECTURE: a visual examination of stone used in buildings does not allow a distinction among the different varieties

The use has been documented in the Roman period with stelae and architectural elements, now exhibited at Civico Museo Archeologico of Milan: capitals of the Republican age (catalogue A0.9.27991 and A0.9.27992, A0.9.1084; catalogue 86 from via Bocchetto, 1901 excavations) made by two different elements (cylindrical shape for the lower part – 80 cm high and 80 cm in diameter; squared shape for the upper part – 100 cm side and 40 cm thick); ionic capitals of the Augustean age (catalogue A0.9.14537 from via Monte di Pietà, 1868 excavations). The Roman villa on the tip of Sirmione peninsula (Lake Garda) features a "Criptoportico" (hidden porch) with piers made of stone coming from

Zovencedo (base and capital) and of stone coming from Montòrio Veronese (shaft): the identification was made through optical microscopy on a thin section in comparison with samples of the different quarries.

Palladio, in the second half of the 16th century, reported about the two bridges of Vicenza ("ponte delle Beccarie sul Retrone" and "ponte San Pietro sul Bacchiglione", both demolished and rebuilt many times): "Both the one and the other of these bridges are made of Costozza stone, which is a soft stone and is sawed like wood" [Palladio, book 3, ch. 15].

Soft stones of Veneto were completely ignored in the Lombard architecture of the Renaissance and Baroque, thanks to the availability of local stones with the same features (i.e. the dolomites from Angera). Afterwards, in the neo-classical period, soft stones of Veneto were employed in some decorative elements (i.e. vases on the top of the façade of Teatro alla Scala – G. Piermarini, 1775–78) or on the top of Villa Reale (L. Pollack, 1790). Soft stones were used to replace the decorations on the pilasters of the façades of Galleria Vittorio Emanuele, after World War II.

DECAY MORPHOLOGIES: disaggregation, crumbling, sulphate skin formation

Pietra Gallina – Matòn (soft limestone of monti Lessini)

COMMERCIAL NAME: this name generally refers to a soft limestone coming from Monti Lessini, in the western part of Veneto. Nevertheless, it is possible to distinguish two kinds of soft limestone: "pietra Gallina" and "Matòn" (from the Venetian dialect, meaning "brick"; in fact, it is easy to cut a soft limestone in a brick-like shape). Another name is "tufo" according to the texture and the lightness of this kind of stone.

REFERENCES

Scamozzi: part 2, book 7, chap. 11, Pietre trattabili (p. 208, line 43)

Jervis: Provincia di Verona, Distretto di Verona; §465 Affi (p. 136), 469 Negarine, §470 Prun (p. 139), §471 Negrar, §472 Parona all'Adige, §473 Quinzano Veronese, §474 Avesa (p. 140), §475 Verona (p. 142)

Salmojraghi: §556 Provincia di Verona; Calcari teneri; Mattone e Pietra Gallina (p. 352)

Marmi italiani: Veneto; Pietra gallina, Pietra mattone (p. 55)

PETROGRAPHIC CLASSIFICATION: clastic sedimentary rock (biosparite; grainstone)

MACROSCOPIC FEATURES: homogeneous surface colour featuring fossil shells both complete or fragmentary (up to 1–2 cm in size). Each lithotype shows a different colour and grain size: white yellowish colour (2.5Y 9/2), fine grain size, homogeneous texture ("pietra Gallina"); yellow-brownish colour (10YR 6/4 or 10YR 7/4), microfossils shells (nummulites, discocyclina, lepidocyclina) clearly visible at naked eyes ("Matòn").

MINERALOGICAL COMPOSITION: calcite

MICROSCOPIC FEATURES: three different varieties are described

- Fragments of tests (0.1–0.5 mm) of foraminifera, echinoids; mosaic cement and syntaxial overgrowth; rounded cavities (0.1 mm) (Àvesa)
- Packed nummulites, discocyclinids (up to 10 mm) and lepidocyclinids, echinoids; microcrystalline calcite matrix, (Quinzano Veronese – monte Ongarine)
- Very packed nummulites, discocyclinids (up to 5 mm) and lepidocyclinids; microcrystalline calcite matrix; mosaic cement and iron oxides; irregular cavities (Montorio Veronese)

GEOLOGICAL SETTING: soft limestones of Veneto pertain to Tertiary formations outcropping in a wide area between the Monti Berici (province of Vicenza) and the eastern shore of Lake Garda (province of Verona), through the southern part of Monti Lessini. Both in the Berici and Lessini areas each term of the stratigraphic succession often shows a great number of *facies* in the same geological stage with different features for different geographic sites.

The southern Lessini area shows particularly the "Calcari nummulitici" formation (middle and upper Eocene) featuring nummulitic and nulliporic limestones together with marly limestones ("pietra Gallina" and "Matòn"); these limestones are in alternating with tuffs and basalts. A sequence of yellowish and fine-grained limestones containing foraminifera are located at the base of the Eocene series (pietra Gallina at Àvesa).

GEOLOGICAL REFERENCES: CGI, Foglio 49 Verona (1968)

QUARRY: Eocene lithotypes were only exploited in Monti Lessini area ("pietra Gallina" – Àvesa, north-east of Verona; "Matòn" – Quinzano and Montorio Veronese, north of Verona)

STONEWORK CATEGORIES: block, small, shaft slab, carving

SURFACE FINISHING: honed

USE IN ARCHITECTURE: a macroscopic examination of stone used in buildings do not allow a distinction among these soft stones (see Chapter "Berici"). The architectural use was large in the medieval times, because the stone was easy to work. The Romanesque buildings of Verona show a wide use of soft limestones together with bricks and river pebbles: i.e. basilica of San Zeno and Duomo.

DECAY MORPHOLOGIES: disaggregation, crumbling, sulphate skin formation

Part III

Foreign countries

Chapter 19

Stones from Europe, Africa, the Americas

Stones coming from the whole Mediterranean world were diffused in Roman times. In the following centuries, only two coloured stones (Rosso di Francia – see *infra*; Broccatello di Spagna – see Chapter 36) were imported from abroad due to their great availability in each Italian region. The use of foreign stones increased considerably in the second half of the 20th century, after World War II and after the abolition of the Fascist policy ("Autarchia"). The number of stones from abroad increased considerably during the 21st century.

IGNEOUS ROCKS

Larvikite

COMMERCIAL NAME: the name comes from the most important town in the quarry area of southern Norway (Larvik); this stone, in Italy, was also named "Labrador", despite the Norwegian provenance

PETROGRAPHIC CLASSIFICATION: igneous intrusive rock (augite-syenite to augite-monzonite)

MACROSCOPIC FEATURES: colour varies from dull grey to silver blue to warm brown to green, including crystals with marked iridescence. Very coarse grain size

MINERALOGICAL COMPOSITION: anorthoclase; titan-augite, dark amphibole (former "barkevikite"), iron rich biotite (lepidomelane)

MICROSCOPIC FEATURES: rhomb-shaped ternary (K, Na, Ca) feldspar crystals with perthite, prismatic crystals of titan-augite

GEOLOGICAL SETTING: "Larvik plutonic complex" of the Oslo rift (Permian – Carboniferous). The outcrop of igneous rocks occupies a narrow territory (about 200 km) of the southern Norway, from Lake Mjøsa (north, Akershus County) to Larvik (south, Vestfold County). Monzonites form the southern part of the pluton.

QUARRY: the main sites are located near Larvik and have operated on a large scale since the last decades of the 19th century. Commercial varieties are blue pearl, eilver pearl and emerald pearl.

STONEWORK CATEGORIES: mainly slabs for flooring or, sometimes, for cladding.

SURFACE FINISHING: polishing with distinctive effect of "labradorescence" (Schiller effect)

USE IN ARCHITECTURE: the stone was used to make buildings in the southern part of Norway since prehistoric times. The use in Italy started in the first decades of the 20th century, and in the period 1920–40 this stone was one of the only materials coming from

DOI: 10.1201/9781003361008-22

abroad and used in Italian architecture. Another important use was devoted to tombstones (i.e. Edicola Bernocchi, Necropoli, Comitero Monumentale, A. Minali, 1936).
DECAY MORPHOLOGIES: scaling

SEDIMENTARY ROCKS

Rosso di Francia – Rouge du Languedoc (Incarnat, Turquin)

COMMERCIAL NAME: in Italy the term "Rosso di Francia" was used to identify this marble matching the ancient Cipollino rosso and Rosso Brecciato (Iassos, southern Aegean coast of Turkey).

The French term is not uniform: different varieties were named after the region of origin (Languedoc, south-western France) or after the carnal colour (Incarnat) or after the presence of bluish-grey shades of irregularly shaped spots (Turquin).

REFERENCES

The Italian writers, despite the large use of this marble in the Baroque architecture of Italy, are lacking in quotations about it. The treatises on ancient marbles (Garofalo, Belli, Corsi) also paid no attention to the French marble, because of its modern use or its potential confusion with the "Cipollino rosso" and "Rosso brecciato".

Two writers reported the marble as "alabaster".

Milizia: "Marmi Moderni. 19. Alabastro . . . 8. Di Roquebrun in Linguadoca di un grigio scuro, con gran macchie di rosso bruno".

Amati: "Marmi Moderni. Alabastro . . . il Roquebrun, che si cava nella Linguadoca, ha il fondo di grigio scuro, con grandi macchie di rosso bruno".

The first quotations were made by Antolini, in his review of Milizia's treatise. [Osservazioni e aggiunte ai Principj di architettura civile, Milano, 1817], and by Pullen, in his survey on the materials used in the churches of Rome.

Milizia- Antolini: "Ordine e gradazione de marmi moderni. . . . Rosso . . . 8. Roquebrun in Linguadoca. 9. Rosso di Francia".

Pullen, in the late 19th century, reported the "Rosso di Francia" as modern and disagreeable. The marble comes from Caunes (south-eastern France), showing red colour with large bluish spots of irregular boundaries. The marble is visible in the churches of Rome: St Anastasia (between Palatino and Circo Massimo), St Maria dell'Anima (close to piazza Navona) and St Peter (Vatican).

Earlier reports, related to the marble of Languedoc, are shown in some contracts concerning the construction of church altars. A "Breccia di Francia rossa et bianca" is reported among other coloured marbles about the altar of the chapel of Sant'Antonio da Padova in the church of St Lorenzo Maggiore (Naples), made by Cosimo Fanzago (1638).

Rouge du Languedoc has been quoted by several French authors since the 17th century.

Alselme Boece de Boot: "Marbres Rouges" do not report the marble of Languedoc [Le parfait joaillier, book 2, ch. 281; Lyon 1644]

Félibien [Des principes de l'architecture . . ., 1676, Paris]: "Book 1, chap. 12. . . . Il se trouve en Languedoc proche la ville de Cosne un marbre incarnat et blanc dont l'on a fait aussi des pieces de 20 pieds de long, pour faire des colonnes. . . . Auprés de Roquebrun à six lieuës de Bezíers, il y a une autre carrière d'un marbre rouge et blanc qui a une grande dureté et un beau lustre: l'on en peut tirer des pieces de plus de 30 pieds de long".

A.C. D'Aviler [Cours d'architecture, Paris, 1691, Première partie]: "Le marbre de Languedoc a le fond rouge, véné et taché de blanc; et celui de Narbonne est d'un rouge pâle, couleur de chair, mêlé de taches blanches".

A.C. D'Aviler [Dictionnaire d'architecture, Paris 1691]: "Marbre. . . . Marbre de Languedoc. Ce marbre, qu'on trouve près de la ville de Cosne, a le fond d'un rouge vif, avec des grandes veines et des taches blanches. Il est assez commun. De ce Marbre sont les deux colonnes ioniques et l'architrave de l'Autel de Notre Dame de Savone, dans l'Eglise des PP. Augustins déchaussés, à Paris; les pilastres et les quatorze colonnes ioniques du château de Trianon. Il y a du Languedoc d'un blanc bleuâtre et gris; iln'est pas si estimé que l'autre".

"Marbre de Roquebrun, à sept lieues de Narbonne. Marbre qui ne diffère du marbre de Languedoc, qu'en ce que ses taches blanches sont toutes comme des pommes rondes. . . . On en voit quelqeus blocs dans les magasins du Roi."

J.F. Blondel [Architecture françoise, tome premier, livre premier, Paris, 1752]: "Des Marbres moderns. Le marbre de Languedoc est le plus commun; aussi l'employe-t-on en general dans les revêtissements de la décoration intérieure, ainsi qu'on le voit aux retraites de l'Eglise de St. Sulpice; ou bien l'on en fait les colonnes des decorations extérieures, telles que sont celles du bâtiment de Trianon. Ce marbre a le fond d'un rouge de vermilion un peu sale et il est mêlé de grandes veines ou taches blanches entremêlées de quelqu'unes bleuâtres et de grises. L'Autel de Notre-Dame de Savone dans l'Eglise des Pères Augustins déchaussés, est de ce marbre."

J.F. Blondel [Cours d'architecture, livre 3, première partie, chapitre 2, article 2, Paris, 1777]: "Des marbres modernes. . . . Le marbre de Languedoc est un des plus communs; il y en a de deux espéces: l'une se tire de la ville de Cosne en Languedoc, sa couleur est un foind rouge de vermlion sale, entremêlé de grandes veines et de taches blanches; on en a revêtu le socle des piliers de la nef de l'Eglise de Saint-Sulpice à Paris, les colonnes Ioniques de la cour du Château de Trianon sont aussi de ce marbre. L'autre qui se tire de Narbonne, est de couleur blanche, grise et bleuâtre et est beaucoup plus estimé".

Quatremère [Encyclopédie méthodique, Architecture, Tome second, Paris, 1801–20]: "Marbre. . . . Marbre de Languedoc. Ce marbre, qu'on trouve près de la ville de Cosne, a le fond d'un rouge vif, avec des grandes veines et des taches blanches. De ce marbre sont les pilastres et les quatorze colonnes ioniques du château de Trianon. Il y en a qui est d'un blanc bleuâtre et gris; on l'estime beaucoup moins que l'autre".

"Marbre de Roquebrun, à sept lieues de Narbonne. Il ne diffère du marbre de Languedoc, qu'en ce que ses taches sont toutes comme des pommes rondes".

Rondelet [1802 – livre 1, sec. 2, art. 10]: "Des marbres modernes. Marbres de France . . . Marbres bleus et autres où le bleu domine. 21. Bleu turquin de Caunes (département de l'Aude). . . . Marbres rouges et autres où cette couleur domine. . . . 99. Marbre incarnat et blanc, de Caunes (department de l'Aude), très beau marbre qui était réservé pour le roi. 100. Il se trouve una carrière de ce même marbre incarnat veiné de blanc, proche de la ville de Narbonne, même département, qui est aussi fort beau. . . . 102. Marbre appelé de Languedoc dont le fond est pâle, avec des grandes taches blanches, est commun dans les départements de l'Aude, de la Lozére et de l'Hérault".

The Italian edition of this treatise, made by B. Soresina (Napoli, 1840), reports the correct translation of the French words.

PETROGRAPHIC CLASSIFICATION: homogeneous limestone with stromatactis

MACROSCOPIC FEATURES: two main varieties are used. One dark red ground with white stromatactis (Incarnat); another orange to red ground with calcite veins and white to grey

stromatactis (Turquin). Stromatactis are calcite formations with a flat base and a dentate top, and they are aligned according to the bedding planes.
MINERALOGICAL COMPOSITION: calcite
MICROSCOPIC FEATURES: mudstone made of crystalline calcite matrix with sponge spicules and crinoids (red ground); wackestone made of fragments of crinoids, bryozoans, ostracods, trilobites, sponges etc. (red ground); non-homogeneous texture of sparry calcite (white stromatactis)
GEOLOGICAL SETTING: marine calcareous sediments of Montagne Noire (lower Devonian) forming four tectonic nappes and pertaining to a mountain range of the Hercynian orogeny; stromatactis are related to sponge-dominated frameworks and differential diagenesis
QUARRY SITES: quarries are located on the southern slope of Montagne Noire, a mountain range of the Languedoc pertaining to two departments: Aude (Caunes en Minervois); Hérault (Saint Nazaire de Ladarez, Roquebrun). The most important is the so-called Carrière du Roy, a few kilometres north-east of Caunes en Minervois.
CATEGORIES AND SURFACE FINISHING: architectural element (column shaft), slab; polishing
USE IN ARCHITECTURE: monolithic column, cornice, panel. The main examples are the peristyle of Grand Trianon (Versailles, 1687, architect Hardouin-Mansart) with coupled column shafts (fourteen columns and four pilasters before and eighteen pilasters behind); the Arc de Triomphe du Carrousel (Paris, 1808, architects Percier and Fontaine) with eight column shafts.
DECAY MORPHOLOGIES: differential erosion

NOTE
Similar red limestones (Rouge Royale) were quarried in southern Belgium (region of Namur) and used as building materials in the whole Belgian territory.
A red hematitic limestone (late Devonian) from Languedoc was also employed for decorative purpose (Palace of Versailles): Griotte, a deep red limestone with nodular textures (Goniatites – Molluscs), quarried near Félines-Minervois (Hérault).

METAMORPHIC ROCKS

Castione

COMMERCIAL NAME: the name comes from a village of Val Leventina (Canton Ticino, Switzerland); the common name "Granito nero di Bellinzona" refers to an erroneous petrographic classification
REFERENCES
Salmojraghi: §634 Canton Ticino; Marmi liasici (p. 450)
PETROGRAPHIC CLASSIFICATION: calcareous contact-metamorphic rock
MACROSCOPIC FEATURES: streaked dark grey ground with reddish spots, medium grain size
MINERALOGICAL COMPOSITION: calcite, plagioclase, garnet (grossularite), pyroxene (diopside)
MICROSCOPIC FEATURES: fairly aligned calcite crystals, twinned plagioclase crystals, rounded aggregates of garnet and pyroxene

GEOLOGICAL SETTING: “Calciphyres of the type *Castione nero*, locally rich in garnets” together with “Calcite marbles” and “Silicate marbles”, included as “Marbles, calciphyres and calcitic micaschists” inside “Carbonate rocks of Alpine metamorphism of Lepontino” (Pennidic).
GEOLOGICAL REFERENCES: Atlas Géologique de la Suisse au 1:25000, #152 Mendrisio-Como (2018)
QUARRY: the main site is located in the lower Val Leventina near Castione, a few kilometres north of Bellinzona. The quarry is placed on the southern slopes of Pizzo di Molinera, at the fork between Val Leventina (west) and Val Mesolcina (east).
STONEWORK CATEGORIES: block, slab
SURFACE FINISHING: polishing
USE IN ARCHITECTURE: the use was mainly local for masonries; nevertheless the architect L. Beltrami used this stone in the early 20th century, because the owner of some Milanese buildings (i.e. Banca Commerciale or Case Bernasconi) was also the owner of the quarry at Castione
DECAY MORPHOLOGIES: scaling

Onsernone

COMMERCIAL NAME: the name derives from the Alpine Valley, a western tributary of the Valle Maggia (Locarno, Canton Ticino, Switzerland)
REFERENCES
Salmojraghi: §634 Canton Ticino; Gneis (p. 449)
PETROGRAPHIC CLASSIFICATION: regional metamorphic rock (gneiss)
MACROSCOPIC FEATURES: dark grey colour with a whitish ground, medium grain size
MINERALOGICAL COMPOSITION: alkali feldspar, biotite, plagioclase (oligoclase)
MICROSCOPIC FEATURES: cloudy crystals of potash feldspar, subhedral crystals of plagioclase, biotite of tabular habit
GEOLOGICAL SETTING: “Complesso di Falde Antigorio – Mergoscia” (lower Pennidic) including gneisses with “augen” texture of biotite, alkaly feldspar and plagioclase (oligoclase)
GEOLOGICAL REFERENCES: Atlas Géologique de la Suisse au 1:25000, #159 Locarno (2018)
QUARRY: a still-working site is located near Vergeletto (elevation 900 m) in a tributary valley of Valle Onsernone
STONEWORK CATEGORIES: block, slab
SURFACE FINISHING: polishing
USE IN ARCHITECTURE: the use is mainly local for different purposes; in Milan the stone was employed for the cladding of the base of Centro Svizzero (piazza Cavour, arch. A. Meili, 1949)
DECAY MORPHOLOGIES: scaling

STONES OF THE 21ST CENTURY

A gradual disappearance of natural stones from the buildings took place in the last years of the 19th century according to the progress of artificial stones. This trend was inverted in the period between the two world wars (1920–40) when the use of building stones was boosted by the great availability, deriving from dozens of quarries throughout the Italian territory and also from Fascist politics, called “Autarchia” (Autarky) or an economic system of self-sufficiency and limited trade. In fact, since 1934, the exploitation of raw materials of the national country against the foreign ones was privileged.

After the end of World War II and in the second half of the 20th century, the use of building stones was mainly devoted to the interior flooring; at the same time the use of “new” materials (concrete, metal, glass, ceramic, synthetic substances, etc.) strongly increased.

This period was a critical point for the Italian quarries, for the Italian stones and therefore for the quarrymen: in particular, dozens of Lombard quarries were closed because of the unbearable level of labour and trading costs. The opening of markets to foreign goods of the whole world and the improvement of transport by road, rail, sea and air, together with the availability of new varieties of stones, favoured an intense use of stone in architecture beginning in the last two decades of the 20th century. A crucial factor, leading to the return of stone on the buildings, was the introduction of the “rainscreen ventilated façade” or “back-ventilated rainscreen”. This kind of façade involves a load-bearing insulated wall (the inner wall) protected by an outer skin made of slabs of stone or of other material (i.e. ceramic tile) supported by a metal grid. The outer skin protects against rain and moisture (chimney effect) and, maintaining a space between the wall and the skin, prevents water from infiltrating the building structure. Rainscreen-ventilated façades allow for the use of very thin slabs of stone, quickly replaced if necessary. The stones involved in rainscreens mainly are igneous rocks of different compositions and colours (granites, diorites, gabbros, etc.), but limestones, travertines, sandstones and marbles are also present. The stonework shops in Italy were able to produce dimension-stone, slab or other different artefacts starting from rough stone blocks, imported from Europe and overseas. The places of origin are spread throughout the world (mainly Africa and South America); nevertheless the origin of stones is almost always hidden by ambiguous labels (i.e. commercial names referred to continental areas) and the simple identification of the quarry sites is very difficult.

Finally, the second decade of the 21st century shows another decrease of the use of stone for façade cladding in comparison with a strong increase of the use of glass, concrete panels, metal sheets, etc.

A brief description of the main characters of these stones is reported, together with the most important use in Milan’s buildings. In this case, their identification is based on macroscopic features and on archive documents in the absence of scientific analyses given the impossibility of taking samples.

Stones from European countries

These stones were already used in their countries; the occurrence in Milan is mainly for flooring and cladding and devoted to commercial shops or restaurants and cafés.

Balmoral: a medium-grained granite (Precambrian) featuring a red ground with black spots. The mineralogical composition involves quartz, potash feldspar, plagioclase and biotite. The main quarry site is located at Taivassalo, a coastal town east of Helsinki (Finland, region of Turku).

NOTE: a Swedish red granite was sometimes used in the first decades of the 20th century (i.e. Milan, portals of Santa Croce, C. Arpesani, 1913–17). This stone came from Solberga-Nässjö (Jönköping province, south Sweden).

Baltic brown: a "rapakivi" granite (Precambrian) featuring brown rounded spots with a dark rim. The mineralogical composition involves a "rapakivi" texture made of large round potassic feldpars mantled by sodic plagioclase (albite), biotite and hornblende. Different quarry sites are spread in the surroundings of Ylämaa (Lappeenranta-Villmanstrand), a town in south-eastern Finland (region of Etelä Karjala – Södra Karelen – Southern Karelia), close to the Russian border.

Rosa Porriño: a medium-grained granite (Carboniferous) featuring pale pink crystals with black spots. The mineralogical composition involves quartz, alkali-feldspars and biotite. The main quarry site is located at Porriño, a village a few kilometres south of Vigo (Spain, Comunidad of Galicia).

Negro Marquiña: a fine-grained limestone (Carboniferous) featuring a black ground with white veins and small streaks (fossils). Calcite is the only mineralogical component. The main quarry sites are located at Marquina-Markina, a few kilometres south of the coast of the Bay of Biscay (Spain, Comunidad of País Vasco, province of Vizcaya).

Nero del Belgio (Belgian black): a very fine-grained limestone (Devonian) featuring uniform dark black ground. The mineralogical composition includes only calcite. The main quarry sites are located at Tournai (Belgium, Wallonie, province of Hainaut), a town close to the French border or in other localities of the adjacent province of Namur. A specific use of this stone was as a black base of coloured inlay work (commesso alla fiorentina) on tables, consoles, cabinets, etc. created by Opificio delle Pietre Dure in Florence (19th century).

Stones from overseas

These stones were exploited in the last several decades in Africa, South America, etc., but the use in local architecture is not relevant. Their use is mainly developed in Europe, North America and China; trade is based on rough blocks then cut and sawn in Italian workshops. The occurrence in Milan, like the aforementioned European stones, is mainly for flooring and devoted to banks, shopping centres, restaurants and cafés.

Azul Macaúbas: a fine-grained quartzite (middle Proterozoic) featuring pale blue ground with dark blue veins. The mineralogical composition involves quartz and dumortierite (aluminum-silicate of boron). The main quarry site is located at Vaca Morta, a village of the Sierra Macaúbas (Brazil, state of Bahia, Boquira region).

Juparanà: a granitoid (Neo-Proterozoic) featuring yellowish-brown ground, swirled and banded by dark brown biotite. The mineralogical composition includes granite and biotite. The main quarry site is located a few kilometres west of Nova Venécia (Brazil, Nordeste, Espìritu Santo) a town about 200 km north of Vitoria, the capital of the state.

Nero Zimbabwe: a medium- to fine-grained dolerite (Pre-Cambrian) featuring a hard black ground. The mineralogical composition includes amphibole, augite, plagioclase and magnetite. The main quarry site is located at Mount Darwin in northern Zimbabwe (province of Mashonaland). It is worth it to note the use as cladding slabs on the ground floor of the Deutsche Bank headquarters (piazza del Calendario 3, Valle Architetti Associati, 1997–2005).

Kashmir white: a coarse-grained leucocratic granite (Proterozoic) featuring a whitish ground with alignments of small pink and brown crystals. The mineralogical composition includes quartz, feldspar, almandine garnet and biotite. The main quarry site is located at Melur, a village near the southern town of Madurai (India, state of Tamil Nadu).

Verde Guatemala: a fine-grained and sometimes brecciated serpentinite featuring a green ground with various shades and dark green veins. The mineralogical composition includes serpentine. The main quarry site is located at Pasabien (Guatemala, department of Zacapa); moreover, other quarry sites supply similar lithotypes (i.e. Kasariyaji near Udaipur, Rajasthan, India).

Kandla: a fine-grained sandstone of two different colours (grey and brown). Quarried in different sites of Rajasthan (India).

Part IV

Stone materials

Part IV

[illegible]

Chapter 20

Coloured marbles (*opus sectile*, inlay)

INTRODUCTION

The marble is a crystalline rock that takes a high polish and originated from the metamorphism of a carbonate rock.

The use of marble marked the birth of the Western civilization in Greece and in the Aegean Sea: the Cycladic schematic figurines, made of white marble and found in graves, date back to the Early Bronze Age (about 3000 BCE). The use of marble was supported by the availability of this material thanks to numerous marble outcrops in the Aegean area both on the mainland (Greece and Anatolia) and on the islands. The white marble has been considered since Antiquity as a metaphor of whiteness, hardness, coldness and purity.

The use of white marbles, on the other hand, was not very substantial in Egypt, because of the shortage of this kind of material in the country. Some coloured marbles were already quarried in Egypt, but they were employed in different way compared to the use made by the Romans: big blocks of Aswan granite are present in Giza's Third Dynasty pyramids (burial chambers or exterior facing), and small vessels of red porphyry were made in the Early Dynastic period. On the contrary, Egyptian alabaster was widely employed (i.e. pavement of the valley Temple of the Fourth-Dynasty king Khaphra at Giza; canopic jars and vessels since the pre-dynastic period) [Aston et al., 2000]. Furthermore, cladding made of coloured marbles were present in the alexandrine buildings of the Ptolemaic period, as indirectly witnessed in the painted reproductions decorating the Hypogea of Alexandria (3rd century BCE) [McKenzie, 2007]. The use of coloured marbles in the Ptolemaic period is revealed by the description, made by the poet Lucan, of the richly decorated Royal Palace of Cleopatra at Alexandria, with porphyry, alabaster and other precious materials [Lucan, 1905; book 10, line 106].

The term "coloured marbles", in the archaeological field, indicates a group of stones of various petrographic origins (marble or monochrome or spotted or veined, alabaster, limestone, ophicalcite, granite, porphyry, breccia, etc.) coming from different sites of the whole Mediterranean basin and widely employed with ornamental purposes.

White and coloured marble artefacts were involved in a massive reuse already during the last decades of the Roman Empire and then in the Middle Ages and beyond. The Roman artefacts were re-employed in the same function or reshaped in other architectural items or cut into small pieces (i.e. stone tesserae for flooring).

At first, it is very interesting to be acquainted with the application, reputation and prestige of marbles through the quotations of ancient and modern authors; moreover, particular

DOI: 10.1201/9781003361008-24

attention was paid to specify the different terms used over the centuries to indicate the same material. The different kinds of marble unearthed in archaeological sites of Milan are then reported together with their significant features and in comparison with the marbles unearthed in other Roman sites in Lombardy.

MARBLES THROUGH GREEK AUTHORS

"The Greeks called *marmaros* (origin of the Latin term *marmor*), the material now called *marble*. The origin of the term is Greek, and this name is due to the brightness and the splendour of this material, as the verb *marmarein* means "to shine". The words of the French architect A. Quatremère de Quincy [Quatremère, 1801; *s.v.* Marbre] perfectly illustrate the most regarded stone material in the whole world: the white marble.

The Italian architect Vincenzo Scamozzi ascribes the presence of marble in many buildings, ornaments and statues in Greece to the large availability of marble in the Greek territory, and he also mentions that these marbles were taken into possession by the Romans when the Greece fell [Scamozzi, 1615; part 2, book 7, ch. 4]. In fact, marbles, both white or coloured, were widely spread in Roman architecture in Greece: i.e. temple of Zeus Olimpios (c. 130 CE) at Athens and Library of Celsus (c. 115 CE) at Ephesus.

Theophrastus was the first Greek author writing about marbles together with other stone materials; his petrographical text, now fragmentary (4th century BCE), reports: "And among such stones the are the Parian [Cyclades], the Pentelic [Attica], the Chian [Aegean Sea], and the Theban [Egypt], and these stone quarries have become widely known" [Theophrastus, 1866; book 6]. Nevertheless, the author used the term *marmaros* about a stone material (limestone or marble) transformed in lime after baking [Theophrastus, 1866; book 9]. Homer, on the contrary, used the term *marmaros*, meaning "stone": the stone pulled up from the walls and employed by Aias, son of Telamon, to kill Epicles [Homer, 1924; book XII, 380] or the block thrown by the Cyclops against Odysseus and their comrades [Homer, 1919; book IX, 499]. Curiously, the most ancient remains about historical chronology (early 3rd century BCE), and now held in two museums, is the "Parian Chronicle" or "Parian marble", a 2 m high stele containing a long and minute inscription; this stele is, obviously, made of marble of the island of Paros. Strabo, among other Greek authors, described accurately the marbles of different quarries spread in the Greek territory: from Pentelikon and Hymettus [Strabo, 1903; book 9, ch. 1, sect. 23] to Paros [Strabo, 1903; book 10, ch. 5, sect. 7]; from Carystus [Strabo, 1903; book 10, ch. 1, sect. 6] to Scyrus [Strabo, 1903; book 9, ch. 5, sect. 16]; from Synnada (Docimia) [Strabo, 1903; book 12, ch. 8, sect. 14] to Proconnesus [Strabo, 1903; book 13, ch. 1, sect. 16]. Pedanius Dioscorides included some stones (alabaster, jasper, gypsum, serpentine) at the end of his treatise on plants and medicine [Dioscorides, 2000; book 5, ch. 87/93]. Pausanias, detailing the regions of Greece, correctly reported, for example, the small pebbles featuring the outcrop of green porphyry near Krokées in Lakonia (Peloponnese) [Pausanias, 1918; book 3, 21, 4]. Paulus Silentiarius scribing the marbles of Hagia Sophia (Proconnesium, Iassos, Carystos, Phrygia together with Egyptian and Greek porphyry, etc.) selects a comparison in the human world (Cipollino rosso from Iassos "show oblique streaks of blood-red and livid white" – Bianco e nero antico from Pyrénées "spreads a milky-white on the marble surface glittering of black poured at random") [Paulus Silentiarius, 2005; lines 590–646].

MARBLES THROUGH LATIN AUTHORS

The use of marbles in Rome started in the Augustean period; as reported by Suetonius about the life of Augustus:

> The city (Rome), which was not built in a manner suitable to the grandeur of the empire, and was liable to the inundations of the Tiber, as well as to fires, was so much improved under his administration that he boasted, not without reason, that he found it of brick, but left it of marble.
>
> [Suetonius, 1889; book 28, 3]

Some older examples are present, such as the Temple of Jupiter Stator, located by the Tiber and later indicated as "Portico di Ottavia", referred to the victor of Macedonians Q. Cecilius Metellus (146 BCE) and known for the magnificence or luxury of marbles [Vitruvius, 1931; book 3, ch. 2]. The use was gradually developed according to the territorial expansion of the Roman Empire towards Greece, Asia Minor, North Africa and Spain, the territories of the quarry sites.

Latin authors held the Greek marbles in great esteem; firstly, Vitruvius reported the marbles from Paros, Proconnesus (island of Marmara), Heraclea (Asia Minor) and Thasos (northern Aegean Sea) [Vitruvius, 1931; book 10, ch. 2, 15], about the construction of the Temple of Artemis at Ephesus.

Plinius wrote, in his encyclopaedic book on natural history, about marble suitable for columns (Hymettus) [Plinius, 1962; book 36, 7], for sculpture (Cyclades – Paros; Thásos, Lesbos) [Plinius, 1962; book 36, 44]; for cladding (Proconnesus, slabs on the brick walls of the palace of Mausolus at Halicarnassus) [Plinius, 1962; book 36, 47]. Plinius's text contains also the detailed description of marbles used in architecture, such as Carystium, Chium, Hymettium, Lacedaemonium, Luculleum, Numidicum, Porphyrytes, Proconnesium, Synnadicum, Taenarium, etc. and the list of the related buildings [Plinius, 1962; book 36, 49–63]. Plinius also reported the possible invention of the art of cutting marble into thin slabs and, according to Cornelius Nepos, reported the first man in Rome to have marble veneers on the whole walls of his house. On one hand, the marble from the island of Marmara (Proconnesium) to decorate the brick walls of the palace of Mausolus at Halicarnassus (Caria, southwestern Turkey, mid-4th century BCE) [Plinius, 1962; book 36, 47]; on the other hand, a Roman knight called Mamurra, chief engineer of Julius Caesar in Gaul, was also the first to have a house including marble columns only (from Carystos and from Luna) [Plinius, 1962; book 36, 48]. Finally, Marcus Lepidus (about 78 BCE) was the first to lay down door-sills of Numidian marble (Giallo antico) in his house [Plinius, 1962; book 36, 49].

About the exploitation of the marble outcropping in the Italian territory, Strabo described in detail the quarries of the Apuanian Alps from which the marble called Lunense was supplied; the name of the marble came from the nearby port of Luna (a few kilometres southeast of Sarzana – Liguria), the port of marble shipping. Strabo reported in the same chapter the white marble and a green veined one (Cipollino apuano), suitable to make monolithic columns or finely carved elements to be employed in Rome and elsewhere [Strabo, 1853; book 5, ch. 2, sec. 5]. Plinius exalted the marble of the recently discovered quarries of *Luna*, suitable to make firm columns [Plinius 1962; book 36, 48] and whiter in comparison with the marble of the island of Paros [Plinius 1962; book 36, 14].

The use of white or coloured marbles spread in every corner of the Roman Empire and, in particular, the use of coloured ones became a peculiar expression of the Roman art, reaching an incomparable perfection, from both a technical and artistical point of view, in the decoration of imperial palaces, religious temples and private houses and villas [Pensabene, 2002]. These marbles were used for decoration and for cladding or paving (polygonal slab), but some of them, thanks to their mechanical characters, were also employed as upright vertical support. The quarries were owned by the emperors, and huge stores of marbles were created in Rome (*Marmorata*, between Aventinus and Testaccio) and of the mouth of river Tiber (*Portus*) [Maischberger, 1997]. The pervasion of the coloured marbles in the Roman architecture was indirectly witnessed by the disapproval expressed by the men of letters celebrating the modesty of the ancient customs compared to the contemporary ostentation of wealth and luxury. Examples come from Horace about the values of the honest life: "Carved ivory I have none, no golden cornice in my dwelling shines; pillars choice of Lybian stone upbear no architrave from Hymettian quarries" [Horace, 1882; book 2, 18]. From Seneca on the superficial blessings:

> Children are pleased by the smooth and variegated pebbles which they pick up on the beach, while we take in tall columns of veined marble brought either from Egyptian sands or from African deserts to hold up a colonnade or a dining-hall large enough to contain a city crowd. We admire wall veneered with a thin layer of marble what defects the marble conceals.
>
> [Seneca, 1917; Letter 115, 8–9]

From Plinius, about the decadence of painting:

> And first we shall say what remains to be said about painting, an art that was formerly illustrious, at the time it was in high demand with kings and nations. . . . But at present time it has been entirely ousted by marbles . . ., we have begun even to paint on the masonry.
>
> [Plinius, 1962; book 35, 1]

Again Plinius, about the use of stones and marbles:

> It remains for us to deal with the nature of stones or, in other words, the prime folly in our behaviour. . . . For everything that we have investigated up to the present volume may be deemed to be created for the benefit of mankind. Mountains, however, were made by Nature for herself to serve as a kind of framework for holding firmly together the inner parts of the earth, and at the same time to enable her to subdue the violence of rivers, to break the force of heavy seas and so to curb her most restless elements with the hardest material of which she is made. We quarry these mountains and haul them away for a mere whim; and yet there was a time when it seemed remarkable even to have succeeded in crossing them. Our forefathers considered the scaling of the Alps by Hannibal and lather by the Cimbri to be almost innatural. Now these selfsame Alps are quarried into marble of thousand varieties.
>
> [Plinius, 1962; book 36, 1]

The stone pieces were mainly used as thin rectangular slabs, set instead painted plaster over walls made of Peperino (a local volcanic stone) or made of concrete; thick rectangular slabs, set instead brick or tile for pavements. In both cases, the arrangement of coloured veins and patches was always taken into the highest consideration. Mosaics also were made using tesserae of coloured marbles (i.e. the floors of Palestrae made of cubic tesserae of red Egyptian and green Greek porphyries, Giallo Antico, white marble – Baths of Caracalla, early 3rd century CE).

Some coloured marbles were also useful as upright structural members (monolithic shaft of columns and pillars or separate drums). In this case, it the quotation about the villa of Gordiani, located at via Prenestina, reporting fifty columns of Cipollino, fifty of Pavonazzetto, fifty of Giallo antico and fifty of Granito del Foro, is pertinent [Scriptores Historia Augusti; Gordiani tres, 1924; ch. 32, 2].

These precious and expensive marbles came from inland Egypt (alabaster, granite, red porphyry), from the Aegean Sea coasts (Cipollino, Cipollino rosso, Fior di Pesco, Portasanta, Settebasi), from the inland of Asia Minor (Occhio di Pavone, Pavonazzetto), from the mainland of Greece (green porphyry, Rosso antico, Verde antico), from northern Africa (alabaster, Giallo Antico). Later on, western European regions also supplied marbles, such as southern France (Pyrénées – Bianco e Nero antico) and eastern Spain (Catalonia – Broccatello). Particularly appreciated were the polychrome breccias showing different coloured patterns for each slab.

Coloured marbles were capillary spread in the provinces of the Roman Empire as witnessed by the archaeological excavations: from *Corduba* (Córdoba) at west till *Hadrianopolis* (Palmyra) at east, from *Colonia Agrippina* (Köln) at north till *Letpis Magna* at south and, obviously, in the whole Italian territory.

The wide variety of marbles traded during the Roman Empire is well documented by the "Edictum Diocletiani" (Diocletian Edict), an edict promulgated in 301 CE in order to properly regulate and honestly organize the Imperial State. Nineteen are the marbles (marble, granite, porphyry, limestone, breccia) classified by a name related to the geographical location of the quarry site; fifteen marbles were now identified (Porphyrites, Lacedaemonius, Numidicum, Luculleum, Pyrrhopoecilus, Claudianus, Alabastrum, Docimium, Tripônticum, Thessalicus, Carystium e Scyreticus as coloured marbles; Lesbium, Thasium e Proconnesium as white or grey marbles) and four still remain unidentified (Euthidemianum, Anacastenum, Heracleoticum e Potamogallenum). The marbles reported by the edict are partially coincident with Plinius's list.

A comparison relative to the price of each marble is very interesting: 250 *denarii* for cubic feet – Porphyrites (Porfido rosso antico) and Lacedaemonius (Porfido serpentino verde); 40 *denarii* for cubic feet of white or grey marbles (Lesbos); intermediate prices for the other ones [Edictum, 1974; 31]. But more interesting is the comparison with other kinds of items: the salary of an agriculture worker (25 *denarii* per day), the salary for stone mason (50 *denarii* per day), for mosaic craftsman (60 *denarii* per day), for a figure painter (150 *denarii* per day) [Edictum, 1974; 7]. Finally, a male slave (16–40 forty years old) was priced thirty thousand *denarii*, a female slave (16–40 forty years old) twenty-five thousand *denarii* and a learned slave sixty thousand *denarii* [Edictum, 1974; 29].

MARBLES THROUGH LATIN POETS

Many Latin authors represented the marbles in a poetical way. Virgil outlined the quarry works to make huge columns to the Carthage's theatre, [Virgil, 1910; book 1, lines 427–29]. Ovid reported the marble as a synonym of stiffness [Ovid, 1922; book 5, Perseus and Gorgon, lines 183, 206, 214 and 234]. Papinius Statius described the room of the upper floor of the villa of Polliux Felix, at Surrentum (Sorrento, Naples), enhancing the colour of the marbles from Aswan (lined by veins), from Phrygia (white background with purple streaks), from Numidia (shining yellow), from Karystos (compared with marine waves), from Lakonia (compared with soft grass) etc. [Statius, 1928; book 2, 2, lines 85–93]. In the poem were also recorded coloured marbles and the white ones from Luna, used to make columns, in the palace of Domitianus Augustus Germanicus [Statius, 1928; book 4, 2, lines 26–29]. Sidonius Apollinaris, in the 5th century, still mentioned the lack of marble as Parian, Cipollino, Proconnesos, Pavonazzetto, Giallo antico, Greek porphyry, Aswan granite for cladding the walls of the palace of Domitius [Sidonius, 1936; to Domitius, 2, 2, §7].

MARBLES THROUGH THE REUSE

The reuse of white marbles was made in three different ways: the first one takes into account the reuse of the artefacts for the same architectural purpose, the second one the reuse with reworking for different purposes; the third one the loss through the burning of marble elements in a kiln to make lime. This last way was blamed by famous architects studying the antiquities, such as Alberti, Bramante and Raphael, and it was witnessed by the Dutch painter Pieter van Laer (1599–1642) called "Bamboccio": his painting *Szene um eiten römischen Kalkofen* (Bayerische Staatsgemäldesammlungen, Alte Pinakothek, Munchen) reproduced a so-called *calcara*, a lime-kiln used also in the Roman Forum. On the other hand, the use of marble to make lime was suggested by Vasari: "white lime made either of marble chips or of travertine" [Vasari, 1907; ch. 4]; by Milizia: "il marmo bianco, dove ne è abbondanza, è preferibile a qualunque pietra per farne ottima calce" [Milizia, 1785; part 3, book 1, ch. 4] or by Rondelet: "Dans presque toute l'Italie la chaux dont on fait usage est fort bonne, parce que la pierre qu'on y emploie est presque toujours une espéce de marbre très-pur" [Rondelet, 1802; book 2, sec. 2, art. 1].

The reuse of coloured marble was a topic in the late-antique and medieval architecture of the Mediterranean basin, and it followed different ways. A structural element may conserve their role (i.e. in a porch) or may be cut into a slab for flooring (*opus sectile*) or small geometrical pieces for decorative purposes. Other artefacts (basins) were transformed as baptismal fonts or holy water stoups. Finally, coloured marble elements, both in the original shape or as irregular fragments, were included in standard masonries together with other stone pieces (block, ashlar).

Some examples allow one to appreciate the size of this architectural phenomenon. The cladding, together with a particular study to the reciprocal arrangement of veined slabs (open-book style) of different building in Ravenna (i.e. Baptistry of Orthodoxes, 5th century) [Krautheimer, 1965]; in Thessaloniki (i.e. church of Agios Demetrios, 5th century); in Istanbul (i.e. Hagia Sophia, 6th century; Chora-Kariye, 1315) [Taylor, 2007]. The architectural elements of Romanesque churches in Milan (i.e. four shafts of red porphyry of the ciborium, Sant'Ambrogio, 12th century); the pavements (*opus sectile*) and the encrustations (together with coloured glass) made by Cosmati or Vassalletto families in Rome (i.e. San Clemente, 12th century), in Latium (i.e. Duomo, Civita Castellana, 13th century), in Umbria (i.e. Collegiata, Lugnano in Teverina, 12th century), in Campania (i.e. Duomo, Salerno, 12th century) [Conant, 1959]. The cladding of the Islamic building in northern Africa (i.e. mosque of Sultan Hassan, Cairo, 14th century) [Williams, 2002]. Finally, the *opus sectile* works motivated also the Baroque architects to decorate floors, walls, columns and pillars with superb coloured patterns, for almost two centuries in Rome (*Marmorari* workshops), Naples (i.e. Cosimo Fanzago worshop), Palermo (*lapidum incisores*) etc. In these cases, the pieces made of coloured ancient marbles were always put together with pieces made of other marble never used in Roman times [Hopkins, 2002]. The practice of the reuse developed by skilful craftsmen, especially in Rome: the ancient buildings and the blocks stored by the Romans were easily available for this purpose. The words of M. de Montaigne, going to the mouth of Tiber, are clear with regard to the reuse of roman marbles: "Il y a quelques ruines et colonnes, comme il y en a plusieurs en ce lieu de Porto, où était cette vieille ville de Trajan, en en fait le pape désenterrer tous les jours et porter à Rome" [Montaigne, 1983; Rome, 15 March 1581].

The marbles stored were also used by *Marmorari* to produce new artefacts, and their workshops had a great role in the repair and reconstruction of ancient monumental buildings. These marble workshops were reported by different authors, such as Amati: "Una serie poi ben officine di marmi ho potuto osservare in alcune officine di Marmorari in Roma" [Amati, 1829; book 2, ch. 7, note]. Since the early 17th century, the enthusiasm for archaeology and for this particular detail (coloured ancient marbles) gave rise to many collections of uniformly shaped specimens, called *Studi di marmi*, picked up from Roman remains and gathered by wealthy connoisseurs and antique lovers in Rome and in other European capitals. This enthusiasm was witnessed by J.W. Goethe during his Italian Journey:

> a collection of specimens which Tischbein had selected for me. . . . There are opportunities enough here for my collecting many more specimens. In our way to the ruins of Nero's palace . . . the temptation to cram our pockets full of the granite, porphyry and marble slabs which lie here by thousands.
>
> [Goethe, 2010; Rome, 18 November 1786]

These collections were blamed by Francesco Milizia pointing out the uselessness of a catalogue of coloured ancient marbles, as the craftsmen often fabricated new "varieties" by cutting the fragments according to different patches or colours.

> Questi sono i marmi principali, di un numero certamente molto inferiore a quello, cui li fanno ascendere i marmorari in quelle loro serie, che chiamansi *Studj di marmi*, dove ordinariamente se ne contano fin duecento specie. Con la stessa industria essi potrebbero moltiplicarli ancora di più, non avendo da fare altro, come già fanno, che dividere e suddividere un pezzo dello stesso marmo in più pezzetti di varie macchie e di vari colori.
>
> [Milizia, 1785; part 1, book 4, ch. 12, 3].

Finally, an outstanding case of use of coloured ancient marbles, through the "commesso alla fiorentina" technique (inlay or thin pieces of porphyry, jasper, coloured marbles, etc. specifically cut and fit into carved-out recesses on a main stone material), was provided by "Opificio delle Pietre Dure" in Florence. The workshop was established in 1588 by Ferdinando I de'Medici, the grand duke of Tuscany, on the basis of artistic ateliers already present, and it was effective until the last decades of the 19th century; some famous artists, such as Buontalenti, Giambologna, Dosio and Ligozzi, lent their designs to make awesome artefacts. Similar workshops were established in Naples by Carlo VII, king of Naples and

Sicily, under the guidance of the Florentine craftsman F. Ghinghi ("Real Laboratorio di Pietre Dure", 1737–1861). The king moved to the throne of Spain (1759) as Carlos III and established in Madrid a new workshop ("Real Laboradorio de Mosaicos y Piedras Duras del Buen Retiro", 1759–1808). Other workshops were installed in Prague, under Rudolph II, king of Bohemia (early 17th century); in Paris under King Louis XIV (Manufacture des Gobelins, late 17th century).

MARBLE IMITATIONS THROUGH PLASTER, STUCCO AND PAINTING

The use of other materials, mainly plaster and stucco, became established in architecture, on column shafts and moulded elements, to save costs or to supply difficulties of coloured marbles; the imitation involves colours, veins and spots featuring a marble surface (see also Chapter 21, Volume 2). Plaster and stucco are generally applied in two superimposed layers made of lime as a binder and sand or crushed marble as aggregate (plaster) or made of lime and sand or gypsum without aggregate (stucco). The imitation of marble is made by a painted layer (plaster) or by a pigment blended into gypsum (scagliola). The painter used different colours and different coats of painting to reproduce ground and veins or spots featuring the original marble. The gypsum craftsman uses a finely ground plaster of paris mixed with glue and incorporating the pigments needed to obtain the marble imitation.

The painted imitation of marble patterns can be very similar to the original; on the contrary, the "scagliola" imitation is less accurate.

A significant example of the use of plaster, applied on a core made of stone or brick, are the granite-like columns (the imitation of pink granite of Baveno is exemplary) of the nave of the churches of San Giuseppe (1607–30) and San Fedele (late 17th century): the masonry shafts were covered with plaster and then by a ground of uniform pink colour, the finish were little black or white spots made of tiny drops of colour. Other examples are the columns and semi-columns in the "Sala da pranzo" or the Room XVII of the Galleria d'Arte Moderna (first decade of the 19th century) at Villa Reale (Milan): the painted layer imitates the green ground and the white spots of Verde antico from Thessaly.

The making technique of gypsum-work (stucco) is clearly described by Francesco Milizia:

> Burnt pure gypsum imitates the marble; it is enough to grind the gypsum as very fine powder, to sieve it, to moisten it with glued water and to blend a colour suitable to form stains and veins as requested by the true marble; this kind of blend reaches a very strong consistency and a very fine lustre.
>
> [Milizia, 1785; part 3, book 1, ch. 6]

Some interesting examples of coloured marble made of stucco are visible in Milan, mainly for columns and pillars (variegated red pattern on a Serlian motif – left side of Cappella Trivulzio, church of Santo Stefano; four column shafts as Cipollino in the "Sale Napoleoniche" of "Pinacoteca di Brera"). Altar frontals made of scagliola are present in different churches of Milan (i.e. church of San Giuseppe, right altar; church of San Fedele, second altar on the right).

An important role was also played by the reproduction of veneer of coloured marbles on the walls. This painting activity was established in the Roman Republican age, probably for a budgetary basis, in order to decorate the interior walls of buildings following Hellenistic examples (Alexandria, Pergamon), as reported by Vitruvius [Vitruvius, 1931; book 7, ch. 5]. In Lombardy there is significant evidence at the Sanctuary (built in the first half of the 1st century BCE), unearthed in 1993 at Brescia (*Brixia*) beneath the Imperial Capitolium (built in the last decades of the 1st century CE): the walls of this Sanctuary were painted simulating panels of breccias and cornices of alabasters. It is remarkable the good reproduction of rounded clasts and dark cement of breccia and of convoluted bands of alabaster: there are real-life copies, although the colours are slightly different from the real ones.

Marble reproduction on walls was diffused everywhere in the Roman world, and Pompeii is full of examples: house of Sallust (*Tablinum*, first style); house of Nozze d'Argento (Room Y, second style); house of Vettii (Art gallery, fourth style) [Ward Perkins, 1981].

Painted walls ("fresco" technique) were present in the medieval buildings or in the Renaissance ones: the imitation mainly concerned the porphyries both from Egypt and from Greece. Two outstanding examples were made by Giotto and A. Mantegna: the painted marble panelling, lining the base of the walls or the thin mouldings, and Cosmati-like frames to the individual scenes at "Cappella degli Scrovegni" (Padua, 1304–13); the red and green porphyries painted behind the Court on the west wall of "*Camera Picta*" (Palazzo Ducale, Mantua, 1465–74). In some cases, such as "Cappella dei Magi" at Palazzo Medici in Florence (1459–60), the red and green porphyries reproduced on wall paintings (Benozzo Gozzoli) are in agreement with the original ones in the floor of Michelozzo.

Examples of this kind of decoration are well represented in Milan. Bernardino Luini painted (1529) slabs of Egyptian and Greek porphyries on the central wall of the church of St Maurizio at Monastero Maggiore. Reproductions of coloured marbles were painted on the naves of Santa Maria delle Grazie (late 15th century) and on the side columns of Santa Maria della Consolazione (green ophicalcite and Macchiavecchia, 1836). In other cases the painted marbles were inserted in geometric spaces surrounded by moulded cornices made of stucco (i.e. Santa Maria delle Grazie, Sauli chapel, fifth on the right nave).

The reproduction of marble, mainly on the walls of the staircases of Milan's apartment buildings, was spread in the first half of the 20th century: the reproduction was made using a painting on plaster (veined marbles) or using a coloured gypsum plaster (stained or dotted marbles).

MARBLES THROUGH EASEL PAINTING

The easel paintings on panel or canvas from the Middle Ages to the Renaissance show accurate reproductions of coloured marbles. An outstanding example is the *opus secitle* pavement reproduced by Hans Holbein the Younger in the painting *The Ambassadors* (oil on oak, 1533, National Gallery, London). Many Italian painters, until the last decades of the 16th century, reproduced ancient marbles in their works. The venetian painter Cima da Conegliano made several reproductions of coloured marbles set in well-proportioned architectures on his religious paintings, now spread in different Italian galleries (Brera, Milan; Galleria, Parma; Accademia, Venice, etc.; oil on panel, 1459–1517). On the other hand, other painters reproduced imaginary marbles through the mixing of different colours (i.e. the throne of the Virgin in *Madonna della Candeletta*, by Crivelli, tempera on panel, Pinacoteca di Brera, Milan, about 1490). The marble reproduction disappeared on paintings in the beginning of the 17th century. It is relevant to report another kind of painting made on a stone used as support: the polished slabs of "Pietra Paesina" of Tuscany (calcareous siltstone with ferrugineous staining among internal fractures) supplied an extravagant background of painted religious scenes.

Finally, coloured marbles and hard stones were used in particular "paintings" reproducing sights of towns, landscapes, religious scenes, "capriccio", still lifes, allegories, etc. made by the Opificio delle Pietre Dure with the technique of "commesso alla fiorentina".

MARBLES THROUGH MEDIEVAL AND MODERN LITERARY SOURCES

Medieval erudites, such as Isidore of Seville, enumerated the marbles following the path of the Latin authors [Isidorus, 1911; book 16, 5]. Still, during the Renaissance, the words of ancient authors were reiterated, sometimes mixed with personal experiences, by erudite writers such as Francesco Grapaldo [Grapaldo, 1494; book 1]; Georg Bauer – Giorgio Agricola [Agricola, 1546; book 7]; Girolamo Cardano [Cardano, 1550; book 7], Tommaso Garzoni [Garzoni, 1585, Discorso 70 De' minerari etc.; Discorso 90 De' Scultori etc.] and Ulisse Aldrovandi [Aldrovandi, 1648; book 4, ch. 57].

White marbles were recorded, in the medieval and Renaissance poetical literature of Italy, always for uncommon characters (whiteness, coldness, hardness, solidity). Petrarch wrote: "I only live on hope, remembering I've seen a little water's constant flow wear away marble and the solid stone" [Petrarch, 2002; 265, lines 9–11]. Marble was also a symbol of wealth or a synonym of sculpture; for instance, Torquato Tasso wrote about the castle of Armida the magician: "Amid the herbs, the grass and flowers rare, the falling leaves down pattered from the wood, the birds sung hymns of love; yet speak I naught of gold and marble rich, and richly wrought" [Tasso, 1749; book 10, 63, lines 5–8]. In other cases, the marble features were compared with human body and soul (heart, breast, skin, purity). The same traits were also influenced by translucent alabaster, as written by Petrarch, in comparison with the idealized body of Laura: "the walls were alabaster, the roof of gold, the entrance ivory, the window sapphire" [Petrarch, 2002; 325, lines 16–17].

Coloured marbles were, instead, reported as part of fanciful buildings. For instance, Dante Alighieri wrote about the staircase of Purgatory:

> Where we came, the first step was of white marble, so smooth and polished that I was reflected there, as I appear. The second was darker than a dark blue-grey, of a rough, calcined stone, cracked in its length and breadth. The third, which is massed above them, seemed like red porphyry to me, fiery as blood spurting from an artery. God's Angel kept both his feet on this, seated at the threshold, which seemed, to me, to be of adamantine stone.
>
> [Dante Alighieri, 2012; Purgatory, book 9, lines 94–105]

Lodovico Ariosto wrote about the sepulchre of Brandimarte "Nor out of Sicily the Count [Orlando] departs, till porphyries he procures and alabasters" [Ariosto, 1823; book 43, 182, lines 1–2]. Giovanbattista Marino, writing about the funeral of Adonis, envisaged Apollo playing the zither and calling back some stones, such as Phrygian marble, Africano, Greek green porphyry, Parian marble and again different porphyries; yet hundreds of smiths and masters saw jaspers and manipulate alabasters. ["Corron veloci alla divina cetra La frigia selce e l'africana pietra, E di Sparta e di Paro il marmo corre. . . . Onde si vede in un balen raccorre Gran quantità di porfidi diversi. . . . Già cento fabbri a prova e cento mastri Segan diaspri, affinano alabastri" [Marino, 1623; book 19, lines 335–36].

A very fascinating tale is the *Hypnerotomachia Poliphili*, one of the most famous books in the world, probably written by Francesco Colonna, published by Aldus Manutius in Venice (1499) together with hundreds of xylographies as illustration of the text. This book contains a description of several marbles, based on their colours, in the sphere of a dreamed magnificent palace (Porphyrite or Porphyro, Numidica, Pyropoecila thebaica, Lacedæmonia, Luculea, Alabastrina, Hexaconthalitho, Ophitea, etc.) [Colonna, 1999]. On the other hand, some writers reported a correct description of the presence of coloured marbles in their own town; for instance, Giovanni Boccaccio [*The Decameron*] wrote about Florence: "As he [Guido Cavalcanti] was between the columns of porphyry there and the tombs [of marble] in question and the door of the church [St Giovanni]" [Boccaccio, 1883; Day 6, story 9].

MARBLES THROUGH TREATISES OF ARCHITECTURE

White and coloured marbles were described together with other building materials by some architects, such as Leon Battista Alberti in 1485 [Alberti, 1988; book 2, ch. 8, 9; book 9, ch. 1]; Pietro Cataneo in 1564 [Cataneo, 1564; book 2, ch. 2, 3]; Giorgio Vasari in 1568 [Vasari, 1907; On Technique, ch. 1]; Benvenuto Cellini in 1568 [Cellini, 1568; Second Treatise], Vincenzo Scamozzi in 1615 [Scamozzi, 1615; part 2, book 7, ch. 4]; Francesco Milizia in 1785 [Milizia, 1785; part 1, book 4, ch. 12]; Giovanni Antolini [Antolini, 1817; part 1, book 4, chap. 12] or Carlo Amati in 1829 [Amati, 1829; book 2, ch. 7, note]. Vasari described some coloured marbles together with other stone materials used in Tuscany. First, he pointed out the hard task to work some coloured stones: the first artist to carve and polish the Egyptian red porphyry according to the antique way was Francesco del Tadda in the year 1555, after vain attempts made by other artists, such as Leon Battista Alberti. In addition to porphyry, Vasari listed: "Serpentino" (Porfido Serpentino Verde, Greece), "Cipollaccio" (probably Verde Antico, Greece), "mischio" (brecciated marbles, "Africano" and "Centopietre", Egypt), red or grey granite of Egypt (Aswan or Eastern Egyptian desert), Basalt (Egypt), transparent marbles (alabaster), "Cipollino" (Carrara and Pietrasanta). The Vasari's list included also: "Paragone" (probably a fine-grained black limestone), statuary marbles (Carrara and Garfagnana), white Pisan marble (monte San Giuliano), travertine (Tivoli, Rome), slate (Lavagna, Genoa), grey granite (Elba Island), Peperino (Roman Campagna and Naples), white limestone (Istria); pietra Serena, Macigno and Pietraforte (sandstones, Florence and Tuscany). Cellini, the Florentine sculptor, goldsmith and engraver, reported five degrees of marble grain size, pointing out the different degree of working difficulties (the best is the very fine-grained marble from Paros Island). The author reported porphyries (green and red); granites (red from Egypt and black and white from Elba Island) and sandstones of Tuscany; finally, Cellini detailed accurately the use of tools such as punch, chisel with notch, chisel, file, rasp and drill.

Scamozzi increased the number of marbles and reported the use in several monuments of Antiquity: Basanite, marble of Chios, Lesvos, Thasos, Paros, Charistos (Cipollino), Himettus, Numidian (Giallo Antico), red porphyry, Proconnesian, green porphyry (Greece), Africano, Claudianus (Granito del Foro), Serpentino, Syenite (red Egyptian granite). However, this list is full of repetitions and sometimes disordered and difficult to understand. On the contrary, Scamozzi showed an uneven knowledge about marbles quarried in Italy, introducing also several notices about their use and about their provenance as results of direct observations on site.

The impact of coloured marbles was significant also in France, as witnessed in different treatises written by peculiar erudites such as Anselme Boece de Boot [Boece, 1644; *livre II Des marbres, chap. 265–281,*] or by architects such as André Félibien [Félibien, 1676; livre 1], Augustin Charles D'Aviler [D'Aviler, 1691; chapitre Des marbres et de leurs différentes couleurs], Jean François Blondel [Blondel, 1675; tome I, livre 1, Des marbres antiques], Quatremère de Quincy [Encyclopédie méthodique, 1801–20] and Jean-Baptiste Rondelet [Rondelet, 1802; book 1, sec. 2, art. 5 "Egyptian granite", art. 9 "Porphyries and ancient marbles"].

MARBLES THROUGH SYSTEMATIC STUDIES

Different original studies, since the 17th century, were devoted exclusively to the description of marbles and their reports became more and more accurate. The authors are Agostino del Riccio [Del Riccio, 1996], Biagio Garofalo [Garofalo, 1738], Francesco Belli [Belli, 1842], Faustino Corsi [Corsi, 1845] and Henry William Pullen [Pullen, 2018] from England. These studies were often linked to private collections of coloured marbles, the already quoted *Studi di marmi.*

The Italian terms used today to indicate the coloured marbles were mainly coined by these craftsmen on the basis of their macroscopic features (Cipollino: streaked as onion) or of their use in renowned buildings (Portasanta: from the doors on the right in Christian basilicas of Rome) rather than the topographical origin (Breccia di Settebasi: from the Villa along the via Tuscolana, south-east of Rome).

Finally, different scientists, such as Gabriele Borghini [Borghini, 1998], Harald Mielsch [Mielsch, 1985], Raniero Gnoli [Gnoli, 1988], Enrico Dolci and Leila Nista [Dolci, 1992], Patrizio Pensabene and Mattias Bruno [Pensabene, Bruno, 1998], Lorenzo Lazzarini [Lazzarini, 2004], Monica T. Price [Price, 2007] and Chandra L. Reedy [Reedy, 2008] in the second half of the 20th century, studied the marbles in a proper way, solving some problems of identification and location of quarry sites.

The availability of coloured marbles in Milan and Lombardy

Coloured marbles were extensively used, during the Roman times, in Milan and other regions of northern Italy. Ancient marbles identified in the sites of *Regio XI Transpadana* (*Mediolanum, Ticinum, Novocomum*, etc.) and of *Regio X Venetia* (*Cremona, Brixia*, etc.) are exactly the same employed in the whole Roman Empire. They were used for paving, veneering, mouldings and other carved ornaments (Bugini, Folli, 2005). In Milan, the Roman buildings provided a partial reuse in the early Middle Ages: four shafts (Porfido rosso antico, 3.4 m high) erected in the ciborium of basilica of St Ambrogio (10th century); four shafts (Porfido rosso antico, 3.0 m high) already in the church of San Carpoforo (9th century) and now at the Museum of Castello Sforzesco; two shafts (Fior di Pesco) in the Chapel of Sant'Ippolito (basilica of San Lorenzo). Furthermore, the architectural development of Milan was characterized by the complete destruction of ancient buildings, and this policy cut off the availability of ancient marbles. An interesting example of the use of coloured marbles is present in the church of Santa Maria dei Miracoli (begun in the late 15th century). The architect Galeazzo Alessi built a wall (greyish silicate marble – Boden) to separate the choir from the deambulatory, featuring geometrical inlays of coloured marbles. Coloured ancient marbles (Africano, Sciro, Breccia Corallina) are set together with other marbles, recalling the aspect of the ancient ones (Serpentine, Portoro).

Other towns of Lombardy show examples of reuse: the funerary monument (Verde antico, Giallo antico, Alabaster, Brocccatello, Portasanta) of Vespasiano Gonzaga (church of Incoronata, Sabbioneta, province of Mantua, 1592); the windows (porfido serpentino) of Casa dell'Arciprete (architect P. Isabello, Bergamo, 1520); the eight columns (Cipollino) of Liceo Volta (architect S. Cantoni, Como, 1820; from an Early Christian baptistery) [Bugini, Folli, 2005].

In northern Italy are noticeable the cases of Venice, where the materials stockpiled after the pillage of Constantinople (1204), supplied the columns for the front and the southern side of the basilica of San Marco and, later, for the veneering of St Maria dei Miracoli (1489) and for the decorations of Scuola Grande di San Marco (1490) and Scuola Grande di San Rocco (1549) [Gnoli, 1988]. The coloured marbles to decorate altars and sepulchres of the church of Santa Croce (1570) at Bosco Marengo (near Alessandria, eastern Piedmont) were sent from Rome by Pope Pius V (Antonio M. Ghislieri, 1566–72), born in the same village.

Finally, starting from the Council of Trent onward, papal patronage stimulated a great architectural renovation and an increasing use of coloured marble; the search for new materials as surrogates for ancient ones, however, became urgent: the stones of Arzo (limestone and breccia) near Lake Lugano and the stones of Varenna black limestone with white veins) near Lake Como. These surrogate marbles have a visual resemblance (colour and texture) with some marbles used by the Romans: stones of Arzo (Macchiavecchia) matched Portasanta (Greece); Varenna black limestone matched Bianco e nero antico or Grand antique (Pyrénées, France).

DESCRIPTION OF MARBLES

Coloured marbles used in the Roman architecture in Milan are described on the basis of their features: macroscopic aspect, petrographical classification, forms of use, provenance. Each marble is identified by the terms reported by modern scientific studies; a series of synonyms used since the Roman times (referred to the authors quoted in previous pages) is also listed.

The words of Belli and Corsi were utilized, if possible, in the descriptions of the macroscopic aspect of marbles.

AFRICANO [Borghini, Dolci, Gnoli, Mielsch, Pensabene, Lazzarini, Price]

Historical terms

PLINIUS Luculleum. ED. DIOCLETIANI Luculleum. ISIDORE Luculleum. AGRICOLA Luculleum. DEL RICCIO Affricano. SCAMOZZI Lucullo – Africano. GAROFALO Luculleum. MILIZIA Africano. ANTOLINI Africano o numidico. AMATI Affricano. RONDELET Affricano. BELLI Africano. CORSI Africano (Chium).

Macroscopic aspect

Varicoloured background (whitish, grey, green, brown, black), light clasts, typical pinkish shades. Green colour features a specific variety.

Petrographical classification

Sedimentary rock: slightly metamorphosed tectonic breccia, containing calcareous clasts made of crystalline calcite, sparry calcite cement, hematite colour

Form of use

Slab (cladding, flooring), column shaft, decorative elements

Provenance

Turkey, Aegean coast, several quarries near Sığacık (ancient Teos, province Izmir)

ALABASTRO (TRAVERTINE) [Borghini, Dolci, Gnoli, Mielsch, Pensabene, Lazzarini, Price]

Historical terms

PLINIUS Alabastrites. ED. DIOCLETIANI Alabastrum. ISIDORE Alabastrites. GARZONI Alabastro. AGRICOLA Alabastrites. VASARI Alabastro. DEL RICCIO Alabastro. SCAMOZZI: Alabastro. ALDROVANDI Alabastrites. GAROFALO Alabastrites. MILIZIA e ANTOLINI Alabastro. AMATI Alabastrite. RONDELET Alabastro. BELLI Alabastro. CORSI Alabastro antico (Alabastrum).

Macroscopic aspect

Brown, yellow, pink and white in concentric bands

Petrographical classification

Calcareous alabaster (travertine, composed by fibrous calcite crystals, together with iron oxide)

Form of use

Slab for flooring or for cladding

Provenance

Near Helwan (province of Giza) or different sites scattered in the provinces of Minya and Asyût; (Egypt); near Hierapolis – Pamukkale (province of Denizli, western Anatolia, Turkey); around the province of Ouahran (Algeria)

BASANITE [Dolci, Gnoli, Mielsch, Pensabene, Lazzarini]
PIETRA BEKHEN [Borghini, Dolci, Lazzarini]
BEKHEN-STONE [Price]
BASALTO [Dolci]
BASALTO VERDE [Mielsch]

Historical terms

PLINIUS Basanites. ISIDORE Basanites. VASARI. DEL RICCIO Selce orientali. SCAMOZZI Basalte – Basanite. BELLI Basalte verde. CORSI Basalte verde (Basanites).

Macroscopic aspect

Deep green uniform background, sometimes with little black points

Petrographical classification

Sandstone (meta-greywacke), very compact and slightly metamorphosed, made of quartz, feldspar and different kind of rock clasts in a silicate matrix

Form of use

Statue, sculte decoration, basin

Provenance

Near Uadi Hammamat, midway between Luxor and Ourgada, Eastern Desert (Egypt)

BIANCO E NERO ANTICO [Borghini, Dolci, Gnoli, Mielsch, Pensabene, Lazzarini]
GRAND ANTIQUE [Price]

Historical terms

Garzoni Nero e bianco. Del Riccio Bianco e nero orientale. Garofalo Gallicus. Milizia e Antolini Bianco e nero (antico). Belli Bianco e nero di Francia. Corsi Bianco e nero di Francia (Celticum).

Macroscopic aspect

White background with angular black clasts

Petrographical classification

Calcareous tectonic breccia, mainly composed by calcite

Form of use

Column shaft (i.e. the columns of the Northern Roman baths of Lutetia, now in the Musée de Cluny, Paris)

Provenance

Near Aubert, Lez River valley, department of Ariège, Mid/eastern Pyrénées (France)

BRECCIA CORALLINA [Borghini, Dolci e Nista, Gnoli, Mielsch, Pensabene e Bruno, Lazzarini, Price]

Historical terms

Del Riccio Breccia minuta di Grecia. Milizia Breccia corallina. Amati Breccia corallina. Belli Breccia corallina. Corsi Breccia corallina. Lazzarini Marmor Sagarium.

Macroscopic aspect

Pale or dark pink to brown ground together with pink, yellow or white angular clasts

Petrographical classification

Clastic sedimentary rock: breccia containing micritic limestone clasts with calcite cement with hematite dispersions

Form of use

Slab for venerring, decorative element

Provenance

Near Vezirhan (province of Bilecik), western Anatolia (Bithynia), Turkey

BRECCIA A SEMESANTO [Mielsch]

SEMESANTO [Borghini, Gnoli, Pensabene e Bruno, Lazzarini, Price]

Historical terms

Del Riccio Breccia sagginata – Breccia panicata. Belli Breccia a semesanto. Corsi Breccia a semesanto.

Macroscopic aspect

Purple ground with white, yellow and brown elongated clasts, maximum grain size 4 mm

Petrographical classification

Meta-microconglomerate made of calcite together with hematite

Form of use

Small slab for veneering, sometimes joined to look like a wider slab

Provenance

Southern part of the island of Skiros, northern Sporades, Greece

BRECCIA DI SETTEBASI [Gnoli, Mielsch, Lazzarini, Price]

BRECCIA DI SCIRO [Borghini, Dolci e Nista, Pensabene e Bruno]

Historical terms

Ed. Diocletiani Scyreticus. Agricola Scyrium. Del Riccio Breccia greca de' Settebassi. Scamozzi Pietra mista di Sciro – Sciro. Garofalo Scyrium. Milizia e

Antolini Sette basi. Amati Sette-basi. Belli Breccia di Sette Basi. Corsi Breccia di Sette Basi.

Macroscopic aspect

Purplish to brown ground with brecciated appearance; elongated white, yellowish, reddish or pinkish clasts

Petrographical classification

Iso-oriented meta-conglomerate featuring clasts of recrystallized calcite

Form of use

Slab for veneering, decorative element

Provenance

Southern part of the island of Skiros and other nearby islands, northern Sporades, Greece

BRECCIA VERDE D'EGITTO [Borghini, Dolci e Nista, Gnoli, Pensabene e Bruno, Lazzarini, Price]

BRECCIA VERDE DURA [Mielsch]

CENTOPIETRE – HECATONTALITHOS [Lazzarini]

Historical terms

Vasari, Del Riccio Marmo detto Araceli. Scamozzi, Belli Breccia verde di Egitto. Corsi Breccia verde d'Egitto.

Macroscopic aspect

Deep green ground with rounded varicoloured stone fragments

Petrographical classification

Slightly metamorphosed conglomerate made of granite, porphyry or basalt clasts with silicate cement

Form of use

Column shaft, basin

Provenance

Near Wadi Hammamat, between Luxor and Hurghada (Eastern Desert, Egypt)

BROCCATELLO [Borghini, Dolci e Nista, Gnoli, Mielsch, Pensabene e Bruno, Lazzarini]

Historical terms

Del Riccio Broccatello di Spagna. Milizia Broccatello di Andaluzia. Antolini Broccatello di Spagna. Amati Broccatello di Andalusia. Belli Lumachellla a Broccatello sanguigno. Corsi Broccatello di Spagna.

Macroscopic aspect

Purple ground and yellow irregular spots with many different shades

Petrographical classification

Compact limestone (wackestone, biomicrite) with dolomite intercalations and fossil remains

Form of use

Slab for paving or for inlay with polished surface

Provenance

Different quarries were located along the valley of river La Llet (tributary of river Ebro), a few kilometres east of Tortosa (the Roman town of *Dertosa*; Catalonia, Spain). The most important were Els Valencians and La Cinta.

CIPOLLINO [Borghini, Dolci e Nista, Gnoli, Mielsch, Pensabene e Bruno]
CIPOLLINO VERDE [Lazzarini, Price]
Historical terms
PLINIUS Carystium. ED. DIOCLETIANI Carystium. ISIDORE Caristeum. GARZONI Caristio. AGRICOLA Carystium. DEL RICCIO Cipollino. SCAMOZZI Charistio – Cipollino. GAROFALO Carystium. MILIZIA e ANTOLINI Cipollino. AMATI Caristio. RONDELET Carystium – Cipollino. BELLI: Cipollino verde. CORSI Cipollino (Carystium).
Macroscopic aspect
Whitish ground with twisted green veins
Petrographical classification
Metamorphic rock or fine-grained marble made of calcite, veins made of chlorite and muscovite
Form of use
Slab for cladding and paving, column shaft, decorative element
Provenance
Southern part of the island of Évia (Euboea) between Nea Styra ans Karystos (Mount Ochi, Greece)

CIPOLLINO ROSSO [Borghini, Dolci e Nista, Gnoli, Mielsch, Pensabene e Bruno, Lazzarini, Price]
AFRICANONE [Dolci e Nista, Lazzarini]
JASSENSE [Lazzarini]
Historical terms
DEL RICCIO Marmo detto Fites. GAROFALO Jassense. AMATI Jallense (sic). BELLI Cipollino rosso. CORSI Jassense – Portasanta or Rosso Brecciato – marmor Lydium.
Macroscopic aspect
Deep red ground with thick, twisted, white veins
Petrographical classification
Metamorphic rock or calcite marble with quartz, plagioclase and hematite impurity
Form of use
Slab for cladding, almost always arranged in open-book style
Provenance
Between Iasos and Akbük, province of Muğla, southern coast of Aegean Sea, Turkey

FIOR DI PESCO [Borghini, Dolci e Nista, Gnoli, Mielsch, Pensabene e Bruno, Lazzarini, Price]
ROSSO DI ERETRIA [Lazzarini]
Historical terms
GAROFALO Molossium. MILIZIA Persico o fior di persico. AMATI: Molossio – Persico. BELLI Fior di Persico. CORSI Fior di persico (Molossium).
Macroscopic aspect
Pinkish ground with purple, red or pink veins and spots
Petrographical classification
Sedimentary rock: cataclastic limestone made of microcrystalline calcite
Form of use
Slab for cladding and paving

Provenance
Near Eretria, central part of the island of Évia (Euboea), Greece

GIALLO ANTICO [Borghini, Dolci e Nista, Gnoli, Mielsch, Pensabene e Bruno, Lazzarini, Price]
Historical terms
PLINIUS Numidicum. ED. DIOCLETIANI Numidicum. ISIDORE Numidicum. GARZONI Numidico. AGRICOLA Numidicum. DEL RICCIO Giallo – Numidico. SCAMOZZI Numidico. ALDROVANDI Numidicum. GAROFALO Numidicum – Lybicum. MILIZIA Giallo antico. AMATI Giallo antico. RONDELET Giallo antico. BELLI Giallo antico. CORSI Giallo antico (Numidicum).
Macroscopic aspect
Dark to light yellow or white ground with purple irregular veins
Petrographical classification
Sedimentary rock: compact or often brecciated limestone with sparitic texture
Form of use
Slab for cladding and paving, column shaft, decorative elements, sculpture
Provenance
Near Chemtou (*Simitthus* – province of Jundūbah), northern Tunisia (*Numidia*)

GRANITO ROSSO [Borghini, Dolci e Nista, Gnoli, Pensabene e Bruno]
GRANITO ROSEO [Mielsch]
GRANITO ROSSO DI ASSUAN [Lazzarini]
GRANITO ROSSO ANTICO [Price]
Historical term
PLINIUS Syenites – Pyrrhopoecilon. ED. DIOCLETIANI Pyrrhopoecilus. ISIDORE Syenites. GARZONI Granito. AGRICOLA Syenites – Pyrropoecilon. CATANEO Granito schizzato di rosso. VASARI Granito. DEL RICCIO Granito rosso. SCAMOZZI Sienitide – Pyriopecilon. ALDROVANDI Syenites – Granitum rubrum – Pyrrhopecilus. GAROFALO Syenites. MILIZIA Granito d'Egitto. ANTOLINI Granito rosso d'Egitto. AMATI Granito d'Egitto – Sienite. RONDELET Granito – Pyropoecilon – Sienite. BELLI Granito rosso. CORSI Granito rosso (Pyrhopoecilus).
Macroscopic aspect
Red ground with grey and black grains
Petrographical classification
Alkali grainite made of quartz, potash feldspar, plagioclase, biotite
Form of use
Column shaft, obelisk, sarcophagus, basin
Provenance
Hills south-east of Aswan (Syene) (province of Aswan), Egypt

GRECO SCRITTO [Borghini, Dolci e Nista, Gnoli, Mielsch, Lazzarini, Price]
Historical terms
BELLI Greco scritto
Macroscopic aspect
White ground with irregular, short, dark grey veins
Petrographical classification
Metamorphic rock, medium- to coarse-grained marble made exclusively by calcite

Form of use
Slab for cladding
Provenance
Near Cap de Gard, a few kilometres north of Annaba (former Hippo Regius – province of Annaba), north-western Algeria

LUMACHELLA ORIENTALE O D'EGITTO [Borghini, Dolci e Nista, Gnoli, Pensabene e Bruno, Lazzarini]
LUMACHELLA D'EGITTO [Mielsch, Price]
Historical terms
ALDROVANDI Marmor orientale. MILIZIA Lumachello antico. AMATI Lumachello antico. BELLI Lumachella d'Egitto. CORSI Lumachella d'Egitto.
Macroscopic aspect
Grey to yellowish ground with small, curved, pale grey fragments
Petrographical classification
Fossiliferous limestone (shells of *Ostrea*), made of calcite
Form of use
Slab for veneering
Provenance
Near Henncir El Kasbate (former *Thuburbo Maius*); Tunisia

OCCHIO DI PAVONE [Borghini, Dolci e Nista, Gnoli, Mielsch, Pensabene e Bruno, Lazzarini, Price]
Historical terms
ED. DIOCLETIANI Potamogallenum (attribution by Lazzarini). MILIZIA Occhio di pavone. AMATI Occhio di Pavone. BELLI Lumachella a Occhio di pavone rosso. CORSI Occhio di pavone.
Macroscopic aspect
Red to brown ground with white fossil fragments, often curved
Petrographical classification
Fossiliferous limestone (shells of *Rudistae*), made of calcite
Form of use
Slab for veneering
Provenanche
Valley of river Sakaria (province of Bileçik), western Anatolia, Turkey

PALOMBINO [Borghini, Dolci e Nista, Gnoli, Mielsch, Pensabene e Bruno]
PALOMBINO ANTICO [Price]
Historical terms
PLINIUS Coralliticus. ISIDORE Coralliticus. AGRICOLA Choraliticus. GAROFALO Coraliticus. AMATI Coralitico – Sangario. RONDELET Coraliticus. BELLI Palombino. CORSI Palombino (Coraliticum).
Macroscopic aspect
Uniform white to yellowish ground (like ebony)
Petrographical classification
Microcrystalline limestone, made of calcite
Form of use
Small, moulded slab, often with animal shape

Provenance
Unknown

PAVONAZZETTO [Borghini, Dolci e Nista, Gnoli, Mielsch, Pensabene e Bruno, Lazzarini]
MARMO PAVONAZZETTO [Price]
Historical terms
PLINIUS Synnadicus. ED. DIOCLETIANI Docimium. GARZONI Frigio. AGRICOLA Synnadicum – Docimium. SCAMOZZI Sinadico. GAROFALO Docimenum – Synnadicum. MILIZIA Paonazzetto. ANTOLINI Pavonazzetto. AMATI Paonazzetto. RONDELET Synnadicum. BELLI Pavonazzetto. CORSI Pavonazzetto (Docimenium).
Macroscopic aspect
White ground almost diaphanous with purple veins and spots
Petrographical classification
Metamorphic rock, marble often brecciated fine to medium grained
Form of use
Slab for cladding and paving, column shaft, decorative element, sculpted element, statue
Provenance
Near Işcehisar (former *Docimium* – province of Afyon), western Anatolia (Phrygia), Turkey

PORFIDO ROSSO ANTICO [Dolci e Nista, Lazzarini, Price]
PORFIDO ROSSO [Borghini, Gnoli, Mielsch, Pensabene e Bruno]
Historical terms
PLINIUS Porphyrites – Leptopsephos. ED. DIOCLETIANI Porphyrites. ISIDORE Porphyrites. GARZONI Porfido. AGRICOLA Porphyrites – Leucostictos. CATANEO Porphido (Leucostycto). VASARI Porfido. DEL RICCIO Porfido. SCAMOZZI Porfiride – Porfido – Leucostito. ALDROVANDI Porphyrites – Porfidum. GAROFALO Porphyrites – Leucostictos. MILIZIA e ANTOLINI Porfido rosso. AMATI Porfido – Porfirite. RONDELET Porfido rosso. BELLI Porfido rosso. CORSI Porfido propriamente detto (Porphyrites) – Porfido rosso.
Macroscopic aspect
Deep red to purple ground with very small white, pink, black spots
Petrographical classification
Volcanic rock, andesite-dacite, made of hornblende (phenocryst), plagioclase (ground mass)
Form of use
Salb for veneering, basin, statue
Provenance
Near mount Dukhân (former *mons Porphyrites*), Eastern Desert, Egypt

PORFIDO SERPENTINO VERDE [Mielsch]
PORFIDO VERDE ANTICO DI LACONIA [Dolci e Nista]
PORFIDO VERDE DI GRECIA o SERPENTINO [Borghini, Gnoli]
PORFIDO VERDE ANTICO [Lazzarini, Price]
SERPENTINO [Pensabene e Bruno, Lazzarini]
Historical terms

PLINIUS Lacedaemonium. ED. DIOCLETIANI Lacedaemonium. ISIDORE Lacedaemonium. GARZONI Lacedemonio. AGRICOLA Lacedaemonium. CATANEO marmo verde di Lacedemonia. VASARI Serpentino. DEL RICCIO Serpentino. SCAMOZZI Serpentino. ALDROVANDI Lacedaemonium – Serpentinum. MILIZIA Serpentino. ANTOLINI Porfido verde. AMATI Lacedemonio – Croceate. RONDELET Porfido verde (antico). BELLI Porfido serpentino verde. CORSI Porfido detto serpentino (Lacedaemonius).

Macroscopic aspect

Deep green ground with light rectangular spots, sometimes crossing

Perographical classification

Volcanic rock, diabase – andesite with plagioclase and pyroxene (phenocryst); glass, plagioclase, pyroxene and chlorite (ground mass)

Form of use

Slab for veneering, decorative element

Provenance

Near Krokeés (pebbles spread on the ground), Lakonia, Peloponissos, Greece

PORTASANTA [Borghini, Dolci e Nista, Gnoli, Mielsch, Pensabene e Bruno, Lazzarini, Price]

Historical terms

PLINIUS Chium. GARZONI Luculleo. AGRICOLA Chium. DEL RICCIO Porta Santa – Libico. SCAMOZZI Marmo di Chio. GAROFALO Chium. MILIZIA E ANTOLINI Portasanta o Sirena. AMATI Sirena – Portasanta. RONDELET Portasanta. BELLI. Portasanta. CORSI Portasanta (Jassense).

Macroscopic aspect

Generally red-brown-pink ground with white, grey, purple spots

Petrographical classification

Clastic sedimentary rock, tectonic breccia with large limestone clasts, microcrystalline calcite matrix and cemen.

Form of use

Slab for cladding and paving, column shaft, decorative element

Provenance

A wide quarry (Theatrolatomio), a few kilometres north of Kios town, island of Kios, Greece

ROSSO ANTICO [Borghini, Dolci e Nista, Gnoli, Mielsch, Pensabene e Bruno, Lazzarini, Price]

Historical terms

PLINIUS Taenarium. DEL RICCIO Rosso orientale. SCAMOZZI Tenario. GAROFALO Taenarium. MILIZIA Rosso antico. AMATI Rosso antico. RONDELET Rosso antico. BELLI Rosso antico. CORSI Rosso antico.

Macroscopic aspect

Red to deep red ground, sometimes crossed by thin black veins

Petrographical classification

Metamorphic rock, very fine-grained marble made of microcrystalline calcite and hematite

Form of use

Decorative and sculpted element

Provenance

Close to the end (cape Matapan or Tainaron) of the central peninsula of Peloponissos, Lakonia, Greece

VERDE ANTICO [Borghini, Dolci e Nista, Gnoli, Mielsch, Pensabene e Bruno, Lazzarini, Price]

Historical terms

ED. DIOCLETIANI Thessalicus. DEL RICCIO Verde d'Egitto. GAROFALO Atracium. MILIZIA e ANTOLINI Verde antico. AMATI Tessalico. RONDELET Verde antico (marmo). BELLI Verde antico. CORSI Verde antico (Atracius).

Macroscopic aspect

Green ground with great rounded grey, black and white clasts

Petrografical classification

Ophicalcite made of fibrous serpentine and calcite

Form of use

Slab for cladding, column shaft

Provenance

Near Omorfohóri (province of Làrisa), Thessaly, Greece

WHITE or GREY MARBLE

MARMO BIANCO DI CARRARA [Gnoli]

MARMO DI CARRARA [Borghini, Pensabene e Bruno]

LUNENSE [Dolci e Nista]

WHITE MARBLE OF CARRARA [Price]

Historical terms

PLINIUS Lunense – Luniense. ISIDORE Lunensis. GARZONI Marmo da Carrara. AGRICOLA Lunense – Carrariense. CATANEO marmo candido del monte di Luni. VASARI Marmi bianchissimi e lattati (di Carrara, di Luni). DEL RICCIO Bianco di Carrara. SCAMOZZI Marmi di Massa di Carrara. GAROFALO Lunense. ALDROVANDI Lunense – Carariense. MILIZIA Statuario (da Carrara). ANTOLINI. Statuario di Carrara – Bianco venato di Carrara AMATI Statuario di Carrara. RONDELET Marmo di Carrara (di Luna). BELLI. Marmo Lunense o di Carrara. CORSI Marmo di Carrara (Lunense).

Macroscopic aspect

White diaphanous ground with grey veins and little spots

Petrographical classification

Metamorphic rock, fine-grained marble made of calcite with some quartz crystals

Form of use

Salb for cladding and pavement, column shaft, architectural element, decorative and sculpted element, statue

Provenance

Three distinct quarry basins north of Carrara (Apuanian Alps), Tuscany, Italy. Modern quarry basins are located north of Massa and Pietrasanta.

MARMO PARIO [Borghini, Gnoli, Pensabene e Bruno]

PARIAN MARBLE [Price]

Historical terms

PLINIUS Parium – Lychnites. ISIDORE Parius. GARZONI Pario (Lichnite). AGRICOLA Parium (Lychinites). CATANEO marmo bianco dell'isola di Paros (lichnite). SCAMOZZI

Pario (Lichinite) – di Paro. Garofalo Parium. Aldrovandi Parium. Milizia Pario. Antolini Statuario di Paros. Amati Pario. Rondelet Marmo di Paros (Licnite). Belli. Marmo Pario. Corsi Marmo greco duro (Parium).
Macroscopic aspect
White ground, diaphanous
Petrographic classification
Metamorphic rock, fine-grained marble made of calcite
Form of use
Statue
Provenance
Near Marathi, in the central part of the island of Paros, Cyclades, Greece

MARMO PENTELICO [Borghini, Dolci e Nista, Gnoli, Pensabene e Bruno]
PENTELIC MARBLE [Price]
GRECO FINO, CIPOLLA, CIPOLLINO STATUARIO [Dolci e Nista]
Historical terms
Agricola Pentelicum. Scamozzi Pentelico – Pentelense. Garofalo Pentelicum. Amati Pentelico. Belli. Marmo Pentelico detto Cipolla o Cipollino statuario. Corsi Marmo greco fino (Pentelicum).
Macroscopic aspect
White ground with yellowish shade and thin greenish veins
Petrographical classification
Metamorphic rock, very fine-grained marble made of calcite
Form of use
Statue, architectural element
Provenance
On the south-western slope of mount Pentelikon (near Athens), Attica, Greece

MARMO BIANCO DI TASO [Gnoli]
MARMO DI TASO [Borghini]
TASIAN MARBLE [Price]
TASIO [Pensabene e Bruno]
THASISCHER [Mielsch]
Historical terms
Plinius Thasium. Ed. Diocletiani Thassi. Isidore Thasium. Agricola Thasium. Scamozzi Tasio – Thasio. Garofalo Thasium. Amati Tasio. Rondelet Marmo Tasio. Belli Marmo greco duro. Corsi Marmo greco livido (Thasium).
Macroscopic aspect
White ground
Petrographic classification
Metamorphic rock, calcite marble (Aliki) or dolomite marble (Vathi); coarse grain size
Form of use
Slab for cladding, decorative element, statue
Provenance
Peninsula of Aliki (east coast) or cape Vathi (northern coast), island of Thásos, Greece

MARMO DELL'IMETTO [Borghini, Dolci e Nista]
GRECO FETIDO, GRECO RIGATO [Dolci e Nista]

IMEZIO [Gnoli, Pensabene e Bruno]
HYMETTISCHER [Mielsch]
Historical terms
PLINIUS Hymettium. AGRICOLA Hymettium. SCAMOZZI Imettio. GAROFALO Hymettium. AMATI Imezio. RONDELET Marmo del monte Imetto. BELLI Marmo imezio statuario – Marmo greco rigato o imezio colonnare – Marmo greco dislocato. CORSI Marmo cipolla (Hymettium).
Macrosopic aspect
White greyish ground with parallel black veins
Petroraphical classification
Metamorphic rock, marble made of calcite
Form of use
Column shaft
Provenance
Northern sector of mount Himettus (near Athens), Attica, Greece

MARMO PROCONNESIO [Gnoli, Pensabene e Bruno]
MARMO DI PROCONNESO [Borghini]
MARMO CIPOLLA [Dolci e Nista]
PROCONNESIAN MARBLE [Price]
PROKONNESISCHER MARMOR [Mielsch]
Historical terms
VITRIVIUS Proconnensium. PLINIUS Proconnesium – Cyzicenum. ED. DIOCLETIANI Proconesi. AGRICOLA Proconnesium – Cizicenum. SCAMOZZI Proconesso. GAROFALO Proconnesium. AMATI Proconnesio – Ciziceno. RONDELET (Marmo) dell'isola Proconeso. BELLI. Marmo greco turchiniccio. CORSI Marmor proconnesium, Bianco e nero antico.
Macroscopic aspect
White ground with thick parallel dark grey veins
Petrographical classification
Metamorphic rock, marble made of calcite plus quartz and muscovite, medium to coarse grain size
Form of use
Slab for cladding and paving, sarcophagus
Provenance
Near Monastyr-Kavala and Sarayalar, Island of Marmara, Turkey

BIGIO ANTICO [BORGHINI, DOLCI E NISTA, GNOLI, MIELSCH, LAZZARINI, PRICE]
BIGIO [Pensabene e Bruno]
Historical terms
PLINIUS Lesbium. ED. DIOCLETIANI Lesbium. ISIDORE Lesbius. AGRICOLA Lesbium. SCAMOZZI Lesbio. GAROFALO Lesbium. RONDELET Bigio antico. BELLI Lesbio – Bigio morato. CORSI Bigio antico, Bigio morato – Grigio giallognolo (Lesbium).
Macroscopic aspect
Grey round with white spots and bands
Petrographical classification
Metamorphic rock, marble made of calcite, medium to coarse grain size

Form of use
Slab for cladding, column shaft
Provenance
Near Moria, island of Lesbos, Greece

BARDIGLIO DI CARRARA [BORGHINI]
BARDIGLIO [GNOLI, MIELSCH, PRICE]
Historical terms
ANTOLINI Bardiglio di Carrara. AMATI Bradiglio (sic) di Carrara. CORSI Bardiglio.
Macroscopic aspect
Grey ground, often with darker veins
Petrographical classification
Metamorphic rock, marble made of calcite, fine grain size
Form of use
Slab for cladding and paving
Provenance
Different marble basins north of Carrara, Apuanian Alps, Tuscany, Italy

Among the coloured marbles, two stones coming from the Italian territory were also included: black limestone and Rosso Ammonitico (see the respective cards in the text).

Historical terms of the black limestone: GARZONI Paragone. VASARI Paragone. DEL RICCIO Paragone. SCAMOZZI Paragone. ANTOLINI Nero di Como. AMATI Nero di Varena. RONDELET Paragone.

Historical terms of Rosso Ammonitico: VASARI pietra ne' monti di Verona . . . rossiccia, e tira in color ceciato. DEL RICCIO Mardorlato e Broccatello di Verona. SCAMOZZI pietre vive di Verona . . . di color carnatino, con alcune macchie più rosse, dove scorrono alcune vene infocate. MILIZIA Breccia di Verona. ANTOLINI Rosso di Verona. AMATI Breccia di Verona. RONDELET Rosso S. Ambrogio di Val Policella (marmo rosso bruno del Veronese).

Another coloured marble was used exclusively in the Baroque (Rosso Francia) and is described in full in Part 2 (Stones from abroad).

REFERENCES

Agricola G., De Natura Fossilium libri X, Basel, 1546.

Alberti L.B., De re aedificatoria – On the Art of Building in Ten Books (J. Rykwert, N. Leach, R. Tavernor eds.), MIT Press, Cambridge, 1988.

Aldrovandi U., Museum Metallicum, Bologna, 1648, Libro IV, cap. 57.

Amati C., Dell'architettura di Marco Vitruvio Pollione Libri dieci, Pirola, Milano, 1829.

Antolini G., Osservazioni ed aggiunte ai Principi di architettura civile di Francesco Milizia, Stella, Milano, 1817.

Ariosto L., The Orlando Furioso (W. Stewart Rose ed.), Murray, London, 1823.

Aston B.G., Harrel J.A., Shaw I., Stone, *in* Ancient Egyptian Materials and Technology (P.T. Nicholson, I. Shaw eds.), Cambridge University Press, Cambridge, 2000.

Belli F., Catalogo della collezione di pietre usate degli antichi per costruire ed adornare le loro fabbriche, Roma, 1842.

Blondel J.F, Cours d'architecture, Paris, 1675–83.
Boccaccio G., Decameron (J. Payne ed.), The Villon Society, London, 1883.
Boece de Boot A., Le parfaict joaillier ou histoire des pierreries, Lyon, 1644.
Borghini G. (ed.), Marmi antichi, De Luca, Roma, 1998 (3rd edition).
Bugini R., Folli L., Sull'uso di marmi colorati antichi in Lombardia (Italia settentrionale), *in* Marmora, 1, 2005, 145–168.
Cardano G., De Subtilitate, Nürnberg, 1550.
Cataneo P., I quattro primi libri di architettura, Venezia, 1564.
Cellini B., Due Trattati – Uno intorno alle otto principali arti dell'oreficeria. L'altro in materia dell'arte della scultura, Panizzi & Peri, Firenze, 1568.
Colonna F., Hypnerotomachia Poliphili: The Strife of Love in a Dream (J. Godwin ed.), Thames & Hudson, London, 1999.
Conant K.J., Carolingian and Romanesque Architecture 800–1200, Penguin, London, 1959.
Corsi F., Delle pietre antiche, Roma, 1845 (3rd edition).
Dante Alighieri, La Divina Commedia (R. Kirkpatrick ed., The Divine Comedy), Penguin, London, 2012.
D'Aviler A.C., Cours d'architecture qui comprend les ordres de Vignole, Paris, 1691.
Del Riccio A., Istoria delle pietre, (manuscript 1597 – printed edition R. Gnoli, A. Sironi eds.), Umberto Allemandi, Torino, 1996.
Dioscorides Pedanius, De medica materia (T.A. Osbaldeston ed.), IBIDIS Press, Johannesburg, 2000.
Dolci E., Nista L., Marmi antichi da collezione, Museo Civico Marmo, Carrara, 1992.
Edictum Diocletiani et collegarum de pretiis rerum venalium (M. Giacchero ed.), Istituto Storia Antica Università, Genova, 1974.
Félibien A., Des principes de l'architecture, Coignard, Paris, 1676.
Garofalo B. [Caryophilus], De antiquis marmoribus, Wien, 1738.
Garzoni T. La Piazza Universale, Venezia, 1585.
Gnoli R., Marmora romana, Edizioni dell'Elefante, Roma, 1988 (2nd edition).
Goethe J., Italian Journey (W.H. Auden, E. Mayer eds.), Folio Society, London, 2010.
Grapaldo F., De partibus ædium, Parma, 1494.
Homer, The Iliad (A.T. Murray ed.), Harward University Press, London, 1924.
Homer, The Odyssey (A.T. Murray ed.), Harward University Press, London, 1919.
Hopkins A., Italian Architecture from Michelangelo to Borromini, Thames & Hudson, London, 2002.
Horace, The Odes and Carmen Saeculare of Horace (J. Conington ed.), Bell and Sons, London, 1882.
Isidorus Ispalensis Episcopus, Etymologiarum sive Originum (W.M. Lindsay ed., Etymologies), Loeb Classical Library, Oxford, 1911.
Krauthaimer R., Early Christian and Byzantine Architecture, Penguin, London, 1965.
Lazzarini L., Pietre e marmi antichi, CEDAM, Padova, 2004.
Lucan M.A., Pharsalia (E. Ridley ed.), Longmans, Green and Co, London, 1905.
Maischberger M., Marmor in Rome, Reichert, Wiesbaden, 1997.
Marino G.B., Adone, Paris, 1623.
McKenzie J., The Architecture of Alexandria and Egypt 300 BC – AD 700, Yale University Press, New Haven, 2007.
Mielsch H., Buntmarmore aus Rom im Antikenmuseum Berlin, Staatliche Museen Preußicher Kulturbesitz, Berlin, 1985.
Milizia F., Principj di architettura civile, Remondini, Venezia, 1785.
Montaigne M., Journal de Voyage (F. Garavini ed.), Gallimard, Paris, 1983.
Ovid P.N., Metamorphoses (Brookes More ed.), Cornhill, Boston, 1922.
Paulus Silentiarius, Descrizione della Santa Sofia (M.L. Fobelli ed., Description of Hagia Sophia), Viella, Roma, 2005.
Pausanias, Ellados perieghesis (W.H.S. Jones, H.A. Omerod eds., Description of Greece), London, 1918.

Pensabene P., Il fenomeno del marmo nel mondo romano, *in* I marmi colorati della Roma imperiale (M. De Nuccio, L. Ungaro eds.), Venezia, 2002, 3–68.
Pensabene P., Bruno M., Il marmo e il colore. I marmi della collezione Podesti, L'erma di Bretschneider, Roma, 1998.
Petrarch F., Rerum vulgarium fragmenta (A. Mortimer ed., The Complete Canzoniere), Penguin, London, 2002.
Plinius, Naturalis Historia (D.E. Eichholz ed., Natural History), Harward University Press, tenth volume, 1962.
Price M.T., Decorative Stones, Thames & Hudson, London, 2007.
Pullen H.W., Handbook of Ancient Roman Marbles (F. Crocenzi ed.), Gangemi, Rome, 2018.
Quatremère de Quincy, Encyclopédie méthodique – Architecture, Paris, 1801–20.
Reedy Ch. L., Thin-Section Petrography of Stones and Ceramic Cultural Materials, Archetype, London, 2008.
Rondelet J., Traité théorique et pratique de l'art de bâtir, Paris, 1802.
Scamozzi V., L'idea dell'architettura universale, Venezia, 1615.
Scriptores Historiae Augustae, Gordiani Tres (D. Magie, A. O'Brien-Moore, W. Heinemann eds.), Putnam's and Sons, New York, 1924.
Seneca L.A., Ad Lucilium Epistulae Morales (R.M. Gummere ed.), Harward University Press, Cambridge, 1917–25.
Sidonius Apollinaris, Epistulae (W.B. Anderson ed.), Loeb Classical Library, Harward University Press, Cambridge, 1936.
Statius P.P., Silvae (J.H. Mozley, W. Heinemann ed.), Putnam's and Sons, New York, 1928.
Strabo, Geografika – Γεωγραφικά (H. C. Hamilton, W. Falconer eds., Geography), Bell and Sons, London, 1903.
Suetonius C.T., De Vita Caesarum – Divus Augustus (J.E. Reed, A. Thomson eds.), Gebbie & Co., Philadelphia, 1889.
Tasso T., Jerusalem delivered (E. Fairfax ed.), Purser, London, 1749.
Taylor J., Imperial Istanbul, Tauris Parke, London, 2007.
Theophrastus, Peri lithon – De lapidibus (F. Wimmer ed., On Stones), Paris, 1866.
Vasari G., Le vite de' più eccellenti Pittori, Scultori e Architettori – Introduzione (L. S. Maclehose ed., The Lives of the Most Excellent Painters, Sculptors and Architects – On Technique), J.M. Dent, London, 1907 (1st ed. 1568).
Virgil P.M., Aeneid (Th.C. Williams ed.), Mifflin, Boston, 1910.
Vitruvius, De Architectura (F. Granger ed., On Architecture), Harward University Press, Cambridge, 1931.
Ward Perkins J.B., Roman Imperial Architecture, Yale University Press, New Haven, 1981.
Williams C., Islamic Monuments in Cairo, The American University Press, Cairo, 2002.

Chapter 21

Mortar and plaster

Mortar is a mixture of binder, aggregate and water. The binder is mainly made of lime coming from the burning of carbonate rocks; the aggregate is added to prevent shrinkage and is mainly made of sand. The use of mortar has been strictly connected with building since prehistoric dwellings, and the importance of the functions carried out by a mortar (coat a wall against weathering, hold together stone or brick in a wall, coat a basin to contain liquids, support for pavements, etc.) was well known by ancient authors. The mortar was reported in the Greek language as "μάλθη" meaning soft stuff; Latin reported "maltha" [Plinius, 1962; book 36, 181] or "materia ex calce et harena" (material made of lime and sand) [Vitruvius, 1931; book 2, ch. 8]; Italian reports "malta" from the Latin and the Greek terms. On the other hand, the English term "mortar" (mix of lime and sand) comes from the Middle English "mortier", the French language reports "mortier"; German reports "mörtel" and Spanish reports "mortero": these terms are always connected to "mortar", meaning "a hard vessel in which substances are pounded with a pestle" (Latin "mortarium"). The Russian language uses the term "раствóр"; Turkish "harç" and Arabic "ملاط".

The description of the features of binder and aggregate as components of a historical mortar precedes the description of the features and uses of different kinds of mortar.

BINDER

The main function of a binder is to hold other materials together and to provide the required strength to the building. In Lombardy the most important binder for mortar is lime, both pure or hydraulic; other binders are Portland cement (employed since the late 19th century) and gypsum (mainly used for decorative elements).

Lime

The ancient Greek term is "κονία", but the term "χάλιξ", meaning "broken stone", was transformed by the Latins as "calx". This term was included in languages derived from Latin: Italian "calce", French "chaux" and Spanish "cal"; German also uses the term "Kalk". On the contrary, English uses the term "lime", coming from the Middle English "lym" meaning "viscous substance"; Russian uses the term "ѝэвесть"; Turkish "kireç" and Arabic "جير" (jir).

The first quotations about this building material pointed out the whiteness and the different characteristics coming from different raw materials. In fact, **Vitruvius** wrote:

DOI: 10.1201/9781003361008-25

> with regard to the lime we must be careful that it is burned from a stone which, whether soft or hard, is in any case white. Lime made of close-grained stone of the harder sort will be good in structural part; lime of porous stone, in stucco.
>
> [Vitruvius, 1931; book 2, ch. 5]

Plinius quoted Cato about the "white stone producing a better quality of lime" and "Lime made from hard stone is more effective for walling, while that made from pporous stone is more suitable for plastering" [Plinius, 1962; book 36, 174].

The same kind of description was reported by Renaissance architects in their treatises: the choice of raw materials was crucial, and several recipes were suggested to make mortar useful for different purposes: bedding, plaster, etc. About the binder (lime) **Palladio** distinguished:

> Lime stones are either dug out of hills or taken out of rivers. . . . The best lime is made of hardest, heaviest and whitest stones. . . . There is also a sort of spungy stone, which makes very good lime for rendering of walls. . . . Pebbles, especially white ones, that are gathered in rivers and rapid streams, make excellent lime . . . commonly used in finishing of walls
>
> [Palladio, 1715; book 1, ch. 5]

Scamozzi gives an accurate report about the lime making in Lombardy and in northern Italy: "the best stones to make lime are . . . the white ones with bright colour . . . with marble-like size as those from river Ticino or from other rivers around the Lake Maggiore". Later, Scamozzi described the loose deposits of the middle course of river Adda: "In these regions at the mouth of the mountain valleys and still 12 to 15 miles in the plain . . . there are stones to make lime . . . and more along the river Adda in the Duchy of Milan" [Scamozzi, 1615; part 2, book 7, ch. 17].

Lime consists of calcium oxide with different proportions of magnesium oxide; it is obtained by burning carbonate rocks (limestone, dolomitic limestone and also marble) in a kiln at a high temperature (>900°C). Calcium carbonate decomposes into *quick lime* (calcium oxide) with the release of carbon dioxide. Quick lime is slaked (addition of water) to obtain *slaked lime* (calcium hydroxide). Slaked lime is the raw material to make mortar and it sets by the evaporation of water becoming a hard material (binder); setting is a slow process, completed by the fixation of carbon dioxide from the atmosphere. Finally, slaked lime is transformed into calcium carbonate.

The burning of a dolomite (double carbonate of calcium and magnesium) involves the production of slaked lime containing magnesium oxide (periclase) and calcium oxide. The process (hydration and carbonation) produces the hydroxides (brucite and portlandite) and then the carbonates (magnesite and calcite).

Lime in Lombardy

"Per la faclità con cui si possono trasportare per via di acqua le calci ai diversi centri di consumazione della Lombardia, ne vediamo fatto un grande consumo in tutte le costruzioni anche dei casolari" [Curioni, 1877, ch. 5].

The use of lime was ubiquitous in Milan and Lombardy, and it has been widely diffused since the Roman period. The scientific evidence pointed out a prevailing use of dolomite to make lime in western Lombardy, but there is no evidence of Roman lime quarries or kilns in this area. The Romans probably exploited the same dolomite outcrops, located along the eastern shore of Lake Maggiore, where lime was produced from the Middle Ages onwards.

The glacio-fluvial deposits of the middle course of the river Adda (cobbles, pebbles of siliceous limestone together with limestone and dolomite outcropping in the river basin) [Carta Geologica d'Italia, sheet 097, 2014] were another medieval and modern source of raw material, mainly to make a moderately hydraulic lime (called "calce forte"), but this feature was not verified in Roman plasters.

The lime of the lakes

> La pietra da calce più in uso fin da tempi immemorabili è la dolomia. . . . Essa ha la proprietà di abbandonare, più facilmente di molte altre pietre da calce, l'acido carbonico, e di fornire calci grasse, che si rigonfiano molto coll'idratarsi.

The eastern shore of Lake Maggiore is interested, both in the northern and in the southern part, by outcrops of dolomite rocks (Triassic) suitable for the production of lime; some outcrops (dome-shaped hills) are also evident in the lake shore morphology. The rocks involved pertain to the Sedimentary series of the Southern Alps: grey dolomites and dolomite-limestone ("Dolomia del Salvatore" formation, Norian) in the northern part (Brezzo di Bedero, Portovaltravaglia, Caldè); grey dolomites often with cyclothemes ("Dolomia Principale" formation, Ladinian-Anisian) in the southern part (Ispra, Angera). About 20 km east from the lake shore, there are other dolomite outcrops located at Rasa and Arcisate (Dolomia del Salvatore) or at Brenno Useria (Dolomia principale) and exploited since the 17th century [Fieni, 2000]. Curioni suggested a hypothesis about the prevailing use of dolomite lime in Lombardy: the ease of transportation on the lakes as the lakes of Lombardy run through very thick dolomite deposits [Curioni, 1877; ch. 5].

The quarries and kilns along the lake shore were capable of using the waterways (Lake Maggiore, river Ticino, Naviglio) to easily reach the building yards in Milan.

Towards the east, there are other quarry sites exploiting dolomite rocks (Dolomia principale formation) on the shores of Lake Como (Lecco branch): Vassena-Oliveto (central part); Paré-Valmadrera (southern part). These quarries are located on waterways (the lake, the river Adda, and the waterway called Martesana) allowing to reach Milan's yards. The use of magnesian lime from the burning of dolomites is documented in Lombardy from the 12th to 20th centuries [Bugini, Folli, 1996].

Finally, this is another site at Vello on the shores of Lake Iseo (north-eastern part), but in this case the lime reached the territory of Brescia.

References on lime quarries in Lombardy

PROVINCES OF VARESE AND COMO

Curioni: 138. Angera. Dolomia gialliccia e rosea . . . prolungamento dei banchi di Arona, ove sono escavati abbondatemente per calci grasse. . . . 139. Ispra. Dolomia quasi candida . . . molto lavorata per calci grasse smerciate nella bassa Lombardia. . . . 140. Caldè. Dolomia color bianco perlaceo. . . . 141. Porto. Dolomia bianco-grigiastra . . . verso Lavello, ove esistono varie fornaci. . . . 142. Porto Val Travaglia. Dolomia bianco-perlacea. . . .

143. Madonna del Monte. Dolomia grigia. . . . 144. Cascina Rasa. Questa dolomia è la più pura di ogni altra dei contorni di Varese. . . . 145. Arcisate. Dolomia . . . in banchi eretti, inclinati a sud-ovest, molto lavorata per calce grassa. . . . 148. Parè (Valmadrera). Dolomia bianchiccia, escavata abbondantemente per calci grasse ad uso edilizio, specialmente per Milano. . . . 149. Vassena [Oliveto Lario]. . . . escavata per calci grasse. [Curioni, 1877]

Jervis: 262. Arona. . . . Calcare compatto. Dalla parte del Nord evvi una cava donde si ottiene della pietra da calce grassa. 265. Germignaga. Calcare. Cava al Lago Maggiore, presso la Punta Lavello. Si cuoce in fornaci esistenti in vicinanza per ottenere la calce. . . . 266. Porto Val Travaglia. Dolomia. . . . Dà calce grassa. 267. Castello Val Travaglia. Calcare dolomitico. . . . Fornisce calce grassa magnesiaca. Cava, poco distante dal centro comunale, presso il Casale di Caldè, proprio sulla sinistra sponda del Lago Maggiore. Si cuoce in fornaci esistenti sui luoghi, ottenendone della calce dolce. 269. Ispra. Dolomia. Adoperasi per far la calce grassa. . . . 270. Angera. Dolomia compatta. . . . Se ne fa pure qualche uso per far la calce grassa. 274. Cabiaglio. Dolomia. Serve come pietra da calce. Cava. . . . 278. Arcisate. Dolomia. . . . Cava. Si cuoce sui luoghi, ottenendone della calce grassa. 279. Induno Olona. Dolomia. Si adopera per stipiti e lavori ornamentali, come pure per far la calce. Cava. 284. Brenno Useria. . . . Calcare biancastro, che è suscettibile di acquistare un discreto pulimento e di cui si possono avere delle grandi saldezze. . . . Cava alle pendici meridionali del monte. Serve per far la calce. 285. Bisuschio. Calcare biancastro. Adoperasi come pietra da calce. Cava. . . . 342. Onno . . . Calcare argilloso. Cava. Viene scavato e cotto sui luoghi per fare la calce idraulica ed il cemento. . . . 351. Valmadrera. Dolomia. Cava nella località detta Gessina: viene cotto per fabbricare calce grassa. Le fornaci sono vicine al villaggio di Parè, distante circa 6 Km da Lecco. [Jervis, 1889]

Salmojraghi: "[Varese] Dolomie. Pietra Grezza e per calci grasse. . . . Porto Val Travaglia; Castello Val Travaglia, Caldè; Ispra; Velate, Rasa; ecc. Triassiche. Arcisate, ecc. Liassiche" [Salmojraghi, 1893].

PROVINCE OF BRESCIA AND SONDRIO

Curioni: 146. Dubino . . . viene escavata per calci grasse. . . . 150–151. Vello [Lago d'Iseo] . . . escavata abbondantemente per calce grassa ad uso edilizio nella provincia di Brescia. . . . 153. Valle Sarezzo. Dolomia escavata alle Tese. . . . 154. Nave. Dolomia subcristallina, di color bianco perlaceo. . . . Viene escavata in grande abbondanza ad uso dei paesi contigui.

Jervis: 411. Vello. Dolomia. Cava. La dolomia di questi monti è adoperata per far la calce forte ed è cotta in una fornace esistente presso Vello. . . . 412. Pilzone. . . . Calcare argilloso nerastro. Cava sulla sinistra del Lago d'Iseo, che serve ad alimentare la fabbrica di calce idraulica di Palazzolo all'Oglio, impiantata nell'anno 1857 dalla "Società ferroviaria dell'Alta Italia" per uso proprio, acquistata in seguito dalla potente "Società Italiana dei Cementi e delle Calci idrauliche", sedente in Bergamo. . . . 413. Paratico. Calcare marnoso. Adoperasi per calce idraulica. . . . 415. Adro. Calcare compatto, atto alla fabbricazione della calce idraulica. . . . 419. Borgo di Terzo. Calcare argilloso, atto a fare il cemento idraulico. . . . 432. Sarezzo. In questo territorio si estrae molta pietra da calce. . . . 441. Brescia. . . . Calcare argilloso giallognolo, in prossimità alla città. . . . Dà calce idraulica. Esiste pure del calcare presso il villaggio di Fiumicello Urago. Si scava per far la calce grassa, la quale viene cotta in fornaci erette sui luoghi. . . . 446. Virle Treponti. . . . Calcare compatto bianco, conosciuto dai cavatori sotto il nome di Corna. Cave nel monte dietro il villaggio di Virle, da cui sono distanti un chilometro. Se ne fa della calce. . . . 454. Vobarno. Calcare, che serva come pietra da calce. . . . 457. Manerba. Calcare. Serve per far calce.

Salmoiraghi: "Dolomie. Diverse: Bagolino; Anfo; Collio; Vello; Marone; Cimmo; Lumezzane; Sarezzo; Sabbio Chiese ecc. Triassica; Pietra Grezza e per calce. Nave ecc. Liassica, candida; Pietra da Taglio e per calce". "[Sondrio] Dolomia. Bormio, Monte Reit, Premadio, Dubino, Bocca d'Adda. Per lo più subsaccaroide; per calce".

The lime from pebbles

The courses of different rivers flowing from the Alps to the Padania plain, in particular the river Adda between Brivio and Trezzo d'Adda, are interested by loose detritic materials referred to as the glacial and glacio-fluvial transport. The clastic materials are cobbles, pebbles and gravels (Pleistocene – Olocene), and their composition reflects the geology of the catchment basin of the river Adda: silicate metamorphic rocks (gneiss, serpentinite); silicate igneous rocks (granite, granodiorite, diorite) of the Alps; carbonate sedimentary rocks (dolomite and limestone of the Triassic, marly limestone of the Jurassic, marly limestone of the "Selcifero Lombardo" formation of the Cretaceous) of the Prealps. These loose materials were burned to obtain the so-called calce forte or calce di Gera d'Adda [Breislak, 1822; chap. 1, sec. 9], a kind of lime with some hydraulic feature due to a clay fraction: "the so-called *calce forte*, made of non-dolomitic limestones with a certain amount of clay coming from lower Jurassic formations. Calce forte was mainly employed for masonries in damp places" [Curioni, 1877; ch. 5]. Unfortunately, scientific analyses to confirm these assertions are lacking.

References on lime from pebbles

PROVINCES OF MILANO AND BERGAMO

Scamozzi: In questa parte d'Italia di qua dal Po e del Ticino . . . per la copia dei torrenti e dei fiumi che scendono dalle Alpi e monti petrosi vi si trovano tutte le migliori sorta di pietre che si possono desiderare per far calcina, le quali servono a Venezia, a Milano e parimenti alle città vicine e anco non poche se ne conducono a Ferrare e ad altre città scoste da essi fiumi. . . . In queste nostre parti dall'uscita delle valli per 12 e 15 miglia nella pianura . . . si ritrovano le pietre da far calcina . . . e molto più lungo l'Adda e il Ticino nel Ducato di Milano, ma le più grosse e migliori sono ad alto. Perché, come abbiamo osservato, a basso divengono ghiara minuta e calcoli e finalmente si riducono in sabbia grossa. [Scamozzi, 1615; part 2, book 7, ch. 17)

Breislak: Dopo che l'Adda è uscita dall'emissario di Lecco, tre volte si dilata formando tre bacini nei quali . . . deve fare nuove successive deposizioni. . . . Allorché il Brembo giunge all'Adda . . . le materie trasportate fino a questo punto si arrestano presso Vaprio dove formano quelle grandi deposizioni. . . . nelle alluvioni straordinarie molti ciottoli giungono al letto dell'Adda . . . la maggior parte di rocce carbonatiche . . . perciò si raccolgono con molta diligenza . . . e servono ad alimentare alcuni forni da calce che sono presso le sponde dell'Adda. Questa è quella calce che in Milano è indicata col nome di "calce forte" o di calce di "Gera d'Adda", e di cui l'esperienza ne ha dimostrato la bontà. [Breislak, 1822; ch. 1, 9]

Curioni: "Negli anni andati si faceva molto uso di calci fatte col cuocere i ciottoli calcari argillosi dell'Adda, presso Cassano, procedenti in specie dal Brembo" [Curioni, 1877, ch. 5]. "147. Bormio. . . . La Valtellina, sotto Bormio, manca quasi generalmente di buona calce; vengono quindi raccolti i ciottoli di questa pietra ad alimento di molte fornaci, estraendoli dall'Adda, lungo buon tratto del suo corso".

Hydraulic lime

The specific features of marly limestones (a limestone with a clay fraction) were exploited in the Renaissance in order to obtain mortars suitable for wet places but without scientific awareness. **Palladio** reported a case from Euganean Hills: "In the hills of Padua, they dig a certain rugged and scaly stone, whose lime is very good for works exposed to the weather and in the water, because it hardens immediately and lasts a long time" [Palladio, 1715; book 1, ch. 5]. **Scamozzi** described some greyish, heavy and flaky stones coming from the territory west of Padua and producing a white lime useful underwater and in a damp place, but useful also to build over the ground; this lime was mainly employed at Venice and Ferrara [Scamozzi, 1615; part 2, book 7, ch. 17].

Both the authors refer to the outcrops of the "Scaglia veneta" formation, made of reddish marly limestones (lower Eocene – upper Cretaceous, Euganean Hills, Padua, region of Venetia).

The burning of marly limestones produces the so-called hydraulic lime, able to set without atmosphere (underwater). During the hydration of the quick lime, silica and alumina combine with calcium hydroxide forming insoluble compounds (calcium silicates and aluminates). Marly limestones are widely present in the Prealps of the central part of Lombardy and Venetia.

The first kind of modern hydraulic lime was named from Palazzolo, a village on the left bank of river Oglio close to the border between the provinces of Bergamo and Brescia; the raw material came from a few kilometres north-east (Lake Iseo, Pilzone, exploiting the limestones of the "Calcari selciferi Lombardi" formation), and its hydraulic lime was employed to build the railway bridge over the river Oglio (270 m long and 9 m wide on the railway line Bergamo-Brescia, 1856–57) and the kilns were assembled at the yard of Palazzolo.

The most important area is located in the lower Val Seriana (province of Bergamo): two important quarry sites (Scanzorosciate and Villa di Serio), located at a short distance in the same hill (a few kilometres north-east of Bergamo), exploiting the marly limestone of the "Sass de la Luna" formation (upper Cretaceous, Albian). In the last decades of the 19th century the quarries were owned by "Società Italiana dei Cementi e delle Calci Idrauliche", with running kilns built into the foot of the hill from which the raw material was obtained.

Another important firm was the "Ditta Fratelli Pesenti fu Antonio" owner of lime furnaces at Alzano Lombardo – Nese (a few kilometres north of Villa di Serio) exploiting the same geological formation ("Sass de la Luna"). Other furnaces were set on Lake Iseo (Riva di Solto and Pilzone – Sulzano). Going north along the Val Seriana, there are quarries at Albino and Pradalunga owned by "Società Italiana dei Cementi e delle Calci Idrauliche" and still exploiting the "Sass de la Luna" formation.

References on hydraulic lime

PROVINCES OF VARESE AND COMO

Curioni: 185. Acquate [Lecco]. Carbonato di calcio argillosi . . . escavato per cementi idraulici. . . . 186. Carabuso [Lecco]. Calcare da cemento romano. . . . 187. Monte Marenzo [Lecco]. Calcare rossiccio molto marnoso. . . . 188. Rova. Calcare molto marnoso. . . . Viene estratto abbondantemente ad alimento delle fornaci della Valle Seriana. . . . 189. Scanzo. Calcare nero molto marnoso, costituente i banchi più bassi visibili della collina detta la Bastia. È escavato abbondantemente per alimento delle fornaci della Società Bergamasca. . . . 194. Capriolo. Calcare molto marnoso.

Jervis: 280. Comabbio. Calcare bianco-giallognolo o cenerognolo a struttura compatta. . . . Cava nella collina tra il villaggio e il lago omonimo. . . . È calcare nummulitico del sistema eocenico. . . . 341. Vassena. . . . Calcare. Si scava per far la calce, la quale è cotta sui luoghi. 342. Onno . . . Calcare argilloso. . . . Cava. Viene scavato e cotto sui luoghi per fare la calce idraulica ed il cemento. . . . 350. Acquate . . . Calcare argilloso . . . è scavato e cotto sui luoghi per la preparazione della calce idraulica e del cemento idraulico.

Salmoiraghi: "Calcare. Acquate, Carabuso. Lastriforme, marnoso, triasico, per cemento". "Calcari. Infraliasici: Onno, per cemento". "Calcari. Eocenici, nummulitici, compatti, talor brecciformi, bianchi ed azzurrognoli; Pietra Grezza e Pietra Concia e per calce: . . . Ternate; Comabbio. Fratturato".

PROVINCES OF BERGAMO AND BRESCIA

Curioni: 164. Vall'Alta [Pradalunga]. . . . calcare marnoso escavato ad alimento della fornace di Pradalunga. . . . 167. Pilzone (sul lago d'Iseo). Calcare nero . . . escavato abbondantemente ad alimento delle fornaci stabilite a Palazzolo. . . . 170. Capriolo. Calcare argilloso. . . . 174. Scanzo. Alla Bastia, trovasi un calcare marnoso nericcio. . . . È escavato abbondantemente ad alimento delle fornaci da calce idraulica.

Jervis: 359. Monte Marenzo. Calcare argilloso, che produce una calce fortemente idraulica. . . . 383. Vallalta. . . . Calcare nero schistoso. . . . Queste varietà di calcare servono alla Società Bergamasca come materie prime a fabbricare la calce idraulica. . . . 388. Desenzano sul Serio. . . . Calcare argilloso. Cava, coltivata per la fabbricazione della calce idraulica e del Cemento. . . . 389. Albino. Calcare argilloso, che si scava per fare calce idraulica e cemento idraulico. . . . 390. Pradalunga. . . . Calcare argilloso. Cava coltivata dalla "Società Italiana dei Cementi e delle Calci idrauliche", avente sede in Bergamo. Le fornaci dove si ottengano questi prodotti sono distanti 1,5 km dalle cave. . . . 393. Villa di Serio. . . . Calcare marnoso giallo brunastro. Cave. . . . Viene estratto per la fabbricazione della calce idraulica e del cemento a presa lenta ed è cotto in fornaci eretto presso le cave stesse. . . . 394. Scanzo. . . . Calcare compatto argilloso siliceo. Cave importanti, coltivate in scala grande, prima della "Società Bergamasca", ora "Società Italiana dei Cementi e delle Calci idrauliche" . . . e che possiede sul luogo delle fornaci a fuoco continuo per la produzione della calce idraulica e del cemento a presa lenta e rapida. Ve ne ha di due colori, ambedue possedenti pregevoli qualità idrauliche. . . . 407. Fonteno. Calcare atto a farla calce idraulica. Cava nei monti sulla riva destra del Lago d'Iseo. . . . 408. Predore. Calcare, detto Maiolica, atto a fare la calce idraulica.

Salmoiraghi: "Calcari. Lias superiore; Pietra Grezza e per calce: Pilzone, Montecolo. Azzurrognolo; ampia cava; dà la calce idraulica di Palazzolo; strati inferiori (Lias inferiore) per calce aerea". "Calcari. Solto, Riva di Solto, Fonteno, Tavernola, Predore, Adrara ecc.; Leffe, Gazzaniga, Vall'Alta, Desenzano al Serio, Albino, Pradalunga, Alzano Maggiore, Nese, Villa di Serio, Scanzo ecc.; Zogno ecc. Per lo più compatti, più o meno argillosi; dall'Infralias al Cretacico; per calci e cementi; Pietra Grezza, talora Pietra Concia".

PROVINCE OF PIACENZA (REGION EMILIA ROMAGNA)

Curioni: "Negli anni andati . . . erano in uso, nei lavori di edilizia idraulica, le calci argillose della Trebbia".

Salmoiraghi: "Calcari. Per lo più marnosi, dell'Eocene. . . . Rivergaro, ecc. Val Trebbia; Bettola, ecc. Val Nure; Val d'Arda ecc. Pietra Grezza e per calce".

Portland cement

The term "cement" is a source of misunderstanding: the Latin term "caementum" refers to "cut stone" (from "caedere" meaning "to cut" or "to break"), and it was employed to indicate the rubble used to make a masonry together with a binder; on the contrary the modern term, in different languages, means "a powder useful to bind together different materials" (English "cement", French "ciment", Italian "cemento", German "zement", Spanish "cemento").

The studies carried out by L. Vicat about hydraulic limes [Vicat, 1818] laid the bases for a new kind of binder with hydraulic features: the so-called Portland cement; the name comes from its resemblance to Portland stone, a Jurassic oolitic limestone quarried in southern England (Isle of Portland, Dorset) and largely used as a building material since the Roman period.

A controlled mixture of calcareous and clayey raw materials reacting in a kiln at high temperatures (about 1450°C) to product the "cement clinker". The clinker is ground with a small proportion of gypsum (calcium sulphate) to control the setting time of the Portland cement.

The raw material, in the beginning, came from natural rocks containing the exact proportion between calcite or dolomite and clay; later on the required proportions were artificially obtained mixing limestones, dolomites and clays. Marly limestones of Lombardy, useful to make hydraulic lime, were also used to make the Portland cement after an appropriate process.

The Portland cement was patented in England by J. Aspidin (1824), but its use became regular in Italy in the last decade of the 19th century, both to make mortar (as a binder) and to make artificial stone (see Chapter 38).

Since the last decade of the 19th century, the use of lime was gradually abandoned and the use of Portland cement was spread in the whole field of architecture, totally changing the methods of building.

As already reported, the first Portland cements in Italy were linked to the outcrop of specific geological formations (marly limestone). The first kilns were located at Casale Monferrato (province of Alessandria, eastern Piedmont), exploiting the "Casale Monferrato" formation (Eocene, marly limestone in alternating with calcareous sandstones, fucoid limestones and dark clay) and then the "Marne di Sant'Agata" formation (Miocene, marls and clays). The production of Portland cement in Lombardy is linked to the plants producing the hydraulic lime (*see supra*) located in Val Seriana (Bergamo). The kilns were installed at Alzano

Lombardo (val Seriana, province of Bergamo) exploiting "Albenza" formation (dolomitic limestones with calcilutites; Hettangian, lower Jurassic) or "Calcare di Moltrasio" formation (limestones with chert and marly joints; Sinemurian, lower Jurassic) or "Calcare di Domaro" formation (limestones with chert and marly-clayey joints; Domerian, lower Jurassic).

Other sites in Lombardy featuring quarries and kilns and exploited in the 20th century are located (from east to west) at:

*Capriolo (south of Lake Iseo, province of Brescia): "Sass de la Luna" formation (marly-limestone, upper Cretaceous, Albian)
*Rezzato (east of Brescia, near Botticino): "Corna" formation (dolomites and limestones; lower Jurassic)
*Tavèrnola Bergamasca monte Saresano (west bank of Lake Iseo, province of Bergamo): "Calcare di Domaro" formation (marly limestones with chert and fossiliferous marls; lower Jurassic)
*Sedrina and Zogno (monte Passata, eastern side of Brembo valley: "Formazione dell'Albenza" formation (oolitic calcarenites and micrite limestone; lower Jurassic); "Maiolica" formation (cherty limestone, lower Cretaceous)
*Ubiale Clanezzo (monte Ubione, western side of Brembo valley): "Formazione dell'Albenza" formation (oolitic calcarenites and micrite limestone; lower Jurassic)
*Palazzago (monte Albenza, western side of Brembo valley): "Formazione dell'Albenza" formation (oolitic calcarenites and micrite limestone; lower Jurassic)
*Carvico and Calusco d'Adda – monte Giglio (Adda valley, province of Bergamo): "Brenno" formation (calcilutites and beds of marl; Maastrichtian, upper Cretaceous) and "Tabiago" formation (calcarenites and clayey marls; Pliocene – Eocene). Monte Giglio is now almost entirely excavated, and the former mountain was transformed in a sunken place
*Comabbio (south of Lake Varese, province of Varese): "Calcari nummulitici" formation (nummulitic limestones with beds of marls; Eocene)
*Caravate (east of Lake Maggiore, province of Varese): "Maiolica" formation (limestone with chert and marly limestone; lower Cretaceous)

References on Portland cement

It is almost impossible to distinguish the quotations about Portland cement, because this material was mainly produced using the same limestone outcrops already used to produce the hydraulic lime. We report only the references about the Casale Monferrato area.

PROVINCE OF ALESSANDRIA (REGION PIEDMONT)

Jervis: 601. Ozzano Monferrato . . . Possiede uno stabilimento di cementi artificiali appartenenti alla Società italiana dei Cementi e delle Calci idrauliche con sede in Bergamo, ed altro della Società anonima per la fabbricazione della Calce e del Cemento di Casale Monferrato. . . . Calcare duro di grana finissima e meno argilloso, distinto dalle seguenti varietà come Marmorino. Calcare di durezza media, di grana ordinaria e mediamente argilloso, distinto dai cavatori come Bastardello. Questi calcari forniscono le calci idrauliche, mentre quello appresso serve a fare i cementi idraulici. Marna molle molto argillosa, di minor durezza delle varietà precedenti e di grana grossolana. Vi sono delle cave numerose in tutto il territorio, comprese quelle di spettanza delle due Società summentovate. Queste rocce appartengono all'Eocene, come pure quelle di San Giorgio Monferrato, Casale Monferrato ecc. ed estraesi in una zona che ha una lunghezza di circa Km 10 con Km 4 di larghezza, tra Pontestura e Casale Monferrato, non lungi dalla sponda destra del Po. . . . 602. San Giorgio Monferrrato. . . . 603. Casale Monferrrato . . . Si coltivano numerose cave, poste nella collina che costeggia la riva destra del Po. . . . Gesso lamellare ossia Selenite in grosse masse cristalline ed in grandi cristalli, che raggiungono una lunghezza di m 0,45, ed avente una decisa tinta giallognola. Questa varietà adoperasi per far del cemento.

Salmoiraghi: Calcari. Terziarii, servono per lo più solo per calce e Pietra grezza, di rado per altri usi: Colline di Tortona, Alessandria, Valenza e Casale Monferrato. Numerosi giacimenti, fra cui notansi i seguenti: Casale, San Giorgio e Ozzano Monferrato. Bigio, giallognolo od azzurrognolo, dell'Eocene, associato ad argille scagliose; dà la calce idraulica, detta di Casale. Due principali qualità: Marmorino, duro, a grana fina (talor con varietà litografiche) e Bastardello, meno duro, a grana media, più argilloso. . . . Gessi. Bianchi, bigi o cinerei, macro o microcristallini, copiosi, per materiali cementizi; Pietra Grezza, raramente Pietra Concia e Pietra da taglio: Colline del Tortonese.

Gypsum

Gypsum is a mineral (dehydrated calcium sulphate) and a building material (after baking of the mineral). English, German, French and Italian languages use terms (gypsum, gips, gypse, gesso) coming from the Greek term "γύψος" (also including lime) and then the Latin term "gypsum". French language discriminates the mineral (gypse) from the building material (plâtre). The similar Arabic term "جبس" (jibs) means lime.

Theophrastus reports this term both for the mineral and for the building material:

> The natural kinds of earth, which are useful as well as superior in quality, are three or four in number . . . the Tymphaic [northern Epirus] or Gypsos. . . . Gypsos occurs in large quantities in Cyprus and can easily be seen; for only a little soil is removed when it is dug up.
>
> [Theophrastus, 1866; 62–64]

Plinius compares lime and gypsum:

> There is an affinity between lime and gypsum, a substance of which there are several varieties. For it can be produced from a heated mineral. . . . The mineral that is heated ought to be like alabaster or crystalline marble. . . . the best kind is prepared from specular stone or from the stone that flakes in the same way. Gypsum, when moistened, should be used instantly, since it coheres with great rapidity. . . . Gypsum is used in plaster and . . . for making moulded figures and festoons in architecture.
>
> [Plinius, 1962; book 36, 182–83]

Scamozzi introduced the gypsum of the hills of Rimini as transparent, like alabaster, "but it is difficult to burn and it assumes after burning a brownish colour unpleasant at the sight". Later the author introduced the "Scaglia di Bologna" as harder than other Italian gypsums. "Each gypsum burns in 20 hours that is the third part of time with regard to the lime of soft stone . . ., the gypsum of Scaglia di Bologna sets more rapidly in comparison with other available gypsums" [Scamozzi, 1615; part 2, book 7, ch. 17].

The production of gypsum ready for architectural use involves the calcination of the dehydrated calcium sulphate; the temperature of calcination is about 150°C and causes the partial loss of water of crystallization and calcium hemihydrate (plaster of paris) is obtained. Plaster of paris is finely ground and slaked (addition of water) before use. A temperature of calcination of about 200°C causes the total loss of water of crystallization, and anhydrite (anhydrous calcium sulphate) is obtained. The setting time of plaster of paris and of anhydrite is very different, and the use of these products is linked to specific purposes and some retarders were used to slow down the rapidity of the setting.

Gypsum as a binder was frequent in ancient Egypt because of the demand of a lesser amount of fuel in comparison with lime making. It was rarely used in Lombardy, despite Plinius's words, except in some decorative mortars (stuccoes). The raw material is the mineral present in evaporitic sediments of Triassic formations of the Lombard Prealps. The most important quarry site was located at Costa Volpino in the lower Valle Camonica (not far from the northern tip of Lake Iseo, province of Bergamo): the quarry exploited a series of beds of gypsum and anhydrite pertaining to a calcium sulphate lens included in the formation of Carniola di Bòvegno (yellowish dolomite with vacuoles, lower-middle Triassic, Skytian-Anisian). Other gypsum quarries, spread in the whole Lombard Prealps supplied only the local use for building or other uses (agriculture, paper mill, etc.). Some significant quarry sites are: Nobiallo (eastern shore of Lake Como, province of Como; Carniola di Bòvegno formation); Lovere (northern tip of Lake Iseo, province of Bergamo) and Auro – Casto (val Sabbia, between Lake Iseo and Lake Garda, province of Brescia) in evaporitic lenses of the upper part of the San Giovanni Bianco formation (marly limestones, upper Triassic, Carnian).

Out of Lombardy, it is worth it to note an important source of gypsum: the "Formazione Gessoso-Solfifera" (upper Miocene) of the northern Apennine (region of Emilia-Romagna) containing gypsum (hydration of anhydrite) together with bituminous marls, dolomitic limestones, etc. The gypsum of some outcrops shows remarkably large twinned crystals (selenite), and it was used as building material (i.e. the "Loggia" of Torre degli Asinelli, 15th century and basal ashlar of Torre Garisenda, late 19th century – Bologna).

The use of gypsum in Milan was scarce and restricted to mortars for decoration as reported in the church of Santa Maria presso San Satiro, planned by Bramante in the late 15th century (see *infra*).

Scagliola

A particular use of gypsum is the so-called scagliola (an Italian word meaning "little flake") used to identify a coloured gypsum. In the late 16th century, this kind of material was experimented in Carpi (province of Modena, Emilia Romagna) close to the Apennine area where gypsum, coming from upper Miocene outcrops, is easily available. A finely powdered gypsum (plaster of paris) was mixed with glue and pigments (both inorganic and organic), then it was polished to make fine decorative elements. The most important use of scagliola was the imitation of inlays made of coloured marbles (*opus sectile*, see Chapter 36): normally a uniform black coloured gypsum, laid with spatula on a wooden table and levelled with a blade and plane, was carved with appropriated tools (chisel) to make elaborated forms and figures; the carvings were filled with pigmented gypsum and glue and finally the whole surface was honed, polished, oiled and waxed. The polishing was an important step in the preparation, involving natural stone such as pumice (volcanic glass rich with cavities, coming from the Aeolian islands, north of Sicily), agate (fine-grained chalcedony) and hematite (iron oxide): the result was a very smooth and highly reflecting surface perfectly reproducing colours, veins and textures, as in a marble [Scagliole intarsiate, 2011].

Altar frontals made of scagliola spread from northern Lombardy to the whole European territory (late 16th–late 18th centuries) following the skilful craftsmen of Valle d'Intelvi, an alpine valley running from the western shore of Lake Como to the mountain ridge towards Lake Lugano. Examples of frontals made of scagliola are present in several churches of Valle d'Intelvi (San Fedele, Casasco, Ramponio), in Lombardy (Busto Arsizio) and in some churches of Milan as Certosa di Garegnano, San Fedele, Sant'Alessandro; Santa Maria della Passione, San Nazaro, etc. A modern use, to make architectural elements as columns, are clearly visible inside the Stazione Centrale (staircases).

Scagliola was also employed as plaster to make plain surfaces simulating a marble surface: this kind of plaster was very widely employed in Milan to cover the walls of staircases. The lower part of these walls (about 150 cm) is characterized by the use of scagliola to simulate a coloured marble cladding: scagliola perfectly reproduces veins, bands, spots, stains, patches, etc. Different products were applied to the masonries: "marmorizzato venato" (veins), "marmorizzato a grandi macchie" (large spots), "marmorizzzato a piccole macchie" (small spots, granite). The veined pattern was prepared by mixing raw materials on a canvas; the gypsum decoration was applied, through the canvas, to the wall masonry prepared with a render coat of plaster; finally, the canvas was gently removed. The spotted patterns were prepared apart inserting quantities of coloured gypsum (marble with large spots) or very small coloured balls (granite) in the ground mass. Examples are visible in hundreds of Milan's apartment buildings, built since the last decade of the 19th century until World War I.

References of gypsum

PROVINCE OF COMO

Curioni: "302. Nobiallo. . . . Questa gessaia è in parte bagnata dalle acque del lago di Como: è di grande importanza per la sua vastità e per la facilità dei trasporti per acqua ad uso di molte città e paesi delle provincie lombarde". "307. Limonta . . . Questa gessaia è molto vasta, ma non presenta la comodità di quella di Nobiallo. . . . 308. Cremeno. . . . Trovasi in potenti banchi raddrizzati". "I gessi del lago di Como, ad uso edilizio e di agricoltura, arrivano a Milano in parte in pezzi ancora crudi, essendovi in questa città forni per cuocerli e macine per ridurli in polvere".

Jervis: 334. Loveno sopra Menaggio. . . . Gesso compatto bianco, con leggera tinta bigia e azzurrognola. È copioso in vicinanza del villaggio di Nobiallo, dove vi ha una gessaia per la sua estrazione. . . . 340. Limonta. . . . Esteso giacimento presso la sponda del lago di Lecco, con gessaia. . . . 341. Vassena. Gesso a struttura cristallina fina. . . . 353. Cassina. Gesso bigio chiaro a grana cristallina fina. . . . 355. Cremeno. Gesso che forma dei banchi considerevoli, associato ad anidrite".

Salmoiraghi: "Gessi. Triassici, bianchi, saccaroidi, talora macromeri, per cemento ed usi agricoli: Moggio; Cassina, Cremeno ecc. Associato ad anidrite. Limonta, Vassena; Loveno sopra Menaggio, Nobiallo".

PROVINCE OF PAVIA

Jervis: 689. Godiasco. . . . Gesso bituminoso bigio azzurrognolo a struttura lamellare e di aspetto perlaceo. . . . 690. Retorbido. Gesso massiccio opaco. 691. Codevilla . . . Gesso bituminoso bigio. . . . Adoperasi tanto quale pietra da costruzione quanto per la cottura. . . . 692. Borgoratto Mormorolo. . . . Gesso. . . . 694. Oliva Gessi . . . Gesso. . . . 695. Mornico Losanna. Gesso. . . . 696. Corvino San Quirico. Gesso compatto bianco opaco di buona qualità. Gessaia. 700. Castana. . . . Gesso. . . . 701. Montescano . . . Gesso lamellare fetido, bigio-scuro o Selenite. . . . 702. Montù Beccaria. Gesso".

Salmoiraghi: Gessi. Montù Beccaria; Montescano; Mornico Losana; Oliva Gessi; Codevilla; Retorbido; Godiasco ecc. Macro e micromeri, spesso bituminosi, bigi, perlacei o nerastri; del Miocene superiore; abbondanti nelle colline; ivi numerose cave (gessaie) pe materiale cementante . . .; in qualche giacimento (Esempio: Montescano) con varietà alabastrine, bianche a marezzature bigio-fumo lucidabili; Pietra per Decorazione all'interno".

PROVINCE OF BERGAMO

Curioni: 289. Volpino. Solfato di calcio (anidrite). . . . Presso la Malpensata questo deposito di solfato di calcio, nelle parti più basse, trovasi allo stato di anidrite. . . . Venne giudicato come una specie particolare, cui fu dato il nome di Volpinite. Vennero coltivati in tutti i tempi tanto il gesso, quanto l'anidrite. . . . 314. Lovere. . . . ad uso di opere edilizie".

Jervis: 370. Dossena. Gesso compatto. . . . 405. Lovere. . . . Gesso candidissimo, con gessaie importanti. . . . 406. Costra Volpino . . . Gesso associato ad Anidrite o Volpinite. Cava. Questa pietra si vende in commercio sotto il nome di Gesso di Volpino ed anche Bardiglione o Bardiglio di Bergamo, a cagione della somiglianza del colore con quello del marmo bardiglio. Essa è impiegata in Lombardia per lavori di scoltura, come il marmo stesso, come pure per certi usi edilizi".

Salmoiraghi: "Volpinite. Volpino, detta Bardiglio di Bergamo, bigia variegata od azzurrognola, associata al gesso; facile a lavorarsi, ma alterabile all'umido; Pietra per decorazione interna; cava antica e grandiosa; Misure Massime 3 metri cubi".

PROVINCE OF SONDRIO

Jervis: "311. Valfurva. Gesso. . . . 312. Bormio. Gesso, costituisce un giacimento esteso sopra i Bagni termali".

Salmoiraghi: "Gessi. Palaeozoici o Triassici: Bormio, Valfurva, Valdidentro, Isolato. Usi locali".

PROVINCE OF BRESCIA

Curioni: "294. Pisogne. . . . La gessaia di Pisogne è la continuazione di quella di Volpino. . . . 299. Pezzaze. Solfato di calce idrato (gesso). . . . 321. Comero . . . Forma un estesissimo deposito".

Jervis: "410. Pisogne. . . . Gesso saccaroide bianco. È abbondante nella valle di Fosio, dove vi sono delle gessaie. . . . 428. Pezzaze. Gesso, presso il villaggio di Lavone, dove vi sono delle gessaie importanti. . . . 448. Idro. Gesso. . . . 450. Comero. Gesso. . . . Gessaia".

Salmoiraghi: "Gessi. Pisogne, Collio, Pezzaze, Comero, Casto ecc. Del Trias, per materiali cementizi e usi agricoli".

AGGREGATE

This term refers to the addition of a non-reacting material to a binder avoiding the excessive shrinkage and increasing the strength of the mortar; only a gypsum binder is commonly free of aggregate. The use of this term is modern, and it is also present in the French (Agrégat) and Italian (Aggregato) languages; the term "filler" (from the verb "to full" or "to complete") is also used instead of aggregate. The term "sand" (sable, sabbia) was used in many cases, because sand is the most employed kind of aggregate.

Latin authors

The aggregates described are either natural or man-made: the sand (*harena, sabulum*) is divided on the basis of colour (black, white, red and *carbunculum*) or origin (quarry, fluvial, marine or sifted out from gravel) [Vitruvius, 1931; book 2, ch. 4]. Vitruvius enhances the quality of a sand aggregate:

> the best sand will be found to be that which crackles when rubbed in the hand, while that which has much dirt in it will not be sharp enough. Again: throw some sand upon a white garment and then shake it out; if the garment is not soiled and no dirt adheres to it, the sand is suitable.
>
> [book 2, chapter 4]

The amount of aggregate to be mixed to the binder is different according to the kind of aggregate [Vitruvius, 1931; book 2, ch. 5]. The use of the powder (marble powder) coming from the lumps containing "transparent grains like salt" is "extremely serviceable" in mortars, and in the case of a lack of these lumps "the bits of marble which marble-workers throw down as they work may be crushed and ground and used" [Vitruvius, 1931; book 7, ch. 6].

The use of volcani-clastic sediment to make mortars is typical of the Italian regions (Latium and Campania) along the coasts of the Tyrrhenian Sea. This material was called "Pozzolana" after the town of Puteoli (now Pozzuoli), located close to Baia and Cuma in the Campi Flegrei volcanic area, near Naples, the Bay and the Vesuvius: "There is also a kind of powder . . . in the neighbourhood of Baiae and in the country belonging to the towns round about Mt. Vesuvius" [Vitruvius, 1931; book 2, ch. 6]. The use of this sediment in a mortar is useful for maritime constructions (harbours):

> Walls which are to be underwater should be constructed as follows. Take the powder which comes from the country extending from Cumae to the promontory of Minerva [now Capo Miseno], and mix it in the vessel through the proportion of two to one.
>
> [Vitruvius, 1931; book 5, ch. 12]

Mortars made of Pozzolana were precisely reported by Strabo in the town of Cuma: "For mixing this sand [pozzolana] with chalk-stones they construct moles in the sea, thus forming bays along the open coast, in which the largest transport ships may safely ride" [Strabo, 1903; book 5, chap. 4, sec. 6].

Other authors such as Plinius in the first century [Plinius, 1962; book 36, 70 and book 36, 177] or Palladius in the fourth century [Palladius, 1988; book 1, sec. 10] reported the Vitruvius quotations. In late Antiquity, Isidorus from Sevilla in the fifth century [Isidorus, 1911; book 16, ch. 3, 11 and book 16, ch. 1, 8] enhanced the peculiarity of Pozzolana (*pulvis puteolanus*) to build the embankments on the sea waters.

Renaissance authors

Palladio distinguished about the aggregate: "Three sort of sand are commonly used in buildings: pit-sand, river-sand, sea-sand", and "There is also in Terra di Lavoro, in the territories of Baia and Cuma, a sand by Vitruvius called Puteolana, which knits together very soon in water, and makes mortar exceeding strong" [Palladio, 1715; book 1, ch. 4]. "To make good mortar, sand is to be combined in such a proportion that one part of lime be put with three parts of pit-sand, and two parts only or river or sea-sand" [Palladio, 1715; book 1, ch. 5].

Scamozzi described the properties of the volcanic ash from Pozzuoli or from around Rome, and also reported other natural materials used in Europe (i.e. yellowish sands at Nuremberg – Bavaria – Germany; coarse-grained sands of the river Danube in Vienna, Austria) [Scamozzi, 1615; part 2, book 7, ch. 21].

The use of "hydraulic mortars" was still enhanced to make mortars for damp places:

> Pozzolana is an earthen volcanic material, famous for the power to make excellent mortar together with lime, a mortar suitable to set rapidly and strongly in water. This substance is largely present into loose deposits spread in the part of Italy laying from the Apennines to Tyrrhenian Sea and from Naples to the southern border of Tuscany.
>
> [Cavalieri, 1831; part 2, book 3, ch. 4, §545]

Pozzolana generally contains lithic fragments, glass shards or droplets and crystals (leucite), and it pertains to the pyroclastic units of both Campi Flegrei (Sintema Vesuviano Flegreo, Campania) and Vulcano laziale (Latium). The first one is a loose, grey trachyte tuff, outcropping from Baia (west), to Averno, to Pozzuoli, to Agnano (east). The second one is a loose deposit of volcanic ashes and lapilli together with bombs and blocks, generally very porous and coloured from grey to reddish brown ("Pozzolane rosse" and "Pozzolane nere" – middle Pleistocene); outcrops are present in the south-eastern part of the region of Latium, and Pozzolana is also available within the Aurelian walls of Rome.

Other "Pozzolanic materials" of volcanic origin were used in regions outside of Italy, such as the volcanic deposits called "trass" (Andernach, Eifel, Rheinland-Pfalz, Germany, between the river Moselle and the Belgian border; Tertiary to Quaternary), the volcanic deposits of the semi-submerged caldera of Santorini (Thira, South Aegean, Greece), the semi-calcined grey limestone used to make lime, called "Cendrée de Tournay" (Hainaut, Wallonie, Belgium), etc.

In other Italian regions, where the true Pozzolana is lacking, the same hydraulic effect was reached using artificial Pozzolana. Under this term were included powder of brick (see Chapter 38), pottery and roof tile; ashes of charcoal from lime kilns; and powder of some burnt stone, such as basalt [Cavalieri, part 2, book 3, ch. 4, §547].

CLASSIFICATION OF MORTAR

Different kinds of mortars are distinguished, each with a particular composition or a particular proportion between the components: plaster (coating of masonry), decoration (decorating of walls, vaults and ceilings), bedding (holding together stone or brick of masonry), paving (supporting a firm surface made of stone, brick or mosaic), special purposes (coating of basins for liquid, sealing, jointing) [UNI-Normal, 2001].

Plaster

Plaster is a mortar useful for coating the masonry and protecting it from weathering. Another purpose is to support painting, mainly executed with a "fresco" technique (inorganic water-soluble pigments applied on a wet plaster).

The Latin term is "tectorium"; the Italian term "intonaco" comes from the Latin "tunicatus", meaning "covered by a tunic"; the French "enduit" comes from the Latin "inducere", meaning "to cover a surface"; the English "plaster" comes from the Greek "ἐμπλάσσειν", meaning "to cover a surface", but the ancient Greek uses the term "κονίασις".

A plaster is typically laid in two or more superimposed coats or layers; only one coat is rarely present: "render" is the coat laid on the masonry, "finish" is the coat of the surface.

The making of the plaster was reported by Vitruvius in the seventh book of his treatise (ch. 3 and 4). Vitruvius reported a render coat (aggregate made of sand – *harenatum*) in different proportions varying according to the environmental conditions (crushed brick or *testa tunsa* is recommended to replace the sand in damp sites) [Vitruvius, 1931; book 7, ch. 4]; a finish coat (aggregate made of marble powder – *marmoratum*), a painted coat (support of pigments – *pictura*) [Vitruvius, 1931; book 7, ch. 3, 4].

Plinius reported the different coats (*harenatum* and *marmoratum*) together with aggregates made of sand (pit, river, sea), of potsherd (crushed brick) or of marble powder [Plinius, 1962; book 36, 175] with different grain size.

The authentic Roman practice, as revealed by scientific analyses on Roman plasters unearthed in different archaeological sites of Lombardy, shows a similar sequence of aggregates: the render coat with quartz sand, the finish coat with crushed calcite (marble powder). Therefore, the aggregates are more heterogeneous, including also potsherd, crushed dolomite or limestone, crushed quartz.

Graffito

This particular kind of plaster is made of two superimposed coats: the finish coat (white) was scratched along a detailed pattern (human figures, landscapes, animals, plants, flowers, etc.) to expose the dark coloured render coat below. This technique, also called "sgraffito") was precisely described by Vasari [Vasari, 1907; ch. 26], and the same architect used them in some buildings of Tuscany (i.e. façade of Palazzo dei Cavalieri, Pisa, 1562).

The graffito in Milan architecture survived in some friezes on the façades of Liberty buildings (early 20th century), but it is perfectly illustrated by the so-called Cascina Pozzobonelli. Now Cascina Pozzobonelli is a modest edifice hidden in the modern quarter of apartment buildings near the main railway station, but in origin this was a great complex of buildings planned by Bramante (around 1498) and located in the northern outskirts of the town. The present-day edifice, the sole remnant after the demolition carried out in 1885, consists of a small portico, with four arches, connected to a "triconch" or a building composed of three semicircular niches (conch) surmounted by a half-dome. The masonry is made of brick, and the inner surfaces of the walls are coated with graffito plasters, featuring architectural themes and perspective views.

Some scientific analyses carried out on the original graffito of Cascina Pozzobonelli allow one to recognize the making techniques and the raw materials [Alessandrini et al., 1989].

The graffito plaster is made of two superimposed coats (total thickness circa 10 mm): the grey render coat contains a magnesian lime binder and a sandy aggregate (river sand, grain size up to 3 mm, metamorphic rock clasts, quartz and muscovite crystals) together with a vegetable-based charcoal; the white finish coat (thickness 0.2 mm) contains only a gypsum binder without aggregate. Regular plasters of the building, always show two superimposed coats (total thickness circa 4 mm): the pinkish render coat contains a magnesian lime binder and a sandy aggregate (river sand, grain size up to 1 mm metamorphic rock clasts, quartz and muscovite crystals) together with crushed brick (0.02–1.0 mm); the white finish coat (thickness 0.2 mm) contains only a gypsum binder without aggregate.

Mortar for decoration (stucco)

The decoration is an addition to the basic structure to enhance its visual appearance; this addition was mainly made using a particular kind of mortar both for the composition and the structure.

The Italian term "stucco" comes from the Old High German term "stuhhi", meaning "plaster", and it is present also in different European languages: English "stucco", French "stuc", German "stuck", Spanish "estuco". According to Vasari (Vasari, 1907; ch. 4):

> Now let us show how the stucco is mixed. Chips of marble are pounded in a stone mortar; no other lime is used for this stucco save with white lime either of marble chips of travertine; instead of sand the pounded marble is taken and is sifted finely and kneaded with the lime in the proportion of two thirds lime to one third pounded marble. The stucco is made coarser or finer, according as one wishes to work coarsely or finely.

Later Giorgio Vasari reports the preparation of the wall underneath the stucco decoration: a skeleton of baked brick or tufa (soft stone) is covered by a rough (coarse and granulated) and then by a finer stucco; besides

> to make mouldings or modelled leafage it is necessary to have shapes of wood carved in "intaglio" with those same forms that you wish render in relief. The worker take stucco that is not actually hard nor really soft, but in a way tenacious, and puts it on the work in the quantity needed for the detail intended to be formed.

The use of iron supports or nails or other armatures is needed to have bolder reliefs in projection. Moreover:

> a bas-relief on a flat wall needs the insertion of numerous nails, here projecting less, there more, according as the figures are to be arranged. . . . While the stucco is hardening the craftsman works diligently, retouching it continually with moistened paint-brushes in such a manner as may bring it to perfection, just as if it were of wax or clay.

Finally, Vasari wrote: "Nor is one to suspect work so done of being perishable; on the contrary it lasts for ever and hardens so well as time goes on, that it becomes like marble" [Vasari, 1907; ch. 13].

Scientific analyses, carried out on mortars of decoration of some historical buildings in Milan, draw attention to a very interesting feature: the use of a mono-mineralic aggregate made of quartz (Santa Maria presso San Satiro) or calcite (Villa Reale).

Church Santa Maria presso San Satiro – stuccoes of the apse

The original church was transformed by Bramante, in the last decade of the 15th century; the architect introduced a fake apse to compensate for the lack of space because of the presence of an urban street. The fake apse (less than 1 m long) includes four fake pillars supporting a fake entablature and a fake barrel vault with coffers: the whole structure is made of a brick masonry, coated with plaster and decorated by a mortar (mouldings of the entablature and coffers): this last mortar shows a binder made of gypsum and a scarce aggregate made of quartz crystals.

Villa Reale – stuccoes of the façades

The Villa was built in 1790 on a plan of the Austrian architect Leopold Pollack (1751–1806) for Count Lodovico di Belgiojoso, but it was purchased (1802) by the municipality and was donated to Emperor Napoleon I, then to Viceroy Eugene de Beauharnais.

The front of the garden features engaged columns with entablature made of skilful stucco-works (moulded profiles, bucrane and garlands).

Entablature profiles show a double-layered lime mortar applied on a core made of bricks: the lower layer of mortar contains a sandy aggregate; the upper one contains a crushed calcite aggregate. Bucranes and garlands show a double-layered lime mortar applied on a core made of roof tile fragments or metallic wires with cloths: the lower layer contains a sandy aggregate with the addition of crushed brick; the upper one contains a crushed calcite aggregate.

An outstanding Renaissance example is Palazzo Te (Mantua, 1535): scientific analyses, carried out on some stucco decorations, are reported to complete the appreciation of this particular kind of mortar.

Palazzo Te (Mantua) – stuccoes of the rooms

The finishing coat of Mannerist stucco-works of three rooms of Palazzo Te shows different features: angular quartz crystals and silicate clasts as aggregate together with a dolomitic lime as binder (cornices and figures of Camera dei Venti and Camera delle Aquile, upper frieze of Camera degli Stucchi); calcite crystals as aggregate and calcium lime as binder (lower frieze of Camera degli Stucchi); limestone clasts as aggregate and calcium lime (and also gypsum) as binder (Hercules statue of Camera degli stucchi).

The use of different aggregate in two similar stucco-works of the same room (Camera degli Stucchi) may be explained by the presence of two craftsmen working at the same time in the room (F. Primaticcio and G.B. Mantovano).

Mortar for bedding

A very important role performed by mortar is to fill the joints of a masonry and evenly transfer the loads downward through the masonry units. The preparation of this kind of mortar seems to be less accurate than the previous mortars. At first, only one coat is sufficient, and the lime binder is always coupled with sandy aggregate.

The composition of the binder and the aggregate remained the same for centuries since the Roman period: lime binder and sandy aggregate. In some cases, a part of crushed brick was added to the aggregate to give a sort of hydraulic function to the mortar. Since the last decade of the 19th century the use of Portland cement as a binder became more and more widespread.

Mortar for pavement (see Chapter 24, Volume 2)

Mortars in pavement are used to create a substrate where the upper surface, made of tiles or slabs, stays firm and fixed to allow regular walking and travelling.

Mortar to contain or to convey water (basin, pipe)

A special purpose performed by a mortar is to coat the inner surfaces of water basins and water channel of aqueducts. It is impossible to have a complete survey of these artefacts in Milan and Lombardy, so it is interesting to report two examples located in Milan and Lombardy (Brescia) and referred to the Roman imperial period.

The first artefact is a water basin (about 5 m long, 2 m wide) unearthed in a suburban house close to the city walls (via Correnti, Milan) and dated to the first half of the 1st century CE. The basin stands in a court below a *compluvium*, and it is connected with a concrete pavement with marble slabs and splinters (see Chapter 24, Volume 2). The external wall of the basin is made of pebbles with lime mortar filling the interstices; the inner coating is made of a lime mortar with crushed brick; the surface is painted with a red pigment. This structure of the water basin is in contrast with the description written by Vitruvius of a cement (*signinum opus*) cistern: the coating of inner walls must be made of a mortar containing a binder of strong lime, an aggregate of pure sand and splinters of hard stone (*silex*, probably "Leucitite" of Latium, see also Italian term "selce" as paving stone) in a proportion of two parts of lime and five parts of aggregate [Vitruvius, 1931; book 8, ch. 6]. These words were later reported by Plinius [Plinius, 1962; book 36, 173].

The second artefact is an aqueduct leading waters from Sarezzo (val Trompia) to Brescia following the course of river Mella for about 14 km. The water pipe (average dimensions: 1 m high, 0.6 m wide) is positioned underground in a dig about 2 m deep, and it is made of masonry of irregular stone pieces with a vault of the same stone materials. Inner surfaces of the water pipe were coated by three red-coloured layers of mortar then covered by a thick layer of calcareous deposit. These layers were sampled and examined using optical microscopy on a thin section and X-ray diffraction. Inner layer: about 1 cm thick; coarse grain size (0.2–2.4 mm); lime binder; two different kinds of crushed brick together with chert, sedimentary and metamorphic rocks fragments, quartz and calcite crystals. Intermediate layer: about 1.2 cm thick; coarse grain size (0.2–5.5 mm); lime binder; aggregate as the inner layer. Outer layer: about 0.4 cm thick; medium grain size (0.1–2.0 mm); lime binder; two different kinds of crushed brick together with sedimentary and metamorphic rocks fragments, quartz crystals. One kind of brick shows a high percentage of temper (quartz, calcite and feldspar); another one shows a low percentage of temper (quartz, muscovite and hematite). The calcareous deposit is about 1.0 cm thick, showing superimposed layers of calcite crystals: the shape of crystals changes from minute crystals (lower part of the deposit close to contact with the mortar) to elongated crystals (central part) to prismatic crystals alternately with fan-shaped very fine crystals (upper part) [Bugini, Folli, 1991].

A similar system of construction was also employed in modern times. Scamozzi praised a water pipe made of glazed earthenware tubes because of the healthiness, the low cost and the duration of this ceramic material. About masonry water pipes, the author described a kind of pipe set around an appropriate wooden cast and made of mortar mixing lime, crushed gravel, crushed brick, marble powder and linseed oil [Scamozzi, 1615; part 1, book 3, ch. 27].

REFERENCES

Alessandrini G., Bonecchi R., Bugini R., Fedeli E., Ponticelli S., Rossi G., Toniolo L., Bramante in Milan: The Cascina Pozzobonelli. Technical examination and restoration, *in* Studies in Conservation, 34, 1989, 53–66.

Breislak S., Descrizione geologica della provincia di Milano, Milano, 1822.

Bugini R., Folli L., Analisi scientifiche sulle malte, *in* Antichi acquedotti del territorio bresciano (F. Rossi ed.), Edizioni ET, Milano, 1991, 61–63.

Bugini R., Folli L., Lommano A., Plaster technologies in Northern Italy from 12th to 19th century (J. Riederer ed.), *in* Proceedings 8th International Congress on Deterioration and Conservation of Stone, Berlin, 1996, 1495–1504.

Carta Geologica d'Italia, Servizio Geologico d'Italia (Progetto CARG), Piacenza, 2014.

Cavalieri San Bertolo, Istituzioni di architettura statica e idraulica, Negretti, Mantova, 1831.

Curioni G., Geologia – Parte seconda. Descrizione ragionata delle sostanze estrattive utili, metalliche e terre, raccolte nelle Provincie lombarde, Milano, 1877.

Fieni L., Calci lombarde – Produzione e mercati dal 1641 al 1805, All'Insegna del Giglio, Firenze, 2000.

Isidorus Ispalensis Episcopus, Etymologiarum sive Originum (W.M. Lindsay ed., Etymologies), Loeb Classical Library, Oxford, 1911.

Jervis G., I tesori sotterranei dell'Italia, Loescher, Torino, fourth volume, 1889.

Palladio A., I quattro libri dell'architettura – The Architecture of A. Palladio in Four Books (N. DuBois ed.), Watts, London, 1715.

Palladius R.T.E., De Re Rustica (R.F. Martin ed., Traité d'Agriculture), Les belles lettres, Paris, 1988.

Plinius, Naturalis Historia (D.E. Eichholz ed., Natural History), Harward University Press, tenth volume, 1962.

Salmoiraghi F., Materiali naturali da costruzione, Hoepli, Milano, 1893.

Scagliole intarsiate (E. Rüsch ed.), Pinacoteca Züst, Rancate, 2011.

Scamozzi V., L'idea dell'architettura universale, Venezia, 1615.

Strabo, Geografika – Γεωγραφικά (H. C. Hamilton, W. Falconer eds., Geography), Bell and Sons, London, 1903.

Theophrastus, Peri lithon – De lapidibus (F. Wimmer ed., On Stones), Paris, 1866.

UNI-NORMAL 10924–2001, Cultural Heritage, Mortars for Building and Decorative Elements, Classification and Terminology, Roma, 2001.

Vasari G., Le vite de' più eccellenti Pittori, Scultori e Architettori – Introduzione (L.S. Maclehose ed., The Lives of the Most Excellent Painters, Sculptors and Architects – On Technique), J.M. Dent, London, 1907 (1st ed. 1568).

Vicat L.J., Recherches expérimentales sur les chaux de construction, les bétons et les mortiers ordinaries, Goujon, Paris, 1818.

Vitruvius, De Architectura (F. Granger ed., On architecture), Harward University Press, Cambridge, 1931.

Chapter 22

Brick and artificial stone

BRICK

"Stones come from the nature or come from the human manufacture. The natural are dug out of the quarry. . . . Man-made stones, called quadrels (brick) because of their shape, are made of chalky, whitish and soft earth" [Palladio, 1715; book 1, ch. 3].

Brick is a solid component of buildings with a prismatic shape; made of clay and soil minerals with other substances; used in walls, in foundations and in pavements. The use of brick is noteworthy in some areas where the natural stones are lacking (i.e. the alluvial plain, called Pianura Padana, in the southern part of Lombardy); nevertheless the use is favoured by some characteristics of brick (lightness, resistance, uniformity of size and shape), and these characteristics influence the regularity, promptness and running costs of a construction.

The raw materials were dug from the alluvial plains, in sediments left by water courses, then they were mixed, moulded, finally dried and, if necessary, burnt at a high temperature, in a kiln, to give strength and weatherproofing.

First of all, a distinction between the sun-dried brick (adobe) and the burnt brick is required. Sun-dried or crude bricks were at first employed because burning involved the availability of expensive furnaces and fuels, then burnt bricks were preferred due to their better resistance in wet climates.

About the terminology, it is necessary to note some differences among the languages. Latin uses one term (*later*) to indicate crude brick and another term (*testa*) to indicate burnt brick, but in some cases the term *later* was used together with different adjectives (*crudus* or *coctus*, *coctilis*, *testaceus*). Greek uses the term πλίνθος, together with ωμός (crude) and οπτός (burnt). In addition to English, also Italian (*mattone*), French (*brique*), Spanish (*ladrillo*), German (*ziegel*) and Arabic (بوط) use one term together with the adjective; Turkish uses kerpiç (crude) and tuğla (burnt brick). Lombard dialects use only one term: *medon* or *quadrell* (Milan), *madòon* or *quadrell* (Cremona), *préda* (Cremona, Mantua). *Quadrell* refers to the square form; *préda* refers to stone because the natural stone is completely lacking near Cremona and Mantua.

Bricks, both crude or burnt, are made using a simple quadrangular mould to produce prismatic elements; crude materials are also used in different manners: rammed-earth or *pisé* (earth and clay mixed together with small pebbles and rammed in a high formwork) and daub or *torchis* (earth and clay mixed with straw). The first one is used as a compact and solid mass, the second one is used as a plaster coat or to make brick.

Finally, the same clay materials are burnt to make roof tiles or floor tiles.

DOI: 10.1201/9781003361008-26

Ancient reports

Some lines of the Bible reported the use of straw to make crude brick:

> The same day Pharaoh gave this order to the slave drivers and overseers in charge to the people. You are no longer to supply the people with straw for making Bricks; let them go and gather their own straw. But require them to make the same number of Bricks as before; don't reduce the quota.
>
> [The Bible, 2011; Exodus, chap. 5, sec. 6–8]

The use of adobe in ancient Egypt was significant since the first dynasty, but an outstanding example is the outer wall of Medinet Habu encircling temples and pavilions (Pharaoh Ramesses III, twentieth dynasty) [Kemp, 2000].

Pausanias in the 2nd century CE, referred to the siege of Mantinea (Arcadia) and pointed out the weakness of the adobe: "Unburnt brick, however, does not suffer so much from engines, but it crumbles under the action of water just as wax is melted by the sun" [Pausanias, 1918; book 8, ch. 8, sec. 8]. On the other hand, burnt bricks were used to build the walls of Babylon along the river Euphrates in Mesopotamia, as reported by Herodotus [Herodotus, 1920; book 1, sec. 180].

Vitruvius wrote about *lateres* and their raw materials, including straw: "Bricks . . . should not be made of sandy or pebbly clay or of fine gravel", but they "should rather be made of white and chalky or of red clay, or even of a coarse-grained gravelly clay". Moreover: "Bricks . . . should be made in Spring or Autumn, so that they may dry uniformly", and they "will be most serviceable if made two years before using" otherwise "cracks make the Bricks weak" [Vitruvius, 1931; book 2, ch. 3]. Finally, brick walls, in order to be perfect and durable, "should be constructed with a structure of burnt Brick on the top . . . to prevent the crude Brick from being damaged" [Vitruvius, 1931; book 2, ch. 8].

Plinius wrote about the earthen architecture describing the rammed-earth and the adobe techniques:

> And then, besides, have we not in Africa and in Spain walls of earth, known as "formacean walls" from the fact that they are moulded, rather than built, by enclosing earth within a frame of boards, constructed on either side. These walls will last for centuries, are proof against rain, wind, and fire, and are superior in solidity to any cement. . . . What person, too, is unacquainted with the fact, that partitions are made of hurdles coated with clay, and that walls are constructed of unbaked bricks?
>
> [Plinius, 1962; book 35, 48]

The earthen architecture was also considered by other Latin authors writing about agriculture and rural buildings, such as Varro (1st century CE) [Varro, 1934; book 1, ch. 4] and Palladius (4th century CE) [Palladius, 1988; book 1, sec. 11].

Modern reports

Among the Italian architects and writers of the Renaissance, Leon Battista Alberti enhanced the better quality of an edifice built using burnt brick: "there is not a better material for any sort of edifice than Brick, not crude but baked" [Alberti, 1988, book 2, ch. 10]. On the other hand, Alberti praised the resistance to fire and the healthiness of the walls made of crude brick: "A building made of crude Bricks is extremely healthy to the inhabitants, very secure against fire, and but little affected by earthquakes. But then if it is not a good thickness, it will not support the roof" [Alberti, 1988, book 3, ch. 11].

Vincenzo Scamozzi accurately described earthen materials and clays to make bricks. Adobe technique was referred to the edifices built in Antiquity [Scamozzi, 1615; book 7, ch. 15]. A particular note reported the edifices of many villages of the Padania plain between Ferrara and Mantua, flooded by river Po and their tributaries: crude bricks were weakened by waters and the buildings fell down, causing a great damage in a land lacking in lime [Scamozzi, 1615; part 2, book 7, ch. 15].

About the burnt brick, Scamozzi lists the features of raw material ("pure, soft and tender clay, without sand"); the best season for digging of clays; the working to remove impurities; the maturing wrapped by a sandy layer; the moulding of the different shapes (prismatic – *mattone, quadrello, laterculo*; tabular – *tavella*; semicylindrical – *coppo, embrice*); the importance of the drying process involving, in a specific site (*aia*), the fitting of long rows of moulded bricks, accurately disposed to allow the circulation of the air and sheltered by small roofs (*rizza*); the burning into a furnace using wood as fuel [Scamozzi, 1615; part 2, book 7, ch. 16].

In the late 18th century, Francesco Milizia exactly reports the words of Palladio and points out six rules to make a good brick: pure clay without gravel or pebble; a long work to purify the clay; rectangular moulds sprinkled with sand to avoid the adherence of the clay paste; two years of desiccation process in a dry place; two days of cooking; another drying process followed by tempering in water and then cooking again at a higher temperature (partial vitrification). The author also lists five properties to evaluate the qualities of a brick: solidity, lightness in relation to the volume, uniform sound if beaten, preservation of colour if wet, without alterations if chilled [Milizia, 1785; part 3, book 1, ch. 2].

Nicola Cavalieri San Bertolo, in the early 19th century, accurately reports, under the heading of "Fake stones" (*Pietre artefatte*): the characters and the general use of a brick, adobe and pisé, raw materials, seasonal working, moulding, shape and size, baking and kilns (bricks placed in different parts of a kiln show different characteristics, and they must be used in different ways), technical tests, awareness of foreign authors, comparisons with ancient authors, defects. In particular, the qualities of a brick can be detected by some signs: a good brick sounds clear and acute when beaten, a bad brick sounds dull and deep; a good brick is compact and fine grained, a bad brick is porous, earthy and crumbles on the edges. Finally, a good brick remains unalterable despite an exposure to rains and chills of a winter season [Cavalieri, 1831; book 3, sec. 1, ch. 3, §514–24].

Adobe features

The brick making was well represented in the translation of Vitruvius's treatise edited by the Venetian architect Giovanni Antonio Rusconi (1520–79): an etching shows some workers mixing wet clay with straw, preparing bricks using a wooden cubic mould before the drying process, but, generally, bricks were made using a prismatic mould [Rusconi, 1590; book 2]. About the size of Lombard brick, Alessandro Capra wrote: "The mould, used in Cremona for building, must be 7 and 10 once long and 3 and seven once wide [oncia was a local measure of about 4 cm] . . ., and the dimensions of bricks are carved on the marble of the Baptistry of

Cremona" [Capra, 1717; book 2, ch. 1]. This rectangular carving is still visible on the ashlar in the lower part of the Baptistry at Cremona: 31 cm long and 15 cm wide.

Adobe in Lombardy

A rural wall of Vho-Piàdena, a small town located on the right bank of river Oglio about 25 km east of Cremona, was examined in order to analyse the crude brick and the mortar used for bedding. Crude bricks form the central part of the wall (about 1.5 m of 3.2 m of total height), the lower part is formed of burnt bricks (about 1.5 m), the upper part (0.2 m) is still formed of burnt bricks covered with roof tiles. The average size of the brick is 28 cm long, 13.5 cm wide, 6 cm thick. The mortar is made of earthen materials, and the joints are 1.0–1.5 cm thick; the presence of a plaster coat is witnessed by some still-visible fragments.

Mineralogical analyses (XRD on powders and optical microscopy on thin section) on the brick show a low percentage of rounded cavities, a composition of illite and chlorite clay, a temper made of quartz (size 40–80 micrometres, irregular surface, conchoidal fractures, sub-angular borders) together with calcite, muscovite and other silicates; the straw is absent. The clay used to make the mortar is different: higher percentage of macroscopic porosity, higher percentage of temper, coarser grain-size; sub-angular quartz crystals show an average size of 150 μm (micrometres) (minimum 10–15 μm and maximum 1 mm) [Bugini et al., 2009].

Clays are easily available in the territory of Piàdena, located between the right bank of Oglio (north) and the left bank of Po (south). The present-day courses of the rivers are filled by recent alluvial deposits; older deposits (*Würm*), made of yellowish clays and sands, are located on the terrace south of Oglio course, superimposed to fluvio-lacustrine deposits (*Interglacial Riss – Würm*) [Carta Geologica, sheet 61, 1969].

Adobe in Piedmont

On the western part of the Padania plain, there are some sites near Novi Ligure and Rivalta Scrivia (eastern Piedmont, Alessandria province) where the use of crude material was boosted by the lack of stone materials: rural building (farmhouses, storehouses, stables, etc.) were built using rammed-earth in a wooden form. The raw material is a clay (illite-chlorite) containing quartz and limestone gravels, up to 6 cm in length. The territory around Novi Ligure shows three water courses (Tanaro, Scrivia and Orba) and in this case also were dug alluvial deposits (Pleistocene). These continental deposits are superimposed to a clastic marine formation (*Sabbie di Asti*): altered sand, silt and clay (*Fluviale medio*, south-western area); non-altered gravel, pebble, clay (*Fluviale recente*, central area); mainly clays (*Post-glaciale e recente*, northern area) [Carta Geologica, sheets 69–70, 1970].

The wooden forms were 25 cm thick and were about 2 m long and 1 m high; the walls were made using four forms. The lower part of the walls (about 60 cm) was made of mortared pebbles (size 20–30 cm), then four superimposed courses of rammed-earth. Burnt brick form the opening frames on the wall. The raw material includes two grain-size fractions: the fraction above 1.4 mm was about 40%, the fraction below was about 60%. The coarser fraction (from 1.4 to 60 mm) sub-rounded clasts mainly made of quartz with limestone and serpentine; the finer fraction (size from 1.4 to 0.04 mm) contains illite and chlorite with quartz and feldspars crystals showing rounded shape and conchoidal surface modelled by the fluvio-glacial environment. Rammed-earth was employed for noble palaces (palace Migliazzi, Frugarolo), churches (Madonna delle Ghiare, Pozzolo Formigaro, with burnt brick), rural houses (Pozzolo Formigaro, Frugarolo and Rivalta Scrivia, two storeys, with burnt brick), rural buildings (Frugarolo and Rivalta Scrivia, one storey, with burnt brick), enclosing walls (Rivalta Scrivia and Bosco Marengo, with clay mortar).

Other Italian towns and areas with earthen architecture are Ferrara (Emilia Romagna), Corridonia (Marche, province of Macerata), Casalincontrada (Abruzzo, province of Chieti) and the Campidano plain (Sardinia, province of Sud-Sardegna) [Bertagnin, 1999; Gilibert and Mattone, 1998; Scudo and Sabbadini, 1997].

Burnt brick features

The shape and size of brick changed many times over the centuries. In the Roman period the bricks showed a squared shape, then sub-divided along diagonal or vertical lines, and the size varied from *pedalis* (one Roman foot long) to *sesquipedalis* (one and a half feet) to *bipedalis* (two feet). Since the Middle Ages, prismatic shape was widely used, and the size was determined by public regulations, so it was slightly variable according to the chronological period and the geographical site. Bricks were also esteemed for their uniformity of shape and size, so they were more suitable for the regularity, rapidity and cost of a construction. The type of burning and the heat control influenced the quality of the brick: a simple clamp burning produces non-homogeneous materials (i.e. the outer bricks are underburned whilst the inner bricks may crack down for the excess of heat). Kilns made of two chambers (a lower chamber for the fuel and an upper chamber for the brick) produce more uniform materials.

The microscopic features seem to remain unchanged over the centuries, but the scientific analyses carried out on Lombard bricks are very scarce and they do not allow one to identify a particular trend. Bricks of the early Christian bell tower of San Lorenzo show a temper made of fine crystals of quartz (equant, angular, 0.05–0.1 mm), with a small fraction of coarser crystals (equant to elongate, angular, 0.2–0.5 mm) and very few poly-crystalline aggregates (0.8–1.0 mm). Bricks of the first decades of the 20th century (Palazzo Comunale, Rho, province of Milan, 1930) show a temper made of quartz crystals (equant to elongate) with a bad sorting (grain size 0.05–0.5 mm).

The use of hollow tile (containing hollow vertical cells) to build interior masonry partitions was developed together with the use of reinforced concrete structures (early 20th century); this kind of brick was made by extrusion (clay mix forced through a shaped casting mould), and the available shapes are various, each adjusted to specific purposes. A furnace for hollow tiles is installed at Borgonato di Cortefranca (Lake Iseo, province of Brescia), exploiting the clay deposits of glacial origin at Provaglio d'Iseo, a few kilometres north.

Burnt brick in Lombardy

A group of Roman kilns (first half of 1st century CE – early 2nd century) was unearthed a few kilometres south of Lonato (province of Brescia): the best-preserved one has a circular chamber (heating space, diameter 6 m) with a base of brick placed vertically (soldier) and a dome-shaped top (disappeared). Two furnaces dating back to the 15th century were installed very close to the Roman furnaces. The area of Lonato involves glacial and glacio-fluvial deposits (Quaternary, Riss) deeply altered to form a reddish clay.

The use of burnt brick was ubiquitous in central and southern Lombardy, as in other regions of Italy far from mountains and stone quarries. The use in Lombard architecture was very important: the medieval Romanesque period shows building where the stone was confined to few architectural elements; then the bricks were employed to build bearing walls or partition walls among the structures. Since the Renaissance brickworks were always coated with painted plasters; the medieval ones were also plastered, but it is impossible to find a witness of this affirmation: the medieval buildings were always scratched during the restoration works started in the last decades of the 19th century.

The scientific analyses on burnt brick in Milan and in Lombardy are lacking; some personal analyses were made on samples coming from buildings of different periods (early Christian to early 20th century). It is possible to note a constancy of brick making both for the prismatic shape (average size about 4:2:1 = 25–12.5–6 cm) and for the recipe: a clay rich of temper (river sand) made of sub-angular quartz crystals (two classes of size: 0.05–0.1 mm and 0.4–0.5 mm) sometimes together with polycrystalline clasts of quartz of metamorphic origin (size 1.0–1.2 mm); the temper percentage varies from 25% to 30%.

The average open porosity is elevated (30–40%) in ancient bricks, and it decreases in modern ones (19th and 20th centuries) according to different types of bricks.

Alluvial deposits pertaining to the Alpine rivers were used in Milan as raw material to make bricks. The Po plain around Milan contains three defined subsurface units (Quaternary), the total thickness increases moving southward, from the Prealps to the plain. Clays, gravels and sands are present in different proportions: gravels are predominant in the upper part of the plain sediments (Middle Pleistocene – Holocene; glacio-fluvial deposits; about 50 m thick); sands are predominant in the middle part (Calabrian – Middle Pleistocene; fluvial deposits, about 50 m thick); clays are predominant in the deeper part (Calabrian; marine, transitional and continental deposits, more than 50 m thick) [Carta Geologica, sheet 118, 2016]. Clays (illite, kaolinite and montmorillonite) available for digging come from loessic deposits ("Ferretto", Middle Pleistocene) located several kilometres north of Milano ("Terrazzo delle Groane"; between Saronno and Meda) or from fine-grained sediments (upper Pleistocene) located few kilometres south of Milano. Around Milan there are many sites named with the term "Fornaci", meaning the presence of a furnace for burning local deposits suitable to make brick (i.e. north: Fagnano Olona, Garbagnate Milanese, Briosco; south: Borghetto Lodigiano, Terranova dei Passerini). In other parts of the Italian territory the name "Fornaci" is still present together with other similar names, such as Fornace, Fornacella, Fornacette, Fornase.

The Lombard architecture has been a burnt brick architecture since the Roman times (*opus testaceum*); bricks were also coupled in a masonry with pebbles or cut stones (*opus vittatum*) (see Chapter 23).

Crushed brick

Everybody knows the properties of Pozzolana (see Chapter 21) in order to obtain a material able to set without atmosphere (i.e. underwater constructions for harbours). To fill the gap created by the lack in volcanic sediments in northern Italy and abroad, different materials were experimented with in order to replace the Pozzolana and to obtain the same goal. Crushed brick, containing aluminium silicates, was one of the most significant surrogates, and it was added, as aggregate, to mortar recipes. This kind of hydraulic mortar has been diffused in northern Italy since the Roman times: i.e. mortars for Roman mosaic (see Chapter 24); bedding mortars of Castelseprio religious buildings (7th century) [Bugini, Folli, 2014]; plasters of the dome of the church of San Satiro (late 15th century).

About the pavements Vitruvius described an intermediate layer made of crushed brick or powdered pottery [Vitruvius, 1931; book 7, ch. 1]; about the plaster, Vitruvius wrote on the use of powdered earthenware instead of sand for a better resistance in damp places [Vitruvius, 1931; book 7, ch. 4]. The argument seems to be neglected by Italian architects; in many cases they simply translated the words of Vitruvius.

Terracotta

The Italian term (literally *baked earth*) was used since early times for sculptured figures and architectural elements.

Terracotta differs from brick and other ceramic bodies (pottery, tile) because it is fabricated using a carefully selected clay and taking heeds during the preparation and the burning.

The Raccomandazione UNI-Normal reports:

> Porous and coloured material obtained by the baking of iron-bearing clay at temperature of about 1000°C. An oxidant atmosphere produces reddish colours, the presence of calcium and its reaction with iron cause the colour to turn yellowish tones. The reducing atmosphere gives a more or less dark greyish tone.
>
> [UNI-Normal, 1998]

Clay was mixed with sand or fired clay to reduce the shrinkage; this plastic mixture was pressed in moulds (made of baked earth or of gypsum plaster of paris). Finally, the firing process was carried out in open kilns with accurate control of the temperature (about 1000°C), mainly preferring a long heating period at a low temperature.

Scamozzi reported the use of "pietra cotta" to make architectural decorations, pointing out the reduced cost and the short time for preparation in comparison with decorations made of natural stone; the author remarked also on the long life of these artefacts if kept in proper conditions [Scamozzi, 1615; part 2, book 8, ch. 12].

The unglazed terracotta was used by sculptors as the first step of their work or to make reproductions of famous sculptures for sale. Terracotta was highly spread in Lombardy since the Middle Ages because of the lack of natural stone; unglazed terracotta shows different colours, textures and qualities according to the clay used and to the temperature of firing.

Moulded elements made of terracotta were employed as a frame around the openings or as cornices on the façades. One of the most significant Milanese examples are the moulded decorations of windows of the oldest parts of the former Ospedale Maggiore (called Ca' Granda) made by Richino (early 17th century) and by Filarete (15th century).

Litoceramica (Klinker)

A particular kind of brick had been produced in Italy since 1925 (Piccinelli S.A. Bergamo, with three different registered trademarks *Klinker*, *Porfilit* and *Porfiroid*); this material was made of a mix of finely ground kaolin, silicates and volcanic rocks, baked at a high temperature to generate a glazing on the surfaces: the result was a kind of ceramic grès with high mechanical and chemical resistances. The usual shape was brick-like, sometimes with moulding on the external surface; the colour varied from red to purple to green. The use reached a maximum in the 1930s as a reaction to the plain plastered surfaces of the architecture of rationalism. Litoceramica was largely employed for cladding in dozens of buildings in Milan and its surroundings: an outstanding example is the Palazzo dell'Arte (architect G. Muzio, 1933), the edifice built to house the architecture exhibition called "Triennale di Milano" [Materiali del moderno, 2017].

NOTE: the term "klinker" is misleading. The same term (clinker) indicates the product of burning of clayey limestone, then ground to obtain Portland cement.

Ceramic tile and glass sheet

These materials are reported in the descriptions of Milan's buildings, but their scientific examination is beyond our field of study.

Ceramic tiles and glass sheets were largely employed in architecture in the 20th century, in fact for some decades after World War II these materials were the leading actors among building materials. Hundreds of apartment buildings, corporate headquarters, public buildings, churches, villas were covered by ceramic tiles of different shapes, from the ground to the roof. Some architects, such as G. Ponti and L. Caccia Dominioni, used a particular kind of ceramic tile in their edifices, showing pyramidal (G. Ponti) or hexagonal shape (L. Caccia Dominioni). In the same way glass covered the building surfaces together with metallic grids, following the International Style.

The definitions of ceramic and glass were fixed by Commissione Normal (Consiglio Nazionale Ricerche and Istituto Centrale del Restauro) [UNI-Normal, 1998].

Ceramica: Materiale inorganico, non metallico, ottenuto da materie prime minerali, foggiato a freddo e consolidato in modo irreversibile mediante cottura.

Ceramic: inorganic, non-metallic material that has been obtained from mineral raw materials; cold formed and then fired to reach an irreversible solidification.

Vetro: Materiale inorganico, non-metallico, ottenuto da materie prime minerali, foggiato allo stato fuso e consolidato per raffreddamento

Glass: inorganic non-metallic material formed by the fusion of mineral raw materials into a mass that cools to a rigid condition without crystallization.

In comparison are reported two technological definitions about ceramic and glass

Ceramic: Pottery. Basic Processes. The clays that most commonly occur in nature are plastic if mixed with an appropriate quantity of water. They can then be modelled into a great variety of shapes, which they will retain during subsequent drying in the air. When this drying has reached a stage at which the clay is "leather-hard" and water content reduced to some 8–15 per cent, the vessel can be further worked by scraping, cutting and turning. After this stage, it is further dried in the air to a "white-hard" state, in which the uncombined water content is reduced to some 3 per cent. The clay is then fired. At a temperature varying from 450°C upward, the clay substance decomposes with the loss of its chemically combined water, becomes incapable of further combination with water, and is then similar to moderately hard stone. If the firing is continued to higher temperatures, the clay substance, according to its composition, will become denser and less porous and will ultimately vitrify and fuse. These latter stages were rarely reached in early pottery, the range of firing-temperature seldom exceeding 1000°C and frequently being much less. [Scott, 1954]

Glass: A rigid but non-crystalline substance, usually transparent. It is made by fusing together alkalis (or their salts), lime and sand (or flint). It is a mixture of silicates of alkalis (soda or potash) and alkaline earths (in practice almost always lime, with some alumina or magnesium. . . . Ideally the material in the crucibles or pots needs to be heated slowly, as well as cooled slowly. [Harden, 1956]

ARTIFICIAL STONE

The use of Portland cement flourished rapidly in the last decades of the 19th century, giving rise to the expression "artificial stone". Artificial stone means a mix of Portland cement, sand, crushed stone and water, moulded and finished to resemble stonework.

In the architectural world, it is possible to consider a strict correlation between a building material and an architectural style: in Italy the correlation between the artificial stone and the Liberty architecture is explicit. In the past, a soft calcarenite (Miocene) widespread in the eastern Apulia (pietra di Lecce) and in south-eastern Sicily (pietra di Siracusa) was connected to the Baroque style which flourished in these provinces; a fine-grained bioclastic limestone (Craie or Chalk, Cretaceous) was linked to the highly carved Gothic style in northern France; a limestone (Portland stone, Jurassic) was connected to the classical architecture of the 19th century in England, together with brick and terracotta.

"Liberty" or "Floreale" ("Art Nouveau" or "Jugendstil" in Europe), flourished at the turn of the 19th century, encompassing architecture, design and decorative arts; the "artificial stone", often considered like a "moulded stone", originated in the same period thanks to the use of Portland cement. The artificial stone was firstly used as a building material and only later, after some improvements to the making techniques, was it used to produce heavily and elaborately decorated elements. The advantage of the use of artificial stones lies in the simplicity of modelling and moulding as one likes, at very low cost. Indeed the different artificial stone pieces to be used in a building were easily fabricated in the same yard, saving time and money. The rich ornamentation of the façades was a typical feature of Liberty architecture, and artificial stone was the key material sometimes coupled with ceramic, mosaic, glass, metal and wall painting.

Early examples are present in some edifices of the last decades of the 19th century (piazza Castello – Foro Buonaparte area, eastern side). The Liberty reached the highest point of achievement between two Exhibitions (Turin 1902 and Milan 1906) in some Italian towns as Milan, Turin and Palermo together with some holiday sites, as Campo dei Fiori (Varese), San Pellegrino Terme (Bergamo) or Salsomaggiore Terme (Parma). The Liberty architecture in Milan was represented in new private houses, as an attitude of wealthy people. Therefore, the lower cost of the artificial stone, in comparison with natural stone, allowed the diffusion of this ornamental style in the workers' houses also.

The years after World War I corresponded to the use of a simplified ornamentation: the so-called Art Decò. In parallel with the Liberty, Art Decò was developed with the use of Portland cement as a main component of the building construction: the framing. The framing involved beams, columns and studs made of reinforced concrete and in few years became the most employed building method completely removing the system of the bearing wall. The use of concrete frame was leading in comparison with the use of steel frames, in the same way the use of wood frame was totally absent.

Raw materials

As reported in Chapter 21, the raw materials to make Portland cement were widely exploited in Lombard territory, mainly in the hill areas of different provinces (Varese, Como, Lecco, Bergamo and Brescia): marly limestones pertaining to different geological formations (Jurassic to Cretaceous) of the Sedimentary series of the Southern Alps.

Artificial stone and cast stone

It is significant to distinguish between the artificial stone (Italian: pietra Artificiale) and the cast stone (Italian: cemento decorativo). In the first case, the material is only a surrogate of the natural stone; in the second case, the moulding properties and the mechanical strength, typical of Portland cement, are employed to realize artefacts, like reliefs and sculptures, with flowing lines and fanciful or extravagant shapes.

Artificial stone is the appropriate definition of an artefact made to imitate the natural stone, such as texture, grain size, colour and surface working. The raw materials are Portland cement, quartz sand, gravel, minute fragments of stone (limestone, marble), pigments; each building element (dimension stone, slab) was prepared and reinforced with metal rods or bars, then worked on-site using the traditional tools (chisel, punch, etc.). Dozens of different recipes are available through the specific literature: the proportions among Portland cement and the aggregate, the different kinds of aggregate, the size of the aggregate, etc. Generally, two beds are present: the first one contains coarse gravel, the second one (very thin) contains fine-grained stone fragments; the proportion of Portland cement and aggregate is variable. A few examples: an igneous rock (granite, syenite, etc.) was imitated through fine-grained fragments of white to pink stones (mainly limestone) mixed with some pigments; a clastic rock (conglomerate) was imitated through the inclusion of some pebbles in a fine-grained matrix; a sedimentary rock (limestone) was imitated through different pigments in a fine-grained matrix; a non-clastic rock (travertine) was imitated through a series of little holes in a fine-grained matrix. Artificial stone surfaces were finished *in situ*, using the same tools as natural ones (point, chisel, etc.). In the case of coating plaster, the imitation of ashlar courses was realized by engraving straight lines like a joint between each course and each block.

Cast stone is the appropriate definition of an artefact made for decorative or ornamental purposes; in this case the imitation of texture or grain size stays behind the outward form. The required shapes (plants or animals) were easily realized with a mould together with the insertion, if necessary, of a reinforcement. The raw materials are always the same, but the marble fragments are predominant as aggregate; in this case also dozens of different recipes are available through the specific literature. Cast-stone decorations were prepared at workshops and then applied to the building façade; the patterns show a great variety (geometry, animals, plants and flowers, fanciful creatures, industrial tools, etc.) and they were exhibited in workshops to facilitate the choice of architects and customers.

The artificial stone and the cast stone are subject to weathering: it is possible to distinguish "hair cracks" and "fractures". Hair cracks develop according to some surface working (punch) or if iron pins and clamps are present inside the artefact (increase of volume for oxidation [rust]). Fractures develop according to static settling of the edifice, and it spread in different artificial stone pieces.

Some petrographic analyses carried out on artificial stone (Portland cement) samples, allow one to know the mineralogical composition of the aggregate. The samples come from different Lombard buildings: Palazzo della Borsa (Milan, architect L. Broggi, 1901) – artificial stone aggregate: crushed calcite crystals. Oratorio Pesenti (Montecchio, Bergamo, 1904) – cast stone aggregate: crushed dolomite together with crushed calcite crystals. Artificial stone aggregate: quartz sand. Villa Gajo Lampugnani (Parabiago, Milan, 1907) – cast stone aggregate: crushed calcite crystals together with crushed limestone and, sometimes, sand. Grande Albergo (Campo dei Fiori, Varese – architect G. Sommaruga, 1909–11) – cast stone aggregate: crushed limestone. Artificial stone aggregate: sand.

A kind of Portland cement, chiefly fabricated for decorative purposes (cast stone), was the "Cemento bianco" (white cement), produced since the late 19th century by Italcementi (Bergamo) and based on a secret recipe, probably using a selected raw material coming from a single quarry together with the addition of some ingredients like kaolin, borax, calcium fluoride, magnesite, etc. Cemento bianco was produced in order to obtain a better imitation of marble works for statues, figures and all kind of ornaments; indeed the grey colour of the regular Portland cement was unable to make marble imitations. In order to know the features of this particular kind of "Cemento", scientific analyses were carried out on private buildings of Montecchio (Alzano Lombardo, val Seriana, province of Bergamo) pertaining to the Pesenti family (owner of Italcementi). These buildings were planned by the architect V. Muzio and built in the years between the 19th and 20th centuries: a villa, an oratory, a stable and a greenhouse.

In particular, the oratory is a small church (about 8 m tall, 6 m wide and 8 m long) with a kind of ornament taken from the typical decorative elements of the Gothic style: pinnacles, railings, splayed arch, statues, reliefs, etc. on the main façade; a simple rusticated motif on the other façades. The ornaments were created assembling pieces of different shape, each with a specific purpose: panels, slabs, open works, ashlars, voussoirs, etc. Some of these pieces with irregular shapes needed reinforcing metallic wires. Scientific analyses were unable to understand their different additions but pointed out the use of two kinds of aggregate: angular fragments of sedimentary rock and calcite crystals (ornaments); river sand (panels). The sedimentary rock pertains to Zandobbio formation, a coarse-grained dolomite with marble-like texture: so the presence of these fragments in the "Cemento bianco" is comparable to the presence of calcite crystals in Roman mortars for plastering or for decoration (stucco). The use of "Cemento bianco" in different kinds or decorative elements of oratory of Montecchio was also useful as advertising about the different items produced by the kilns of Italcementi [Carlessi, Bugini, 2001].

LIBERTY ARCHITECTURE IN MILAN

The importance of Liberty in architecture and its diffusion in the new quarters of Milan, is enhanced by this survey on private houses and apartments built in 1900–20 in the peripheral areas of the town. The natural stone (granite, conglomerate) is present almost always on the baseboards, but in some cases it is substituted by artificial stone. The artificial stone is present in façade claddings imitating the natural stone; the cast stone is present as decoration and ornament (flowers, leaves, fruits, birds, female figures with long and undulating hair, etc.). Plaster coating is mainly present in the upper floors; other materials are brick, ceramic tile, wrought iron and glass.

In order to show the use of artificial and natural stones in this architecture, we chose two northwestern quarters where Liberty buildings are present in a great number: Quartiere Sempione (around corso Sempione) and Quartiere Magenta (around corso Magenta). Obviously, there are other quarters showing Liberty architecture (i.e. Quatiere Venezia, in the north-eastern sector of the town): the most interesting buildings are also reported in some zones (see Part 1).

Buildings of the quarters are listed according to: "Il Liberty a Milano" (Bossaglia, Terraroli, 2003), "Milano Déco" (Bossaglia, Terraroli, 1999), "Milano Liberty" (Melano, Veronesi, 1991). In many cases the name of the architect and the year of construction are lacking. The list follows the address (alphabetical order), and the identification of building materials refers to the façade.

QUARTIERE SEMPIONE

The quarter grown around corso Sempione (the long avenue leading north, to the Simplon pass, Switzerland and north Europe) includes several buildings of the first decades of the 20th century, sometimes mixed with other ones built after World War II.

via Abbondio Sangiorgio 9

Baseboard: artificial stone (fake granite); cladding (ground floor): artificial stone (fake conglomerate); coating (other floors): plaster (rough finish); coating (vertical band): plaster (honed surface); decoration: cast stone

via Aleardi 17

Baseboard: artificial stone; cladding (ground floor and sides): artificial stone; coating: plaster; decoration: cast stone

via Arimondi 12

Baseboard: Granofiro di Cuasso (rough block); cornice: Montorfano (moulded element); cladding (ground floor): Granofiro di Cuasso (rough block); cladding (other floors): brick; coating: plaster (graffito); decoration: cast stone

via Canova 19 (*casa Knappworst*) – Architect: R. Perego (1909)

Baseboard: artificial stone; cladding (ground floor): artificial stone; cladding (other floors): brick; decoration: cast stone

via Castelvetro 5 (*casa Carugo*) – Architect: L. Tenconi (1908)

Baseboard: ghiandone; cladding (ground floor): artificial stone; cladding (first floor): ceramic tile; decoration: cast stone

via Castelvetro 7 (*casa Pardelli Losi*) – Architect: E. Provasi (1910)

Baseboard: Serizzo Ossola; cladding (ground floor): artificial stone; coating: plaster; balconies and windows: cast stone

piazza Gramsci 8

Baseboard: artificial stone; cladding (ground floor): artificial stone; coating: plaster; decoration: cast stone

via Guerrazzi 10 – Architect: A. Manfredini (1910)

Baseboard: Ceppo Brembo; coating (ground floor): plaster (new?); cladding (first floor): artificial stone; coating (other floors): plaster; decoration: cast stone

via Leopardi 19 (*casa Sormani*) – Architect: F. Gallavresi (1905)

Baseboard: artificial stone (fake conglomerate); cladding: artificial stone, brick; decoration: cast stone

via Leopardi 23 (*casa Sormani*) – Architect: F. Gallavresi (1905)

Baseboard: artificial stone (fake conglomerate); cladding: artificial stone, brick; decoration: cast stone

via Leopardi 24 – (1904)

Baseboard: Baveno; cladding: artificial stone, brick; decoration: cast stone

via Mantegna 6 (*Alfieri & Lacroix*)

Address: via Mantegna 6
Baseboard: artificial stone; cladding (ground floor): artificial stone; coating: plaster; decoration: cast stone

via Messina 16

Baseboard: Ceppo Iseo (new addition): coating: plaster; doorway: cast stone; decoration: cast stone

via Messina 24 (*Carminati & Toselli*) – Architects: G.F. Carminati, E. Gussalli (1911)

Baseboard: artificial stone; coating: plaster; decoration: cast stone (railway elements)

via Piero della Francesca 38

Baseboard: artificial stone; cladding (ground and first floors): artificial stone; coating: plaster; decoration: cast stone

via Piero della Francesca 42

Baseboard: Serizzo Ossola; cladding (ground floor): artificial stone; coating: plaster; decoration: cast stone

via Procaccini 24 – (1914)

Baseboard: artificial stone; cladding (ground to second floor): artificial stone; cladding: brick; coating: plaster; decoration: cast stone

via Procaccini 26

Baseboard: artificial stone; cladding: artificial stone (different surface finish); decoration: cast stone

via Procaccini 28

Baseboard: artificial stone; cladding (ground, first floor and central part): artificial stone; cladding: brick; coating: plaster; graffito (upper floor); decoration: cast stone

via Procaccini 41

Baseboard: artificial stone (fake conglomerate); cladding (ground to first floor): artificial stone (two bands); coating: plaster; decoration: cast stone

via Procaccini 47 (*casa Conti-Borbone*) – (1927)

Baseboard: artificial stone (fake granite); cladding (ground floor): artificial stone; coating: plaster; decoration: cast stone

via Rosmini 11

Baseboard: artificial stone; cladding (ground floor): artificial stone; coating: plaster; decoration: cast stone

via Rosmini 15

Baseboard: artificial stone (fake granite); cladding (ground floor): artificial stone; cladding: brick (central part); coating: plaster (side parts); decoration: cast stone

via Rosmini 17

Baseboard: artificial stone; cladding (ground floor): artificial stone; coating: plaster; decoration: cast stone

via Sarpi 3 – Architect: G. Valerio

Baseboard: artificial stone; cladding (ground and first floors): artificial stone; cladding: brick; decoration: cast stone

via Sarpi 5 – Architect: A. Manfredini

Baseboard: artificial stone; piers (ground floor): artificial stone; coating: plaster; decoration: cast stone

via Sarpi 14

Baseboard: artificial stone; cladding (ground floor): artificial stone; coating: plaster decoration: cast stone

via Sarpi 44

Baseboard: artificial stone (fake conglomerate); cladding (ground floor): artificial stone; coating: plaster; decoration: cast stone

corso Sempione 5 (*casa Calegari*) – Architect: T. Magriglio (1899)

Baseboard: Ceppo Brembo (painted); cladding, pilasters (ground and first floors): artificial stone (rustication, painted); coating (other floors): plaster; decoration: cast stone (painted)

corso Sempione 17 – Architect: G. Calegari (1900)

Baseboard: artificial stone; coating (ground and first floors): plaster; cladding (other floors): brick; decoration: cast stone

corso Sempione 60 (*casa Bugini*) – (1908)

Baseboard: Ceppo Brembo; cladding (ground and first floors): artificial stone; cladding: brick; decoration: cast stone

QUARTIERE MAGENTA

This quarter, grown between corso Magenta and via XX Settembre, north-east from the ancient urban walls, maintains a certain uniformity showing series of apartment buildings, of four to five storeys, built in the first three decades of the 20th century, according to the town plan of 1884–85.

via Alberto da Giussano 7 – G. Boni (1922)

Baseboard: Ceppo Brembo; coating (ground floor): plaster; cladding: brick; decoration: cast stone

via Alberto da Giussano 16 (*casa Macchi*) – Architects: G. Macchi, G. Rusconi Clerici (1914)

Baseboard: Serizzo Ossola (lower part), Montorfano (upper part); cladding (ground floor): cast stone; cladding: brick; decoration: cast stone

via Alberto da Giussano 18

Baseboard: Montorfano; doorway: Montorfano (two columns with pedestal); cladding (ground and first floors): artificial stone; coating: plaster; decoration: cast stone

via Alberto da Giussano 22, 24 – Architect: G. Combi (1909)

Baseboard: Ceppo Brembo; cladding (ground and first floors): artificial stone; coating: plaster; decoration: cast stone

via Alberto da Giussano 26, via Pallavicino 27

Baseboard: artificial stone (fake granite); cladding (ground and first floors): artificial stone; coating: plaster; decoration: cast stone

via Ariosto 4 – Architect: O. Bongi (1907)

Baseboard: Montorfano; cladding: artificial stone (fake conglomerate); decoration: cast stone

via Ariosto 8 – (1910)

Baseboard: artificial stone (fake conglomerate); coating: artificial stone; decoration: cast stone

via Ariosto 10 – (1908)

Baseboard: artificial stone (fake conglomerate); coating: artificial stone; decoration: cast stone

via Ariosto 19

Baseboard: Serizzo Ossola; coating: artificial stone; decoration: cast stone

via Ariosto 20 – Architect: G. B. Bossi (1905–06)

Baseboard: Ghiandone; coating: artificial stone (fake conglomerate); decoration: cast stone

via Ariosto 21 (*casa Cavalli Agostoni*) – Architect: A. Menni (1906–08)

Baseboard: Ghiandone; cladding (ground floor): Ghiandone and artificial stone (in alternating bands); doorway: Ghiandone; coating: artificial stone; decoration: cast stone

via Ariosto 22 – Architect: G. B. Bossi (1905–06)

Baseboard: Ghiandone; coating: artificial stone; decoration: cast stone

via Ariosto 23 – Architect: R. Turchetti (1925–26)

Baseboard: Serizzo Ossola; coating: plaster; decoration: cast stone

via Ariosto 24 (*casa Cambiaghi*) – Architect: Zucconi (1907)

Baseboard: Ghiandone; coating: artificial stone; decoration: cast stone

via Ariosto 26

Baseboard: Serizzo Ossola, cladding: artificial stone; decoration: cast stone

via Ariosto 27 (*casa Bottarelli*) – (1904)

Baseboard: Ceppo Brembo; cladding (ground and first floors): artificial stone (rustication); coating (other floors): plaster (graffito); decoration: cast stone

via Ariosto 28, 30

Baseboard: Ghiandone; coating: artificial stone; decoration: cast stone

via Ariosto 32 – Architects: C. Tenca and R. Valabrega (1912)

Baseboard: artificial stone (painted); cladding: artificial stone (fake conglomerate, painted); decoration: cast stone (painted)

via Ariosto 34 (*casa Comini Camozzi*) – Architect: A. Comini (1906)

Baseboard: artificial stone (painted); cladding: artificial stone (painted); decoration: cast stone (painted)

via Belfiore 5 (*casa Ferrandi*) – Architect: A. Malinverni (1915)

baseboard: artificial stone; cladding (ground floor): artificial stone; coating (bow window): plaster; decoration: cast stone

via Bianchi Mosè 12, 14

Baseboard: artificial stone (fake conglomerate); coating: plaster; decoration: cast stone

via Boccaccio 4

Baseboard: Ghiandone; doorway: Ghiandone; cladding: artificial stone; decoration: cast stone

via Boccaccio 7 (*casa Isotta*) – Architects: A. Binda, G. Boni (1901)

Baseboard (two parts): Baveno (below: squared block; above: moulded element); cladding (ground and first floors): Ceppo Brembo; cladding: brick; decoration: cast stone

via Boccaccio 11 – Architect: A. Comini (1901)

Baseboard: Ceppo Brembo; doorway: cast stone; cladding (ground and first floors): artificial stone; coating: plaster; decoration: cast stone

via Boccaccio 33

Baseboard: Baveno; coating: plaster; decoration: cast stone

via Boccaccio 37

Baseboard: Serizzo Ossola; cladding (ground floor): Ceppo del Brembo; coating: plaster; decoration: cast stone; balconies (brackets): Montorfano

via Colonna 15

Baseboard: Ceppo Brembo; cladding (ground floor): artificial stone; coating: plaster; decoration: cast stone

via Colonna 17–19

Baseboard: artificial stone; coating: plaster; decoration: cast stone

via Colonna 23

Baseboard: artificial stone; cladding (ground floor): artificial stone; cladding (first floor): brick; coating (second floor): plaster; decoration: cast stone.

via Colonna 25

Baseboard: Sarnico; cladding (ground floor): artificial stone; coating: plaster; decoration: cast stone

via Colonna 36

Coating: plaster; decoration: cast stone

piazza Conciliazione 1 (*casa Binda*) – Architect: A. Binda (1908)

Baseboard: Ghiandone; cladding (ground and first floors): artificial stone; cladding (other floors): brick; decoration: cast stone

piazza Conciliazione 2–4

Baseboard: Ghiandone; cladding: artificial stone, brick; decoration: cast stone

viale Ezio 9

Baseboard: Serizzo Ossola; cladding (ground floor): artificial stone; doorway: cast stone; coating: plaster; decoration: cast stone

via Gioberti 1 (*casa Donzelli*) – Architect: U. Stacchini (1903–04)

Baseboard: Ceppo Brembo; cladding (ground floor and balcony): artificial stone; cladding: brick; decoration: cast stone, ceramic, metal

via Guido d'Arezzo 5 (*casa Frisia*) – Architect: E. Cattò (1920–22)

Baseboard: Ghiandone; doorway: Ghiandone; cladding (ground floor): artificial stone (fake conglomerate); coating: plaster; decoration: cast stone

via Guido d'Arezzo 7

Baseboard: Ceppo Brembo; doorway: Ghiandone; cladding (ground floor): artificial stone; coating: plaster; decoration: cast stone

via Guido d'Arezzo 9,11 – Architect: S. Lissoni

Baseboard: artificial stone; cladding (ground floor): artificial stone; cladding (central part): brick; coating (side parts): plaster; decoration: cast stone

corso Magenta 96 (*casa Laugier*) – Architect: A. Tagliaferri (1905–06)

Baseboard: Ghiandone; cladding (ground floor): artificial stone (rustication); cladding (first floor): brick; coating (other floors): plaster; decoration: cast stone, ceramic; bracket: cast stone

via Mascheroni 10

Baseboard: Baveno; coating: plaster; decoration: cast stone

via Mascheroni 14 – (1904)

Baseboard: Ceppo Brembo; coating: plaster; decoration: cast stone

via Mascheroni 16

Baseboard: artificial stone; coating (ground floor): plaster; coating (other floors): plaster (graffito); decorations: metal

via Mascheroni 18 (*casa Felisari Carugati*) – Architect: G.U. Arata (1907–08)

Baseboard: Ghiandone (honed, lower part; quarry face, upper part); coating: plaster; balconies: cast stone

via Mascheroni 19 – (1916)

Baseboard: Serizzo Ossola (bottom and top), Occhiadino (middle); cladding (ground and first floors): artificial stone; cladding: brick; coating (bow window): plaster; decoration: cast stone

via Mascheroni 20 – Architects: C. Tenca and A. Chiappa (1914)

Baseboard: artificial stone (fake conglomerate); cladding (ground and first floors): artificial stone; coating: plaster; decoration: cast stone

via Mascheroni 22 – L. Cesa Bianchi (1902)

Baseboard: Ceppo del Brembo; cladding (ground floor): artificial stone; coating: plaster; decoration: cast stone

via Monte Rosa 14 (*casa Citterio – Galetti*) – Architect: A. Menni (1909)

Baseboard: Serizzo Ossola; doorway: cast stone; coating: plaster; decoration: cast stone

via Monti 12 – Architects: A. Bonomi, G. Dugnani (1905)

Baseboard: Baveno (below); Botticino (above); cladding (ground and first floors): artificial stone; coating: plaster; decoration: cast stone

via Monti 54 – Architect: C. Ruggieri (1910)

Baseboard: Ghiandone; cladding (ground and first floors): artificial stone; coating: plaster; decoration: cast stone

via Pagano 37, 39 (*casa Sardi*) – Architects: E. Baroncini, G. Vanini (1912)

Baseboard: Ghiandone; cladding (ground floor): Ghiandone, artificial stone; coating: plaster; decoration: cast stone

via Pagano 63

Cladding (ground floor): Ghiandone, Baveno, Porfido Monumentale, Verde Roja, Serizzo Ossola (quarry faced); cladding: brick; brackets: Ghiandone; balconies: plaster (Graffito); doorway and windows: igneous rocks (quarry face)

via Pagliano 2

Baseboard: artificial stone; cladding (ground floor): artificial stone; cladding: brick; decoration: cast stone

via Pagliano 4

Baseboard: artificial stone; cladding (ground floor): artificial stone; cladding: brick; decoration: cast stone

via Petrarca 2 (*casa Carugati Felisari*) – Architect: P. Ricotti (1903)

Baseboard: Ceppo Brembo; coating: plaster; cladding: brick; decoration: cast stone

via Petrarca 4 (*casa Macchi*) – Architect: L. Macchi (1903)

Baseboard: artificial stone; coating: plaster; cladding: brick; decoration: cast stone

via Petrarca 13 and 15 – (?)

Baseboard: Montorfano (squared block); coating (ground floor): plaster; cladding: brick; brackets: Montorfano; decoration: cast stone

via Petrarca 16 (*casa Vanoni*) – Architect: A. Menni (1903)

Baseboard: artificial stone; coating: plaster; cladding: brick; decoration: cast stone

via Petrarca 22, 22a – Architect: A. Volpi (1906)

Baseboard: Ceppo Brembo; coating: plaster; decoration: cast stone

via Petrarca 24 – Architect: L. Macchi (1903)

Baseboard: Ceppo Brembo; coating: plaster; cladding: brick; decoration: cast stone

via Fratelli Praga 8 (*casa Mazzucchetti*) – Architects: C. and R. Carini (1906)

Baseboard: artificial stone; cladding: artificial stone (red colour, central part); coating: plaster (side parts); decoration: cast stone

largo Quinto Alpini 8

Baseboard: Ceppo Brembo; coating: plaster; cladding: brick; decoration: cast stone

largo Quinto Alpini 12

Baseboard: Urago; coating: plaster; cladding: brick; decoration: cast stone

via Revere 7 (*casa Donzelli*) – Architect: U. Stacchini (1907–09)

Baseboard: Ceppo Brembo; coating: plaster (rough finish); decoration: cast stone

via Revere 11 (*casa Bistoletti Mora*) – Architect: C. Ruggeri (1910)

Baseboard: Ghiandone; coating: plaster; decoration: cast stone

via Revere 15 (*casa Castelli-Croff*) – Architect: D. Castelli (1907–08)

Baseboard: Urago; coating: plaster; cladding: brick; decoration: cast stone

via Revere 18 (*casa Perotto-Malfante*) – Architect: U. Tononi (1903)

Baseboard: artificial stone; coating: plaster; cladding: brick; decoration: cast stone

via Rovani 3 – Architect: F. Portaluppi (1901)

Baseboard: Ceppo Brembo; doorway: Baveno; coating: plaster; cladding: brick; decoration: cast stone

via Rovani 9

Baseboard: Ghiandone; coating: plaster; decoration: cast stone

via Rovani 11 (*casa Tassoni e Turati*) – (1907)

Baseboard: Ghiandone; doorway: Ghiandone; coating: plaster; decoration: cast stone

via Ruffini 3 (*casa Volpi Macchi*) – Architect: L. Macchi (1907)

Baseboard: Montorfano; doorway: Ghiandone; cladding: artificial stone; decoration: cast stone

via Saffi 8 (*casa Bosisio*) – Architect: U. Bosisio (1902–03)

Baseboard: Ceppo Brembo; coating (lower floor): plaster; decoration (upper floors): cast stone

via Saffi 9 (*casa Dugnani*) – Architect: U. Castiglioni (1902–03)

Baseboard: Baveno (lower part); artificial stone (upper part); coating (first floor): plaster; cladding (intermediate floors): brick; cladding (upper floor): ceramic (tile)

via Saffi 10 – Architect: E. Bennati (1902)

Baseboard: Serizzo Antigorio; coating: plaster; decoration: cast stone (fake conglomerate)

via Saffi 12 (*casa Maffioretti*) – Architect: G. Maffioretti (1900)

Baseboard: Ghiandone; coating: plaster; cladding: brick

via Saffi 14 (*casa Maffioretti*) – Architect: G. Maffioretti (1902–03)

Baseboard: Ceppo Brembo; doorway: Ceppo Brembo; coating: plaster; decoration: cast stone

via Saffi 20

Baseboard: artificial stone; coating: plaster (rustication); decoration: cast stone

via Saffi 24 (*casa Binda Castiglioni*) – Architect: A. Binda (1903)

Baseboard: Ceppo Brembo; cladding: brick; coating: plaster; decoration: cast stone

via Saffi 26 – Architect: A. Binda (1904)

Baseboard: Ceppo Brembo; coating: plaster; cladding: brick; decoration: cast stone

via Saffi 32

Baseboard: Ghiandone (lower part), Ceppo Brembo (upper part); doorway: Ceppo Brembo; cladding: artificial stone; decoration: cast stone

via San Siro 33 – Architect: G. Mettler (1921)

Baseboard: artificial stone (fake conglomerate); cladding (ground floor): Beola; coating (other floors): plaster; decoration: cast stone

via Sebeto 1 (*casa Verga*) – Architect: E. Pirovano (1899)

Baseboard: Ceppo Brembo; coating: plaster (corner stone, rustication); decoration: cast stone

via Sebeto 3 (*casa Verga*) – Architect: E. Pirovano (1899)

Baseboard: Ceppo Brembo; coating: plaster (rustication); decoration: cast stone

via Silva 39

Baseboard: artificial stone (fake Baveno); coating: plaster; decoration: cast stone

via Tasso 8 (*casa Donzelli*) – Architect: E. Zanoni (1912–13)

Baseboard: cast stone (fake conglomerate); coating: plaster; decoration: cast stone

via Tasso 10–12 (*casa Apostolo*) – Architect: U. Stacchini (1906–08)

Baseboard: Ceppo Brembo; coating: plaster; decoration: cast stone

via Tasso 14 (*casa Donzelli*) – (1911)

Baseboard: artificial stone; coating: plaster; cladding: brick; decoration: cast stone

via Tasso 17 – Architect: A. Comini (1906)

Baseboard: Serizzo Ossola; coating: plaster; cladding: brick; decoration: cast stone

via Telesio 17

Baseboard: Serizzo Ossola; coating (ground floor): plaster (rough); coating: plaster (honed); decoration: cast stone

via Telesio 19 – Architect: G. Laveni (1921), addition of other floors

Baseboard: artificial stone; cladding: artificial stone; decoration: cast stone

via Telesio 22 (*casa Battaini*) – Architect: A. Menni (1908)

Baseboard: Ghiandone; cladding: artificial stone; decoration: cast stone

piazza Tommaseo 5 – Architect: N. Acerbi (1906)

Baseboard (above and below): Serizzo Ossola; baseboard (middle): artificial stone (fake conglomerate); coating: plaster; cladding: brick; decoration: cast stone

via Toti 2

Baseboard: Ghiandone; pilasters: Ghiandone; coating: plaster; decoration: cast stone

corso Vercelli 53 – Architect: U. Stacchini (?)

Baseboard: artificial stone (fake granite); cladding (ground and first floors): artificial stone (rustication); coating (other floors): plaster; decoration: cast stone

piazza Wagner 4,6

Baseboard: Serizzo Ossola; cladding (ground floor): artificial stone (pilasters); coating: plaster; decoration: cast stone

REFERENCES

Alberti L.B., De re aedificatoria (J. Rykwert, N. Leach, R. Tavernor eds., On the Art of Building in Ten Books), MIT Press, Cambridge, 1988.

Bertagnin M., Architetture di terra in Italia. Tipologie, tecnologie e culture costruttive, Monfalcone, 1999.

The Bible, New International Version, Biblica Inc., New York, 2011.

Bossaglia R., Terraroli V., Il Liberty a Milano, Skira, Milano, 2003.

Bossaglia R., Terraroli V., Milano Déco, Skira, Milano, 1999.

Bugini R., Biondelli D., Folli L., L'architettura in terra della pianura Padana (Italia sett.): gli edifici in mattoni crudi nella provincia di Cremona, *in* Mediterra 2009–1ª Conferenza mediterranea architettura in terra cruda, Cagliari, 2009, 23–32.

Bugini R., Folli L., Medieval rubble walls of Castrum Sibrium (Italy), *in* Proceedings 9th International Masonry Conference (P.B. Lourenço, B.A. Haseltine, G. Vasconcelos eds.), Guimarães, ID1381, 2014, 1–8.

Capra A., La nuova architettura civile e militare, Cremona, 1717.

Carlessi M.A., Bugini R., L'oratorio Pesenti di Montecchio (Alzano Lombardo), *in* Atti Convegno Scienza e Beni Culturali, Padova, 2001, 469–482.

Carta Geologica d'Italia, ISPRA-Progetto CARG, sheet 118 (V. Francani et al. eds.), Piacenza, 2016.

Carta Geologica d'Italia, Servizio Geologico d'Italia, sheet 61 (F. Petrucci, S. Tagliavini eds.), Napoli, 1969.

Carta Geologica d'Italia, Servizio Geologico d'Italia, sheets 69–70 (A. Boni, R. Casnedi eds.), Napoli, 1970.

Cavalieri San Bertolo N., Istituzioni di architettura statica e idraulica, Negretti, Mantova, 1831.

Harden D.B., Glass and glazes, *in* A History of Technology (C. Singer, E.J. Holmyard, A.R. Hall eds), Oxford, 1956.

Gilibert A., Mattone R. (eds.), Terra: incipit vita nova. L'architettura di terra cruda dalle origini al presente, Politecnico, Torino, 1998.

Herodotus, The Histories (A.D. Godley ed.), Harward University Press, Cambridge, 1920.

Kemp B., Soil including mud-brick architecture, *in* Ancient Egyptian Materials and Technology (P.T. Nicholson, I. Shaw eds.), Cambridge University Press, Cambridge, 2000.

Materiali del moderno (L. Cupelloni ed.), Gangemi, Roma, 2017.

Melano O.P., Veronesi R., Milano Liberty, Mursia, Milano, 1991.

Milizia F., Principj di architettura civile, Remondini, Venezia, 1785.

Palladio, A., I quattro libri dell'architettura (N. DuBois ed., De Architecture of A. Palladio in Four Books), Watts, London, 1715

Palladius R.T.E., De Re Rustica (R.F. Martin ed., Traité d'Agriculture), Les belles lettres, Paris, 1988.

Pausanias, Hellados Periegesis (W.H.S. Jones, H.A. Omerod eds., Description of Greece), Harward University Press, Cambridge, 1918.

Plinius, Naturalis Historia (D.E. Eichholz ed., Natural History), Harward University Press, tenth volume, 1962.

Rusconi G.A., Della architettura di Gio. Antonio Rusconi, Venezia, 1590.

Scamozzi V., L'idea dell'architettura universale, Scamozzi, Venezia, 1615.

Scott L., Pottery, *in* A History of Technology (C. Singer, E.J. Holmyard, A.R. Hall eds.), Oxford University Press, London, 1954.

Scudo G., Sabbadini S., Le regioni dell'architettura di terra. Culture e tecniche delle costruzioni in terra in Italia, Maggioli, Rimini, 1997.

UNI-Normal 10739–1998, Cultural Heritage, Ceramic Technology, Terms and Definitions, Rome, 1998.

Varro M.T., Rerum rusticarum de Agricultura (W.D. Hooper, H.B. Ash eds., Agriculture), Loeb Classical Library, Harward University Press, Cambridge, 1934.

Vitruvius, De Architectura (F. Granger ed., On Architecture), Harward University Press, Cambridge, 1931.

Part V

Building structures

Chapter 23

Masonry

INTRODUCTION

Vitruvius tried to describe the attempts of men to have a shelter against weather excesses at the birth of the building technique: "And first, they set up forked stakes connected by twigs and covered these walls with mud. Others made walls, of lumps of dried mud covering them with reeds and leaves to keep out the rain and the heat". "Then . . . they began to build . . ., houses with foundations, having brick or stone walls, and roofs of timber and tiles". [Vitruvius, 1931; book 2, ch. 1]. The importance of walls is perfectly outlined by these words and masonry (vertical structure made of superimposed units so that the weight of a unit bears on the one below) was the most employed building system since Antiquity.

Latin authors took into consideration the masonry (called *opus*, meaning *work*) and proposed a classification based on the shape and size of the single elements and on the materials employed. Vitruvius reported two *opus* about the Romans: one made of regular beds and mortar joints (*reticulatum*); another made of imbricated rubble (*incertum*). Three were the *opus* in Greece: the masonry of equal courses (*isodomos*), the masonry with courses of various thickness (*psudoisodomos*) and the masonry in which only two faces are dressed and the rest is laid at random (*emplecton*) [De arch., book 2, ch. 8]. The typically Roman masonry made of rubble (*structura caementicia*) is reported in relation to the characters of sand [Vitruvius, book 2, ch. 4]; Plinius reported the Greek terms [Plinius, 1962; book 36, 171–72].

Roman masonries

Some remains allow one to consider the masonry used in Milan during the Roman times in Milan. The remains of the towers of the Circus and the remains of the city walls (San Maurizio al Monastero Maggiore) show a masonry made of brick with mortar joints. Each course is laid such that the middle of a brick covers the vertical joint of the underlaying course. The most representative masonry is the so-called *opus caementicium* where the *caementa* are irregularly shaped stone pieces bonded by a lime mortar. In Milan this kind of *opus* is visible in the foundations of the amphitheatre (via De Amicis) or in the foundations of a rounded edifice (near via Brisa), one of the few remains of the Imperial Palace; in both cases the *caementa* (rubble) are represented by well-sorted river pebbles. This kind of structure was always hidden from sight because of its setting below the ground (foundation) or of the cladding made of marble slabs.

DOI: 10.1201/9781003361008-28

Romanesque masonries

Romanesque buildings show unlimited options about masonries and a complete classification would include a too large number of groups and sub-groups [Parenti, 1990]. In fact, masonries change in the same building according to the different architectural parts: the masonry of the apse may be different from the masonry of the sides or a masonry made of squared blocks was employed in the parts of higher structural importance (pier, buttress, etc.), and a masonry made of rough stones was employed in the parts of lower structural importance (non-load-bearing wall).

The classification proposed here ranks four different groups of masonries according to the character and shape of the elements, and each group is strictly connected with some particular lithologies.

1 Masonry
Rubble – pebbles, cobbles and uncut stones, laid in irregular courses or in herringbone work, with thick joints of mortar (western Lombardy, i.e. Agliate, Baptistery)
Lithology
Glacial deposits (morainic or glacio-fluvial) of mixed lithologies connected to the rocks outcropping in the basin of the glaciers or the rivers: igneous rocks (granite, diorite, porphyry); sedimentary rocks (limestone, dolomite, sandstone, conglomerate); metamorphic rocks (gneiss, green stones)
2 Masonry
Rubble – roughly squared blocks of igneous, metamorphic or massive sedimentary rocks with different shapes and sizes and laid in fairly regular courses together with pilfered blocks from Roman artefacts (central and northern Lombardy, i.e. Gallarate, San Pietro)
Lithology
Massive igneous rocks (granite, diorite, porphyry, etc.) or metamorphic rocks (gneiss); thick-bedded sedimentary rocks (limestone, conglomerate, etc.); different lithologies (granodiorite, conglomerate) pilfered from ancient Roman buildings
3 Masonry
Ashlar – dressed sedimentary rocks of different thicknesses laid in quite horizontal courses with thin joints of mortar (central and eastern Lombardy, i.e. Pavia, San Michele)
Lithology
Thin or irregularly bedded sedimentary rocks (limestone, sandstone, etc.)
4 Masonry
Brickwork – bricks of different shape and size laid in regular courses or in herringbone work, with even joints of mortar (southern Lombardy, i.e. Lodivecchio, San Bassiano); sometimes bricks are mixed with rough stones or are set in alternate courses with squared stones or pebbles.
Lithology
Bricks, made of baked clays, together with rounded stone elements or together with limestone, sandstone, etc.

The masonries show almost always stone or brick arranged in parallel courses, less or more horizontal. In particular, the irregular thickness of the courses, if the masonry is made of sedimentary rocks (i.e. limestone, sandstone), is accordant with the different thickness of each natural bed. The courses of stone or brick almost always need the use of a mortar to obtain the suitable solidity. The joints of mortar change according to the shape of the stone: a masonry made of round pebbles requires thick mortar joints, on the contrary a masonry made of squared stones or bricks requires thin joints. Mortars used for bedding were based on a quicklime (magnesian) coming from Triassic dolomites of the Lake Maggiore area or from the loose sediments (siliceous limestone, limestone, dolomite, etc.) collected along the riverbeds at the northern boundaries of the Po plain. The binder, often containing white lumps made of calcium carbonate (5–20 mm wide), was mixed with an aggregate made of siliceous and/or carbonate sand coming from the alluvial deposits, but sometimes the "cocciopesto" (crushed brick) replaced the sand.

The lithology of the region around Milan shows a particular aspect: loose sediments of the alluvial plain (Pianura Padana) made of sand, gravel and clay; loose sediments of glacio-fluvial deposits along the courses of some rivers, made of boulders, pebbles and cobbles. This local aspect strongly influenced the availability of building materials in Milan and surroundings, mainly in the Middle Ages when the supply of stone was hampered by a poor road network and a lack of vehicles suitable for transport. The most employed building materials were brick and pebbles together with the sand and gravel used for mortars: the pebbles pertain to Quaternary glacio-fluvial deposits (matrix-supported gravels) or to Quaternary glacial sediments of moraines (till, unsorted).

Gravel and sand come from quarries dug into the water table and scattered around the Milan urban area. The clastic materials were selected (washing, sieving) in order to obtain the proper aggregate (filler) for different kinds of mortar [Bugini et al., 1996].

The glacio-fluvial deposits of Lombardy contain sediments got together, transported and piled up by the alpine glaciers running southward from the Alps to the plain (Ticino and Adda valleys) and they are widely diffused north of Milan, in the western part of the region (Varese, Como and Lecco provinces). Boulders, cobbles and pebbles are made of igneous rocks (granite and granodiorite, porphyry, green stone, etc.), metamorphic rocks (mainly gneiss) of the Alps; sedimentary rocks (limestone and dolomite; sandstone and conglomerate) of the Prealps. Cobbles and pebbles are useful as building stone, and their morphology follows the lithology: spheric (igneous or massive sedimentary rocks), tabular (metamorphic or thin-bedded sedimentary rocks); disc-shaped (foliated metamorphic rocks). Cobbles and pebbles were set without any kind of working or, at the most, cobbles were split in two parts; tabular and disc-shaped pebbles were arranged in herringbone work. Large, erratic boulders were first cut and transformed into rough squared blocks by Roman craftsmen and then the blocks were taken from the dismantled Roman buildings and set in the lower part of the Romanesque buildings [Bugini, Folli, 2014].

The original Romanesque masonries of Milan and its provincial territory were investigated about stone material and setting. Petrographical identifications refer to the stones employed in the masonry, including piers, buttresses, pilasters and quoins and excluding capitals or bases, portals or lintels, trimmings and porticoes [Bugini, Folli, 2014]. Romanesque masonries of Lombardy were widely described by Arthur Kingsley Porter [Porter, 1915].

MILAN AREA

Chiesa Rossa (late 12th century)
Masonry: fine brickwork
Courses: very horizontal, separated by mortar beds of moderate thickness

Material: regular bricks of large size

Sant'Ambrogio (church; Monaci bell tower, Canonici bell tower 12th century)

Atrium (12th century). Masonry: closely resembles that of the apse; brickwork with occasional blocks of stone introduced seemingly at haphazard into the masonry of the walls, especially in the lower parts of the buttresses. Courses: horizontal or in herringbone style; mortar beds about 2 cm thick. Material and lithology: bricks with great variations in size; Serizzo, Ghiandone, Ceppo, Proconnesian marble, Trachyte, Molera.

Apse (12th century). Masonry: fairly rough brickwork; bricks often broken coming from an earlier edifice; the great majority of bricks have a fairly honed outside finish. Squared stones inserted in the lower parts. Courses: horizontality not well maintained and numerous herringbone and vertical courses; mortar beds average about 2 cm. Material and lithology: bricks with very irregular size and shape; Ghiandone, Serizzo, Ceppo, Angera, Ammonitico.

Monaci bell tower (9th century). Masonry: rough brickwork (some bricks projecting from the surface, other receding); squared blocks of stone (lower part). Courses: seldom horizontal; many herringbone and vertical courses are inserted; very wide mortar beds somewhere. Material and lithology: bricks (average thickness 7 cm); Ceppo (lower part).

Canonici bell tower (12th century). Masonry: brickwork, distinctively more advanced than that of the atrium (lower part). Courses: no herringbone (lower part). Material: bricks.

San Celso (11th century)

Masonry: brickwork

Courses: horizontal with mortar-beds of normal thickness

Material: bricks; wide, but generally of moderate length

Sant'Eustorgio (apse, 11th–12th centuries)

Masonry: brickwork

Courses: approximately horizontal, but much herringbone works; thick mortar beds

Lithology: bricks of all size and shapes

Santa Maria Rossa a Crescenzago (middle 12th century)

Masonry: brickwork

Courses: horizontal with mortar beds of moderate thickness

Material: large and well-squared bricks

Santi Nazaro e Celso (11th century)

Masonry: brickwork

Courses: horizontal for the most part

Material: regular bricks

San Satiro (bell tower) (middle 11th century)

Masonry: extremely rough, brickwork; pilfered squared stone blocks (lower part)

Courses: seldom horizontal, interrupted at frequent intervals by herringbone masonry (upper part)

Material and lithology: bricks of enormous size; Serizzo, Ghiandone, Musso

San Sepolcro (13th century)

Masonry: rough brickwork with occasional blocks of stone especially as quoins (lower part of south side)

Courses: widely departing from the horizontal; layers of herringbone are present at frequent intervals

Material and lithology: bricks greatly varying in size; Angera
San Simpliciano (south side, 12th century)
Masonry: brickwork; very large stone blocks pilfered from Roman buildings (lower parts)
Courses: nearly horizontal for the most part, separated by mortar beds of moderate thickness
Material and lithology: bricks enormously large; Ghiandone, Serizzo, Ceppo
San Vincenzo (9th century)
Masonry: brickwork with occasional blocks of stone
Courses: regular and horizontal, in the main, but broken at times by vertical and oblique, thick mortar beds
Material and lithology: bricks of small and irregular shape; Ghiandone, Serizzo

PROVINCIAL AREA

Agliate Baptistery (11th century)
Masonry: rubble; rounded stones and pieces of brick
Courses: bricks laid in a mass of mortar; stones laid carefully with an approximation towards horizontal and narrower mortar beds
Lithology: pebbles, Tufa; bricks
Agliate, Santi Pietro e Paolo (11th century)
Masonry: rubble, rounded stones and bricks, pilfered big blocks
Courses: irregularly horizontal with a mass of mortar
Lithology: pebbles (gneiss, granite, limestone), Ghiandone; bricks
Chiaravalle Abbey (early 13th century)
Masonry: brickwork
Courses: horizontal with wide beds of mortar
Material: large bricks
Morimondo Abbey, Santa Maria (late 11th century)
Masonry: brickwork of excellent character
Courses: horizontal, separated by wide mortar beds
Material: bricks large, regular and well formed
Vaprio d'Adda, San Colombano (11th–13th centuries)
Masonry: fine ashlar; large and accurately squared blocks
Courses: horizontal, separated by mortar beds of moderate thickness
Lithology: Ceppo
Viboldone Abbey, Santi Pietro e Paolo (12th–13th centuries)
Masonry (apses): brickwork with little changes between choir and nave
Courses: horizontal with wide mortar joints
Material: bricks of regular size

The masonry since the Renaissance

Architects of the Renaissance, such as Alberti or Palladio, described some masonry adding new options. Alberti wrote: "The different sorts of masonry are these: the ordinary [squared stones placed exactly], the chequer sort [small stones placed on their corners] and the irregular [rough stones with their sides answering one to the other]" [Alberti, 1988; book 3, ch. 6]. Palladio wrote:

> The Ancients had six sorts of walls, one of which called *Reticulata* or net-work [*or Chequer-work, it is no more in use at this time*], another of Quadrels or Bricks [walls faced on both sides with bricks and the middle filled up with cement rammed together with brick-bats]; the third of Cement, which is a mixture of flints or pebbles and earth roughly laid with or without mortar; the fourth made of various stones and called Rustick; the fifth of squared stones; the sixth called *Rimpiuta* or Coffer-work [taking planks and placing them edge-way in two rows distant one from another . . .; filling the space . . . with Cement, stones of all sorts, earth and mortar mingled together.
>
> [Palladio, 1715; book 1, ch. 9]

In Milan the masonries became more and more uniform, using regular courses of ashlars mainly made of limestones or sandstones coming from the area of the Prealps (north of Milan).

About the ashlar masonry, Palladio reported:

> Whereas it happens sometimes that buildings are made, the whole, or a good part of marble or of some other great stones; I think it very proper here to explain what the Ancients did on such occasions, because it is to be observed in their works, that they were so nice in the joining of their stones together, that sometimes the joints are difficult to be perceived. . . . As far as I can understand, they first squared and wrought those sides of the stones, which were to be laid one upon the other, leaving the other sides rough.
>
> [Palladio, 1715; ch. 10]

Modern masonries

The traditional brick of prismatic shape was the main component of the masonries since the last decades of the 16th century, masonries covered with coats of plaster. A typical feature was the decrease in the thickness of the wall going upwards. This feature was already described by Palladio:

> It is to be observed, in the erecting of the walls, that they ought to diminish proportionally, as they raised and grow higher. That part therefore which appears above ground, is to be one half thinner than he foundations, and the second story half a brick thinner than the first story; and so successively to the top.
>
> [Palladio, 1715; book 1, ch. 11]

Since the commencement, in the first decades of the 20th century, of the construction based on reinforced concrete framing, the masonries were mainly used for interior partitions using hollow tiles, therefore brick masonries were totally neglected.

Cladding and veneering

> The art of cutting marble into thin slabs may possibly have been invented in Caria. The earliest instance, so far as I can discover, is that of the palace of Mausolus at Halicarnassus, the brick walls of which were decorated with marble from the island of Marmara.
>
> [Plinius, 1962; book 36, 47]

The cladding is defined as any thin layer of stone applied over a structural backing in order to present a more decorative aspect hiding the almost always irregular structure of the wall. A question of cost is a valid reason for cladding: a slab, the most important element of cladding, with its thinness may flawlessly simulate a structure made of squared and finished stone blocks.

This particular expression of masonry embellishment was mainly present in Roman times, when slabs of highly priced coloured marbles were employed (see Chapter 20, Volume 2).

Alberti, writing about wall claddings, reported the use of well-polished stone slab; the slabs were secured to the wall surface in two ways:

> For fastening on the slabs, if they are thick, fix into the wall either pins of iron, or little spars of marble sticking out from the wall, to which you may fasten your slab without any thing of cement. But if the slabs are thin, after the second plastering, instead of mortar, take wax, pitch, rosin, gum mastic, all melted and mixed together, and warm your piece of marble by degrees.
>
> [Alberti, 1988; book 6, ch. 10]

In modern times the stone or marble cladding was mainly employed into religious buildings; wood panels or plaster coating together with stucco were mainly employed into private palaces and into public buildings.

Roof

Roman tiles (*tegula*) were flat, rectangular with turned-up edges, and they were drawn in rows, each row subdued to the following one; the contact edge between two tiles was covered by a semicircular tile (*imbrex*) [Plinius, 1962; book 36, 159].

The artefacts used to cover the roofs in northern Italy were precisely described by Scamozzi:

> A wide part of Italy and chiefly the Padania plain from Piedmont to Lombardy to Venetian and Friuli or from the Apennine to the Alps, where a clayey earth is found and the buildings are covered with imbrex (coppo) made of terracotta, suitable to receive and to take away the rains.
>
> [Scamozzi, 1615; part 2, book 8, ch. 23]

The structure that caps the building was normally a hip-roof exclusively made of hip tiles, or a tile (imbrex) with the shape of hollow, truncated cone cut along its axis, laying alternately overlapping one up and another down. The material was terracotta produced through the baking of clay, coming from sediments spread in the alluvial plain where Milan is located. The framework to support the roof was made of wood with a typical series of trusses. Different kinds of tiles were also employed in the late 19th century and after as Belgic tile (rectangular tile with an S-shape cross-section with concave and convex curves of unequal size, so the large curve overlaps the small curve of the adjacent tile).

The use of stone for roofing was practically absent in Milan architecture, but there were largely employed in the mountain area of Lombardy using local stones as foliated gneisses (western Lombardy) or "porfiroide" (a low-grade metamorphic rock coming from volcanic rocks; central Lombardy). Thick slabs made of Beola or other metamorphic foliated rocks were used in eaves cornices.

The roofs changed in the 20th century with the rising of reinforced concrete structures; in particular, the architecture of "Razionalismo" built flat roofs, and the tiles were completely neglected in favour of terraces on the top floor with insulating systems or roofs made of metallic sheets with low inclination.

REFERENCES

Alberti L.B., De re aedificatoria – On the art of building in ten books (J. Rykwert, N. Leach, R. Tavernor eds.), MIT Press, Cambridge, 1988.

Bugini R., Folli L., Lommano A., Plaster technologies in northern Italy, *in* Proc. 8th Intern. Congr. Deterioration of Stone (J. Riederer ed.), Berlin, 1996, 1495–1504.

Bugini R., Folli L., Medieval rubble walls of Castrum Sibrium (Italy), *in* Proceedings 9th International Masonry Conference (P.B. Lourenço, B.A. Haseltine, G. Vasconcelos eds.), Guimarães, ID1381, 2014, 1–8.

Palladio, A., I quattro libri dell'architettura (N. DuBois ed., The Architecture of A. Palladio in Four Books), Watts, London, 1715.

Parenti R., Una proposta di classificazione tipologica delle murature postclassiche, *in* Proceedings Convegno Scienza e Beni Culturali (G. Biscontin ed.), Libreria Progetto, Padova, 1990, 49–61.

Plinius, Naturalis Historia (D.E. Eichholz ed., Natural History), Harward University Press, tenth volume, 1962.

Porter A.K., Lombard Architecture, Yale University Press, New Haven, 1915–17.

Scamozzi, L'Idea della architettura universale, Venezia, 1615.

Vitruvius, De Architectura (F. Granger ed., On Architecture), Harward University Press, Cambridge, 1931.

Chapter 24

Pavements

INTRODUCTION

The smooth surface of a room or the firm surface of a road are construction systems where the outward, suitable for walking or travelling, is only the upper and visible part of a more complicated structure made of superimposed layers of mortar. A pavement is susceptible to a marked decay and, obviously, it is often repaired or replaced.

ANCIENT AUTHORS

"Paved floors originated among the Greeks and were skillfully embellished with a kind of paintwork until this was superseded by mosaics" [Plinius, 1962; book 36, 184].

The evidences of paved floors are nevertheless more ancient: i.e. the basalt pavement of the pillared court of the Pyramid Temple (Pharaoh Khufu, fourth dynasty, c. 2630–510 BCE); the alabaster pavement of the T-shaped interior pillared hall of Valley Temple (Pharaoh Kaphre, fourth dynasty); [Aston et al., 2000]; the gypsum pavement of the Throne Room of the second palace at Knossos (Crete, c. 17th century BCE) [Preziosi, Hitchcock, 1999].

The treatise of Vitruvius [Vitruvius, 1931; book 7, ch. 1] gives detailed information about the making techniques of mortar layers used as the foundation of mosaic or stone paving; the author reports different sequences of mortar layers according to the location of the floor: on the level ground or on the upper floors; outdoor or frost-proof. The typical sequence consists of four layers: the first and the second made of mortar with rounded stones decreasing in grain size (from 15 cm to 5 cm, called respectively *statumen* and *rudus*); the third made of mortar with crushed brick (*nucleus*) and the fourth made of stone tesserae or stone slabs or bricks; finally, a fine mortar coat fills the gaps among stone elements of the surface. The thickness of the layers is: rudus about 25 cm, nucleus about 12 cm. The proportion of the mortar components are: three or five parts of stone and one or two parts of lime (rudus); three parts of crushed brick and one part of lime (nucleus).

DOI: 10.1201/9781003361008-29

ROMAN PAVEMENTS IN LOMBARDY

During the Roman times, three different kinds of surface finish were employed in Milan and Lombardy:

- concrete (*opus signinum*)
- stone (*opus sectile*)
- mosaic (*opus tessellatum*)

Concrete

"The first pavement, in my opinion, were those now known to us as barbaric and *subtegulam* pavements, a kind of work that was beaten down with the rammer" [Plinius, 1962; book 36, 185]. This kind of pavement is reported as *opus signinum*, meaning a layer of mortar with crushed brick, splinters of terracotta and sand as aggregate, including cubic pieces of stone (tessera) disposed in a simple layout or rod-shaped pieces or slabs with unpatterned disposition.

Some examples of Lombardy, based on scientific analyses, are useful to understand the sequence of layers. Starting from the bottom: a layer made of pebbles, coarse-sized fragments of brick joined together by a lime mortar; a mortar layer made of lime, crushed brick, sand or gravel together with small pieces of stone (mainly black and white limestone) sometimes following a geometric pattern; a painted layer (uncommon) with red pigment.

Milan – *Domus* via Correnti

- Upper layer: lime mortar with sand and a low percentage of crushed brick. This layer also contains coloured marble slabs (maximum side 20–50 mm) with geometric shape disposed in nearly regular rows around a centre, together with black cubic stone tesserae set in quincunx. Coloured marbles are Africano, Breccia Settebasi, Cipollino, Giallo antico, Pavonazzetto, Porfido rosso antico, Portasanta together with white and grey marbles.
- Lower layer: pebbles set in one row and put together by a lime mortar.

Milan – *Domus* via Illica

- Upper layer: lime mortar with sand and crushed brick together with small chips of black limestone. This layer contains also coloured marble slabs (side 20–50 mm) with irregular shape (Porfido serpentino verde) or geometrical shape (white marble) arranged in a regular pattern.

Brescia. Roman republican sanctuary, western room (first quarter of 1st century BCE)
Total thickness: circa 100 mm.
Upper layer: thickness circa 50 mm; mortar made of lime with crushed brick; the surface includes rectangular slabs of white or black limestone.
Lower layer: thickness circa 50 mm; mortar made of lime with gravel (size 20–60 mm) and splinters of stone (until 50 mm).
Cremona. Excavations at piazza Marconi, rubble (US 591), 1st century CE.
Total thickness: circa 200 mm.

Upper layer: thickness circa 100 mm; mortar made of lime with coarse-grained crushed brick and splinters or tesserae (side 10 mm) of black (micritic limestone with sparry calcite veins) or white limestone.

Lower layer: thickness circa 100 mm; coarse fragments (more than 50 mm) of brick together with a lime mortar.

Cremona. Excavations at piazza Marconi, rubble (US 1086), 1st century CE.

Total thickness: circa 150 m.

Upper layer: thickness circa 70 m; mortar made of lime with crushed brick and quartz crystals. Together with splinters (about 10 mm) of black or white limestone (oolitic limestone with sparry calcite veins).

Lower layer: thickness circa 80 m; coarse fragments (more than 50 mm) of brick together with a lime mortar.

Calvatone (ancient *Bedriacum*). "*Domus dei Signini*", 2nd century CE

Total thickness: undetected.

Upper layer: thickness circa 10 mm; mortar made of lime with crushed brick (until 10 mm) and crushed limestone, quartz sand and calcite crystals. Cubic tesserae (side 10 mm) of black or white limestone.

Lower layer: thickness more than 60 mm; coarse fragments (until 30 mm) of brick together with a lime mortar.

Stone

> To fresh rubble should be added a third of its weight in pounded potsherds; and then the rubble, mixed with two-fifths of its weight in lime, should be rammed down to a thickness of one foot. After this, a final coat 4½ inches thick must be added to the rubble and large square stones not less than 1½ inches thick laid on it.
>
> [Pliny 36, 189]

This kind of pavement is called *Opus sectile*, and it means precisely cut stone slabs with a variety of shapes, mainly arranged in geometric patterns and laying on mortar layers. *Opus spicatum* is called a pavement with a walking surface made of brick, mainly disposed in herringbone style.

To describe the Roman *opus sectile*, Vasari used the term "commesso" (past participle of the Italian verb "commettere", meaning "put together two or more things"): "stone pavements diversified with various blending of porphyry, serpentine and granite. . . . They cut the marble into little pieces. . . . As we still see in the Baths of Caracalla in Rome" [Vasari, 1907; ch. 6].

This kind of pavement was probably very diffused in Lombardy as witnesses by thousands of fragments of slabs of coloured marbles unearthed in different excavations around the Lombard territory: these fragments were found as rubble, but for a great part they pertained to pavements no longer conserved.

Few examples of Lombardy, based on scientific analyses, are enough to understand the sequence of layers. Starting from the top: stone slab (limestone, marble); a layer for bedding made of lime and lacking in aggregate; one intermediate layer made of lime, sand and crushed brick and including some stone slabs or bricks used to give a better support to the coloured marbles; sometimes another intermediate layer with the same composition; a lower layer of coarse pebbles joined by a mortar.

Brescia. Roman Basilica (piazza Labus), second half of the 1st century CE.
Total thickness: undetected.
White marble slabs (different provenances), thickness 35–70 mm.
Bedding layer: lacking.
Intermediate layer: lime mortar with crushed brick (size up to 20 mm).

Desenzano. Roman villa (Borgo Regio), sector B, room 35, second half of the 4th century CE.
Total thickness: about 300 mm, excluding the paving slabs.
Stone slabs (*in situ*): rhombus (black, white, red – thickness 35–60 mm).
Bedding layer: thickness 20 mm
Lime mortar without aggregate.
Intermediate layer: layer C_1 – thickness 90 mm.
Lime mortar with sand and crushed brick.
Intermediate layer: layer C – thickness 90 mm.
Lime mortar with gravel (0.5–60 mm).
Lower layer: layer D – thickness 110 mm.
Rounded stones (about 120 mm long) together with fine-grained mortar filling the voids.

Mosaic

"Mosaics (*Lithostrota*) came into use as early as Sulla's regime. At all events, there exists even to-day one made of very small cubes which he installed in the Temple of Fortune at Palestrina" [Plinius, 1962; book 36, 189]. The Greek term *Lithostrotos* (stone laid) or *opus tessellatum* means cubic pieces of stone (*tesseræ*) of different size and arranged to make geometric or figured patterns, supported by superimposed layers of mortar.

Vitruvius explained with many details the preparation of a mosaic, and the indications are comparable to the evidences coming from archaeological excavations.

> Next a layer of stones is to be spread, each of which is not less than a handful. After spreading the stones, the rubble, if it is fresh, is to be mixed, three parts to one of lime; if it is of old materials, five parts of rubble are to be mixed with two of lime. Let it then be laid on, and rammed down with repeated blows by gangs of men using wooden stamps. When the stamping is finished, it must be not less than nine inches thick. Upon this, a hard coat of powdered pottery is to be laid, three parts to one of lime, forming a layer of six inches. On the finishing coat, a pavement of marble slabs or of mosaic is to be laid to rule and level. When it is laid, and the proper fall is adjusted, it is to be rubbed down. . . . After the rubbing down, when they are completely smoothed and finished, marble dust is sprinkled over, and over that coats of lime and sand are to be applied.
>
> [Vitruvius, 1931; book 7, ch. 1].

It is worth it to note the discrepancy between the composition of Roman mosaics in Milan and in Lombardy. Some examples of Lombardy, based on scientific analyses, are useful to understand the sequence of layers. Starting from the bottom: a layer of coarse pebbles joined by a mortar; two intermediate layers of mortar made of lime, sand and crushed brick; a thin layer of mortar, made of lime and lacking in aggregate, where the tesserae are lying; a layer of stone tesserae (mainly cubic shape, side circa 1.0 cm) together with a mortar made of lime and fine sand filling the interstices among the tesserae.

The layer where the tesserae are lying, made of lime without aggregate, is totally neglected by Latin authors. This layer was recently reported in different mosaics of the Roman world (Moore, 1968; Dunbabin, 1999).

ROMAN MOSAICS IN LOMBARDY

Two superimposed mosaics, one lying upon another, located in the area of the Imperial Palace (Milan, via Gorani), were analysed on the basis of optical petrography.

- Upper mosaic (*MIGO10* US *1247*), 4th century CE.
  Total thickness: about 180 mm. From the top: (A) cubic tesserae made of black or white limestone (side 20 mm); (B) a setting bed (5 mm thick) made of lime mortar without aggregate; (C) a mortar bed (*Nucleus*, 40 mm thick) with brick fragments (10–20 mm) in a lime binder with coarse-grained crushed brick aggregate; (D) a mortar bed (*Rudus*, 50 mm thick) with small pebbles (10–20 mm long) in a lime binder with coarse-grained quartz crystals and gneiss clasts aggregate; (E) a bed (*Statumen*) made of one row (50 mm thick) of pebbles (80–100 mm long).
- Lower mosaic (*MIGO10* US *1246*), 1st century CE.
  Total thickness: about 100 mm. From the top: (A) cubic tesserae made of black or white limestone (side 10 mm); (B) a setting layer (5 mm thick) made of lime mortar without aggregate; (C) a mortar layer (*Nucleus*, 30 mm thick) made of lime binder with a coarse-grained crushed brick aggregate; (D) a mortar layer (*Rudus?*, 30 mm thick) with stone pebbles and brick fragments (30–40 mm long) in a lime binder with coarse-grained crushed brick; (E) a bed (*Statumen*) made of one row (40 mm thick) of pebbles (50–80 mm long).

[Bugini et al., 2015]

The use to superimposing mosaics on each other, is also witnessed in the Roman villa of Desenzano del Garda: five different mosaic floors from the late 1st century BCE to the early 1st century CE to the second half of the 4th century CE (room 54, sector C) [Bugini et al., 2012].

The different sizes of stone tesserae allow for some considerations about the raw material needed to make a mosaic, ignoring the raw materials needed to make the mortar layers (lime, sand, pebble, etc.).

Using cubic tesserae with 10 mm sides, as in the via Gorani lower mosaic (1st century), a square metre is covered by 10,000 tesserae, ignoring the thickness of the mortar joints and the lack of regular shape of each tessera. Using cubic tesserae with 20 mm sides, as in the via Gorani upper mosaic (4th century), a square metre is covered by 2,500 tesserae, still ignoring the mortar joints and the shape of tesserae.

One cubic metre of stone, ignoring the loss of material caused by cut, supplies 1 million cubic tesserae with 10 mm side and allows to make 100 sq. m of mosaic. One cubic metre of stone, still ignoring the loss of material caused by cut, supplies 125,000 cubic tesserae with a side of 20 mm and allows one to make 50 sq. m of mosaic. Considering the specific weight of the limestones employed in these mosaics (about 2.7 g/cm^3), a 10 mm per side cubic tessera (1 cm^3) weight 2.7 g; a 20 mm per side cubic tessera (8 cm^3) weight 21.6 g. So a square metre of mosaic will weigh respectively 27,000 g or 27 kg and 54,000 g or 54 kg.

Finally, using 20 mm side tesserae instead 10 mm tesserae, a double amount of stone will be necessary to cover the same surface and the mosaic will have a double weight, still ignoring the weight of the mortars used for the foundation beds.

Stones of tesserae of Roman mosaics

Different archaeological sites of Lombardy were considered in order to identify the stones used for making tesserae of mosaic [Bugini, Folli, 2009, 2013].

Brescia (Brixia)
Late republican Sanctuary near the Capitolium (1st century BCE)
Domus (vicolo Settentrionale, 1st century CE)
Domus (Collegio Arici, via Trieste 17, 1st century CE)
Domus (western courtyard of Santa Giulia, 2nd century CE)
Domus (Ortaglia di Santa Giulia, via Musei 81, 1st century CE)
Domus (Palazzo Martinengo, via Musei 32, 1st century CE)
Calvatone (Bedriacum): *domus* (1st and 2nd century CE)
Desenzano: *villa* (2nd and 4th century CE)
Garlate: *villa* (Santo Stefano, 1st century CE)
Milan (Mediolanum): *domus* via Amedei 2 (2nd century CE)
Sirmione (Sirmio): *villa* "Grotte di Catullo" (1st and 2nd century CE)
Toscolano: *villa* (2nd century CE)

White colour
Ooids limestone (San Vigilio formation – lower Jurassic. Valpolicella, north of Verona)
Biancone limestone (Biancone formation – lower Cretaceous. Valpolicella)
Botticino dolomite (Corna formation – lower Lias. Botticino, north-east of Brescia)
White marbles (coarse-grained, average size 2.8 mm)
Black colour
Black mudstone (Triassic formations of Lombard Prealps)
Red colour
Scaglia (Scaglia formation – upper Cretaceous. Valpolicella)
Nodular limestone (Rosso Ammonitico formation, upper/middle Jurassic. Valpolicella)
Yellow colour
Nodular limestone (Rosso Ammonitico formation, upper/middle Jurassic. Valpolicella)

The quarry site of stone material is sometimes very far from the mosaic: i.e. ooids limestone used in Milan (via Amedei, 2nd century CE) comes from a quarry on the eastern shore of Lake Garda (distance about 140 km).

PAVEMENTS OF THE MIDDLE AGES IN MILAN AND LOMBARDY

One kind of pavement (stone) was used in the Middle Ages, while other kinds of pavement (concrete, mosaic) seem to be abandoned. In particular, pavements, made of white or black stone tiles, are spread in religious buildings of the present-day territory of Lombardy. On the contrary, the Lombard edifices are lacking in stone pavements, called "Cosmati work", made of small pieces of ancient coloured marble; these smooth-running geometric patterns (used also on altars, tombs and pulpits) featured the floors of the early Christian churches (4th–7th centuries) [Claussen, 1998; Guidobaldi Guiglia, 2001]. This kind of stone floor was widely spread in central and southern Italy and the making techniques were handed down from father to son, remaining unchanged for centuries [White, 1993].

The features of pavements made of black tiles, together with white ones, were investigated in some floors of Milan and Lombardy: for each site were reported the size (centimetre) and the shape of the tiles. The provenance of black tile refers to Triassic limestones of the Lombard Prealps, the provenance of the white one refers to white marble from Musso or from the Apuanian Alps or to Mesozoic limestones of Lombard Prealps [Bugini, Folli, 2008].

Early Christian buildings

St Giovanni Baptistry Milan: octagon room, on-site (4th century); rhomb (diagonal 40–20 or 10–5); hexagon (side 15 or 12); square (side 25 or 18)

St Giovanni Baptistry Novara: octagon room, on-site (4th century); triangle

St Tecla church Milan: nave, on-site (4th century); hexagon (side 12 or 15); triangle (12, 15 or 7.5)

St Nazaro church Milan: right transept chapel, on-site (4th century); hexagon (side 15); equilateral triangle (side 15); right-angled triangle (side 20–14–14; square (side 14); rectangle (sides 4–5 to 20–26)

St Giovanni in Conca Milan: crypt, museum (5th–6th centuries); hexagon

St Pietro in Atrio church Como: nave, on-site (5th century); hexagon (side 14)

St Vincenzo church Cantù – Galliano (Co): presbytery, on-site (5th century, relocated 11th century); hexagon (side 15)

St Giovanni Baptistry Riva S. Vitale (Ticino – CH): octagon room, on-site (5th century); hexagon (side 9 or 15); triangle (side 9 or 15); rhomb (diagonal 21)

Medieval buildings

St Giovanni Baptistry Castelseprio, octagon room: on-site (7th century); hexagon, square, triangle, rectangle (side 10 or 15)

St Giovanni Baptistry Lomello, octagon room: on-site (7th century); hexagon (side 12); rhomb (diagonals 19–12)

Oratory Garlate buried below St Stefano church: remains (7th century); hexagon (side 10 cm, thickness: 0.9–4.5)

St Maria Foris portas ch. Castelseprio, nave: on-site (9th century); hexagon (side 10, thickness: 0.9–4.5)

Patterns

Early Christian pavements still *in situ* involve a great number of patterns combining black and white tiles together with different shapes and sizes. The co-existence of different patterns in the same edifice is usual.

MODERN PAVEMENTS

A quotation of Palladio is useful to introduce the differences based on the materials employed: "The superficies of the floors may be made out of mortar, . . . or of square tyles, or of hard stone". These words of Palladio are useful to distinguish three different kinds of paving, according to the materials employed for the surface:

- the first, called "terrazzo" in Venice, is made of lime, crushed brick, coarse sand and stone chips ["Those floors of mortar or plaster, are very good when made out of beaten cement and fine sand, or with river-plebbles"];
- the second one is made of earthenware tiles ["the many forms into which they be made and of the colours they are capable to receive by the different colours of the earth they are made with"];
- the third one is made of stone slabs suitable to polishing, such as marble or limestone ["in winter they would be too cold; but in galleries or other public places, they will agree well enough"].

[Palladio, 1715; book 1, ch. 22]

These kinds of surfaces are well represented in Milan, but it is impossible to make a list of buildings or churches describing their pavements. Moreover, as already reported, the pavements are frequently subjected to partial or total, and they are often disregarded by the description of a building.

"Terrazzo" alla veneziana

This pavement consists of chips of stones of different colour set in a coloured mortar and perfectly polished. The support was not investigated by scientific purposes, but it is probable the presence of different layers made of mortar with an aggregate decreasing in grain size towards the surface.

Example in Milan. Villa Belgioioso – Reale (via Palestro): the pavement of the colonnaded "Sala da Pranzo" (first decade of the 19th century; now Room XVII of the Galleria d'Arte Moderna), show white marble and limestones of various colours (red, pink, black) in a green mortar ground. Palazzo Litta (corso Magenta): different floors show mosaics made of white, grey and black tesserae.

Earthenware

The lack of stone material in the territory of Milan and the contemporaneous abundance of clay material allowed the use of terracotta to make tiles for pavement. Typical of Lombard furnaces was the production of rectangular or square tiles with a particular marbling effect involving red, orange and white colours.

Example: Santa Maria della Sanità (via Senato, C.F. Pietrasanta, 1708): square tiles shiwing reddish or yellowish colours.

Stone

A precise history of the development of the Milan's pavements is very difficult to draw: this kind of architectural element is frequently subject to a partial or total replacement caused by the surface wear or by the breaking, but also the adjustment to new national rules (i.e. heating system in a church). The pavements were also threatened by the mutation of taste, trend or style during the centuries.

Renaissance pavements still showed geometrical-shaped tiles (triangle, rhombus, square, rectangle, trapezium), but the patterns were more complicated, involving three colours (black, white, red) and, in some cases (concentric pattern), the single tiles may show curved boundaries. A peculiar difference distinguishes the Renaissance pavements from the medieval ones: large quadrangular slabs of white stone (Candoglia marble) forming the ground where black or red elements are inlaid.

Original examples are still visible in some churches of Milan, such as Duomo, Santa Maria dei Miracoli or San Giuseppe.

Duomo – Northern Sacristy (architect Marco da Carona, 1404–07): the side parts show a Roman pattern with three coloured rhombi made of Candoglia (white), black Varenna limestone (black), Rosso Arzo (red); the central part shows a series of white squared tiles containing a black six-point star around a hexagonal figure made of white, black and red triangles; each stellar tile is inserted in a white and black square grid (the stone materials are the same of the lateral part). Other kinds of stone were introduced by restoration works of the late 19th century (i.e. Pietra d'Oira instead Nero di Varenna).

Duomo – Naves, transept, deambulatory: "Il pavimento è tutto lastricato di marmi uniti a vari colori con arabeschi, ma non è ancora tutto compito". According to the words of Serviliano Latuada [Latuada, 1738], the pavement was partially achieved between 1585 and 1820, but only in 1940 was it completed. The pavement shows a complicated pattern arranged in quincunx and consisting of lobate black figures on white ground surrounded by red semi-circular figures in alternating lobate black figures. The original pattern was planned by Pellegrino Tibaldi (1527–96) using three different materials: Candoglia (white), black Varenna limestone (black), Rosso Arzo (red). Continuous maintenance works, due to the high rate of wear of white Candoglia marble, compelled to substitute the great part of the pavement, however, preserving the original pattern; in some cases Rosso di Verona was employed instead of Rosso Arzo. Other different patterns, always based on three colours, are present in some areas around the main altar.

Santa Maria dei Miracoli (architect M. Bassi, 1599): the pavement varies according to the different parts of the church; it is based on geometric figures, but more elaborated patterns with curvilinear elements based on floral figures are also present. The pattern featuring the central nave contains rows and lines in alternating round and octagonal figures made of black (black Varenna limestone) and red (Rosso Arzo) motifs on a white ground (Candoglia); the same pattern is present on the ceiling of the barrel vault over the nave (coffers made of moulded mortar). The pattern featuring the side naves is based on the area of each bay with quincunx of round or octagonal black figures (black Varenna limestone) on a white ground (Candoglia) and a few little red discs (Rosso Arzo); the space corresponding to the pillars is occupied by a red-and-black fret on a white ground. A wide stripe separates the pavement of the nave and the pavement under the dome and before the main altar: the floral motif is mainly made of black elements on a white ground.

The central part under the dome features a round pattern with sixteen sectors radiating from a tri-coloured stellar centre; each sector contains five circular white-and-black figures decreasing in size towards the centre; so the pattern shows five concentric circumferences with sixteen round elements each. In this case also the pattern of the pavement is reproduced on the ceiling of the dome (coffers made of moulded mortar).

More complicated patterns, with curvilinear elements, feature the pavement of the deambulatory of the apse, consisting of five squared bays separated by four triangular intervals. Squared bays and triangular intervals show black geometrical patterns on a white ground. The pattern on the church central axis contains four symmetrical figures made of curvilinear red and black elements on a white ground. Finally, floral patterns accompanied by frets, mainly made of red stone are present in correspondence with the pillars of the apse. It is worth it to note, in this deambulatory, the use of Rosso Verona instead Rosso Arzo.

San Giuseppe (1607–30): the pavement shows five patterns according to the different parts of this centrally planned church. The main hall under the dome is occupied by a concentric pattern made of white (Candoglia) and black (Black Varenna limestone) trapezia with eight red stripes (Rosso Arzo) radiating from an octagonal centre; the four parts to be left in the angles show a pattern of white and black triangles. The passage between the main hall and the presbytery shows the Roman pattern with white, black and red rhombi; the presbytery shows a pattern of rows of white and black trapezia in alternating with lines of red rhombi. Finally, the area between the balustrade and the main altar shows a pattern of rows with alternating white and black trapezia.

19TH–20TH-CENTURY PAVEMENTS

Some significant examples, both in apartment and in public buildings, allow to describe the pavements as an important factor in the architecture of the period 1920–40.

The majority of pavements remained invisible, as they are almost always included in private properties.

Together with the use of stone or marble slabs (opus sectile), it is worth it to note the use of mosaic made of stone tesserae (called Palladiana); the size of tesserae (2–4 cm) is bigger than the size of Roman tesserae (1–2 cm) and the modern ones often show an irregular shape. The most employed stones are: Botticino, black limestone, Rosso Verona, Levanto red.

Stone

Some examples of use of stone slabs are:

Cappella Espiatoria (Monza, G. Sacconi and G. Cirilli, 1901–10): the pavement of the chapel and the pavement of the crypt were made by O. Fedeli (Rome) introducing in the Milan area the *opus sectile* tradition, held by the craftsmen of Rome (called Marmorari). The marbles of the chapel are "ancient coloured marbles", such as Africano, Giallo antico, red Egyptian porphyry (Porfido rosso antico), green Greek porphyry (Porfido serpentino verde), Proconnesian, Cipollino, etc.; the marble setting follows the pattern of early Christian churches (interlaced circles).

Banca Commerciale: the floor of the main hall of this former bank (piazza della Scala, L. Beltrami, 1907) shows a curvilinear pattern made of red (Rosso Verona), white (Botticino) and grey (Bardiglio) slabs.

Monumento ai Caduti: this huge monument (piazza Sant'Ambrogio, Muzio et alii, 1927–30) was severely damaged by the aerial bombings of 1943. The pavement shows different patterns based on the use of coloured stones coming from Lombard quarries (Crevola, Zandobbio, etc.).

Stazione Centrale: different pavement according to the different parts of the building. Stones employed are Chiampo, Verde Roja, Occhiadino, black limestone, Abbazia, Rosso Collemandina, Serpentinite, Rosso Verona and yellow, Balma, Repen, Aurisina, Rosso Porfirico, Rosso Levanto, Griotte, Perlato Sicilia.

Villa Necchi: a privately owned urban villa (via Mozart 14, architect P. Portaluppi, 1932–35) with several stone materials employed for flooring different rooms (dining room, bathroom, etc). The list includes limestones (Arabescato Orobico, Grigio Carnico, Botticino, Breccia Aurora); travertine; alabaster; Verde Roja and Verde Patrizia (ophicalcite).

Casa Wassermann: a privately owned urban villa with different stone materials employed for flooring (viale Lombardia 17, architect P. Portaluppi, 1934–38). The list includes limestones (Nero assoluto, Portasanta, Rosso Amiata), marbles (white from the Apuanian Alps, Calacata avorio vena oro, Candoglia/Ornavasso, Fior di Pesco Carnico, Paonazzetto della Rocchetta, Valle Strona); ophicalcites (Rosso di Levanto, Verde Alpi Cesana, Verde Challand, Verde Issorie); Serpentino classico and Verde Roja.

Stone pieces

This kind of floor includes small slabs (few centimetres long) with a quadrangular shape.

Galleria Vittorio Emanuele (G. Mengoni, 1865–77, renovated in 1967 and restored many times): this wide pavement combines areas of stone pieces (Giallo Mori) together with *opus sectile* (Bardiglio; Rosso Verona) and mosaic (tesserae 2 cm) using white, black or red limestones.

former Casa dei Fasci (via Nirone 15, P. Mezzanotte, 1926–27): the pavement of the first floor is made of white or grey marbles together with coloured marbles (Arabescato Orobico, Rosso Levanto, Grande Antico d'Italia, Rosso Verona, Serpentino, etc.); the large joints among the slabs were filled by a grey mortar.

Mosaic

The use of mosaic made of stone tesserae (cubic shape, about 2 cm side) was spread in the first half of the 20th century: hundreds of Milan's apartment buildings show this kind of floor, starting from the ground spaces with direct access from the exterior (entries, vestibules, reception halls, etc.) or spaces linked to the staircase (stair halls, stair platforms, landings). The colours vary from black to white with different shades of grey, from red to pink, yellow and green.

The patterns are various, and they are inspired by geometric motifs or by animals, flowers or human figures. Mainly monochromatic sedimentary rocks (limestone, dolomite) were employed to make the mosaic, together with marbles and ophicalcites. A list of the stones includes Rosso Verona, Rosso Levanto, Verde Roja, Giallo Mori, Rosa del Garda and white or grey (Bardiglio) marbles from the Apuanian Alps. Finally a coloured mortar fills the gaps among the stone tesserae.

Some interesting examples are:

Ca' Brutta (via Moscova, G. Muzio, 1919–22): black, white and grey tesserae set in geometrical patterns.
former Casa dei Fasci (via Nirone, P. Mezzanotte, 1926–27): the mosaic occupies the hall of the ground floor; it is made of cubic tesserae, mainly grey and white marble and black limestone, featuring a geometrical pattern together with military symbols.
Casa Borletti (via San Vittore, G. Ponti; 1927): geometrical lines of black tesserae in a ground made of white tesserae.
Stazione Centrale: mosaics are made of Chiampo, Rosso Verona, Bardiglio, white marble, black limestone.
Palazzo della Triennale (viale Alemagna, G. Muzio, 1932–33): the main hall contains a mosaic of triangular patterns; equilateral triangles made of tesserae of different colours (Rosso Levanto, Rosso Verona, Giallo Verona, Rosa del Garda, black limestone, Bardiglio, serpentine).
Palazzo Civita (piazza Duse; G. Zanini, 1933–34): black, grey and white tesserae, made of white and grey marble and black limestone, arranged in geometrical or figurative (fish) patterns.

A wide survey of the entries of apartment buildings and its stone materials in the period 1920–70, is reported in "Entryways of Milan. Ingressi di Milano" [Kolbitz, Signori, 2016].

STAIRCASES AND BALCONIES

The traditional system to make staircases and balconies of civil buildings until the first decade of the 20th century involved the use of thin and wide slabs of Beola: 3–5 cm thick and up to 4 m long. The employ of this stone also matched its high resistance against deterioration caused by constant wear.

Beola slabs were sustained by thin corbels made of metal or by thick corbels made of Baveno or Montorfano granite; the railings were always made of metal.

Staircases of noble palaces involved, on the contrary, coloured marbles (stones from Arzo) or granites (Baveno or Montorfano) for stairs; coloured marbles for railings. Balconies of noble palaces involved granites (Baveno) for platforms and corbels; limestone (Viggiù) or conglomerates (Ceppo Brembo) for railings.

PARQUETRY

This kind of pavement lies outside of stone materials, but it is very important for the floors of palaces, villas and also apartment buildings. The motifs go from simple arrangements of wood pieces (herringbone style) to complicated patterns involving geometrical or curvilinear figures with wood pieces of different of colours or grains.

A sort of mosaic composed of relatively thick wood (7/8 inch); used for pavements.
Outstanding examples are present in royal villas of Milan and Monza, built in the late 18th century.

- Villa Reale (Milan). The wooden floors are located at the first floor as the ballroom on the centre.
- Villa Reale (Monza): different rooms with wooden pavement are locate on both the first and second floors.

TARSIA

The wood inlay called "tarsia" consists of precisely cut pieces of coloured wood fit together in the same manner with incrustations of coloured marbles ("opus sectile" or "commesso alla fiorentina"). This important use of wood is mainly devoted to the decoration of the furniture, but it is worth it to note the case of the decoration of the stalls arranged in the church choirs to sit down monks or priests. An outstanding example is the choir of Santa Maria dei Miracoli with twenty-two stalls: eighteen are arranged as three sides of a hexagon (six per side), two bigger stalls are arranged at the beginning and at the end of the choir and two other stalls are arranged on the angles. The choir was planned by the architect Galeazzo Alessi (born in Perugia, 1512–72) and it was sculpted and assembled by the wood carvers Paolo Gaza or Paolo de' Gazi (from 1567 to 1573) and by Giovanni Taurini, after the death of Gaza (from 1593 to 1613).

The stalls show a decorated architecture made of caryatids, entablatures, pediments, etc. preceded by a sculpted balustrade: the central part of each stall is occupied by a panel where the use of different kinds of wood allows for the creation of architectural views in perspective with buildings, towers, obelisks, columns, porticoes, loggias, arches, bridges, staircases, etc. Vasari reported this technique as made of: "little, pieces of wood, variously coloured, fitted and joined together in panels of walnut". The author mentioned also the work of Fra Giovanni da Verona (about 1457–1525): "giving various colours to the woods by means of dyes in boiling water and of penetrating oils, in order to produce the lights and shadows with these variously tinted woods, as in the art of painting". Wooden works of Giovanni da Verona are present in the churches of Sant'Anna dei Lombardi (Naples, 1506–10, partially re-assembled in the Sacrestia Vecchia) and Santa Maria in Organo (Verona, 1491–99, choir in the apse and sacresty). Finally, Vasari came to end with a severe opinion: "[Inlay in wood] is a work that soon becomes black . . . and is also of short duration because of worms and fire, it is considered time thrown away in vain to practise it" [Vasari, 1907; ch. 31]. On the contrary, the choir of the Milan's cathedral shows sculpted panels instead of inlays: seventy-one scenes of the life of Sant'Ambrogio and other bishops of Milan. This great wooden choir, with a double order of stalls and a highly decorated structure, was promoted by cardinal Carlo Borromeo, planned by Pellegrini and then by Aurelio Luini and Camillo Procaccini, and executed from 1567 to 1614 by different wood carvers as Paolo Gaza, Giovanni Taurini with Giacomo and Giampaolo Taurini, Virgilio de'Conti (1571–1615).

ROAD PAVING

A particular use of stone was the paving of the roads or rather the structure assembled to give to the ground the capability to bear the movement of vehicles and people and to drain rapidly the waters. The stones were employed as an upper layer of this structure.

Road paving in the Roman period

Road paving was already important during the Roman period: the consular roads (*viae militares*) were perfectly paved with stones to allow the rapid deployment of the legions of the army and the transport of goods. The stones, in the form of thick slabs of polygonal shape, both quadrangular and pentagonal, were set as the upper layer of a composite stratification of different kinds of materials.

A detailed description about the Roman roads and their planning and structure, was reported by Plutarch, the Greek biographer noted for his "Parallel lives" of distinguished Greeks and Romans personages:

> roads were carried straight through the country without deviation, and had pavements of quarried stone, and substructures of tight-rammed masses of sand. Depressions were filled up, all intersecting torrents or ravines were bridged over, and both sides of the roads were of equal and corresponding height, so that work had everywhere an even and beautiful appearance.
>
> [Plutarch, 1921; chap. 7, 1]

On the contrary, the most quoted authors on architecture, such as Vitruvius and Pliny, lack precise descriptions about road construction and materials. The original paving of "via Appia" (linking Rome towards the south-east to Brindisi) was represented by Giovan Battista Piranesi, the famous engraver, architect and archaeologist (1720–78). The etching "View of the via Appia Antica" shows a portion of the paving made of polygonal stone elements with oblique sides (like an inverse frustum of a pyramid) together with the underneath well-rammed layer made of lime, pozzolana and splinters of stone [Piranesi, 1756].

Jean Rondelet in the early 19th century, enhanced the qualities of the Roman road and, according to N. Bergier [Bergier, 1736] tried to list four motives to explain these qualities: to employ of the troops during the periods of peace, to accelerate the movement of the correspondence in the whole territory, to make easy the military expeditions, to make comfortable the commerce and voyages. Roman roads were divided into eight classes: *via* (wide road); *actus* (narrow road); *iter* (pedestrians); *semita*, *trames*, *diverticulum* or *divortium* (different kinds of path); and *callis* (mountain path). After a description of the features of Roman roads, the author considered a comparison between Roman and modern roads or the solidity versus the width [Rondelet, 1802; book 2, sec. 3, art. 11].

The stone used to pave the roads, in Rome and in central Italy, was a volcanic rock, called "*silex*" (meaning "hard stone") or "selce" in Italian. This is a leucitite (leucite, plagioclase, pyroxene) coming from the volcanic complex of Colli Albani (Pliocene-Pleistocene), outcropping about 20 km south-east of Rome.

Other kinds of stone (marble, limestone, etc.) were used to pave some particular kind of urban streets or squares without vehicular traffic (i.e. the slabs of limestone of the Forum at Pompei or the slabs of white marble for the court and the slabs of Pavonazzetto and Giallo antico for the side porticoes at Trajan's Forum).

Road paving in Italy

Four modern architects, Alberti, Palladio, Scamozzi and Milizia, wrote about the roads, taking the roads built by the Romans as a model; the paved roads were unexpected for a traveller such as Michel de Montaigne in Italy in the last decades of the 16th century "A quinze milles nous découvrîmes la ville de Rome. . . . Nous rencontrâmes aucunes contrées de chemins relevés et pavés d'un fort grand pavé, qui semblait à voir quelque chose d'ancien" [30 November 1580; Montaigne, 1983].

Leon Batttista Alberti, in the late 15th century, paid attention to the layout and the width of the roads both inside or outside the towns:

> the ways out of the city should be short, strait and secure . . .; when they come to the town, if the city is noble and powerful the street should be strait and broad . . .; but if it is only a small town, it will be better . . . to have the streets wind about sometimes to the right, sometimes to the left.
>
> [Alberti, 1988; book 4, ch. 5].

The paving of a road is similar to the paving of a house floor: a thick layer of gravel on a well-hardened ground, a layer of sand (fluvial or marine), a layer of stone as surface [Alberti, 1988; book 3, ch. 16]. The Romans accurately chose the stone to avoid wear and slip, and the best one was the already quoted volcanic rock called "selce"; each piece of stone was set with a smoothed surface and perfectly joined with another ones; the preferred size was about 0.5 m (width) and 0.30 m (thickness).

Andrea Palladio in the second half of the 16th century wrote about ways and streets within the city and ways or roads without the city.

> The ways ought to be short, convenient, secure, pleasant and beautiful. (. . .) The ways will be further convenient, if they are made every where equal; that is to say, that there be no places in them where armies may not easily march and that there be no difficulty of passage, either from waters or rivers.
>
> [Palladio, 1715; book 3, ch. 1]

Ahead "the middle of the street will be left for carriages and beasts, and may be paved with flint or any other hard stone" [Palladio, 1715; book 3, ch. 2]. Further ahead "It was paved with uncertain stones, that is, such as had unequal sides and angles" [Palladio, 1715; book 3, ch. 3].

Scamozzi, in the early 17th century, after a description of different kinds of Roman roads (width, drainage of rainwater, etc.) and a quotation of Latin authors, reported the succession of layers required to pave a road: a level of gravel and coarse-grained sand very well hardened before the laying of bricks, suitable for pedestrian use, or of hard stones (volcanic or clastic rocks) suitable for carriages and wagons. The accurate laying of these stones makes strong and long lasting the paving and makes more constant the journey of the vehicles [Scamozzi, 1615; part 2, book 8, ch. 26].

Milizia, in the late 18th century, reported five adjectives featuring the Roman roads: firm, useful, comfortable, attractive and happy. After that, the architect gave a more detailed description of the different materials and work involved in the road construction. Firstly, he paid attention to the different materials according to the raw materials available in each region. In some cases the road was simply cut into the rock, in other cases layers of gravel, sand, mortar, brick, pebble and squared stone were gathered. The typical stratification involved, like the house pavement described by Vitruvius in the seventh book: three superimposed layers (*statumen*, *ruderatio* and *nucleus*) as support, containing lime mortar with aggregates of different composition (stone, brick) and size; a fourth layer was made of a whitish and hard cement; a fifth one was made of strongly hardened pebbles or of thick slabs of stone. These stone were irregularly shaped, as the *opus incertum*, but perfectly joined together; in the presence of rectangular stones, their alignment was always oblique in respect to the parallel sides, to avoid a probable disconnection caused by load of the wheels. The author described also the raised shoulders and ditches for water drainage on each side. Finally, the importance of a constant maintenance was suggested, because the surveillance together with low expenses allowed them to conserve the works; on the contrary negligence produced disadvantages and waste of money [Milizia, 1785; part 2, book 3, ch. 10, §8].

Road paving in Milan

During the Roman period the stones for paving in Milan and northern Italy [Previato, 2015] came from the Euganean Hills, a volcanic Tertiary complex located a dozen kilometres south-west of Padua (region of Venetia). In this area outcrops include a variety of volcanic rocks, such as trachyte, basalt, ryolithe and tuff; the most important are the grey trachytes quarried at Monte Lonzina, Montemerlo, Teolo, Zovon, etc. These alkali-feldspar rocks were employed in thick slab with polygonal shape (average size 50 cm long, 40 cm wide, 25 cm thick – as detected in Milan, via Spadari 1, 1887). The application of trachyte pieces for paving was evinced by the high resistance of this silicate stone to the wearing caused by constant use. Some fragments of Roman paving are still visible (allochthonous location) in the Archaelogical Museum or in the underground station of Missori (Metropolitana linea 3, piazza Missori): the presence of deep ruts caused by the carriage wheels is remarkable. Some findings during archeological excavation allow us to know the width of the streets of the town: from 5 m (via Dante, 1889) to 4.5 m (via Piatti, 1962; piazza del Duomo, 1983) with a convex surface.

Mediolanum was a cluster of routes, mainly in the late Imperial period: *via Aemilia* to Rome through *Ariminum* (Rimini); *via Postumia* to the eastern region of Venetia (*Aquilieia*); different transalpine routes to the north (Spluga pass) or to the north-west (Piccolo and Gran San Bernardo passes).

Modern road paving in Milan

A general road paving in Milan is a matter of the late 18th century, starting from the city centre around the cathedral. In the previous centuries, the town was furrowed by streets paved with rammed-earth, but some examples of stone paving were drawn from the treatises of architecture. Scamozzi reported the use of the "Serizzo", also called "Selce", for street paving together with elevated sideways, designated to the pedestrians and paved with bricks set in herringbone style; Serizzo is a gneiss normally used for buildings and coming from Lake Maggiore (val d'Ossola, Piedmont), [Scamozzi, 1615; part 2, book 7, ch. 8]. Latuada, in the first decades of the 18th century, noted the stone paving of piazza Duomo: the pavement is made of "pietra viva" (hard stone as granite) [Latuada, 1738; vol. 1, #1].

The most important methods of paving involved pebbles or thick slabs.

Pebble paving

The *Vocabolario milanese-italiano* published by G. Banfi (Libreria Ubicini, Milano 1857) reports some terms about the road paving, involving river pebbles.

Rizzàa, rizzadura: Acciottolato, ciottolato, selciato: il pavimento delle strade, stalle, dei cortili e simili.
Rizzadin, rizzin: Selciatore, ciottolaio.
Rizzoeu: Ciottolo, ghiaiotto, ciotto: sasso di fiume bislungo e rotondo da ciotolar (rizzà) le vie.

Pebbles were used first as coming from the loose sediments of the rivers flowing from the Alps. Pebbles are mainly made of metamorphic rocks together with granites, quartz and, to a lesser extent, carbonate rocks. Pebbles of uniform size (about 10 cm long) and oblong shape were fitted together on a sandy substrate; two parallel carriage ways were placed in the centre of the street in order to make smoother the roll of the wheels; two raised sidewalks were also added in favour of pedestrians. These sort of wheel tracks were made from rectangular thick slabs of Baveno granite, but the wheels rapidly dug deep ruts on this stone. Baveno granite was quickly replaced by San Fedelino granite, coming from the upper part of Lake Como (val Chiavenna) and showing a greater resistance to wear. The use as a building stone of San Fedelino (a grey, fine-grained granite) was restrained because of its uniform and dull colour in comparison with the intense and vivid colours of Baveno or Montorfano granites.

New examples of this kind of paving were applied in some areas almost free of traffic (i.e. piazza Borromeo), but the pebbles were laid with a mortar based on Portland cement.

Paving in thick slabs

In the early 20th century the increase of traffic on the urban streets required continuous maintenance of the pebbles and stoneways. A paving completely made of stones was achieved, taking into account some important features: easier washing and sweeping away and reduced amount of mud. The question of the mud was already made clear by Scamozzi reporting the use of spreading some sand above the stone paving and the consequent production of mud [Scamozzi, 1615; book 8, ch. 26].

The shape of each stone is prismatic, with one size more developed than the other two; to avoid the aforementioned alignment of different rows of stones, the rectangular ones were set together with polygonal ones (five sides), thus creating an oblique disposition in respect of the sides of the street. The size of the stone elements was about 0.4–1 m long, 0.25 m wide and 0.15 m thick.

The stones were chosen among the granite of San Fedelino, Sienite di Balma and Granofiro di Cuasso. In many cases it is possible to see a mix of these stones in the same area, probably caused by later maintenance works (substitution of broken pieces). A revival of granite paving was developed in the 21st century, in pedestrian areas without cars.

This kind of paving involved also a convexity of the profile with a maximum in the centre and a depression at the two sides and also the presence of two walkways for pedestrians slightly raised in respect of the road surface. The rim of the walkway was marked by prismatic stone elements, also made of San Fedelino granite; in some cases a sandstone was preferred, coming from the east coast of Liguria (Riviera di Levante).

The first stones are broadly described in this text, but the last one was used only for the purpose of paving.

A Ligurian sandstone, called "Arenaria di Lèvanto", is reported by Marmi Italiani: "Arenarie. In diversi luoghi della Liguria orientale (Riccò del Golfo, Biassa, Monterosso, ecc, in provincia Spezia) . . . si estraggono delle ottime arenarie cementate da silice, molto adatte alla lastricatura perché di difficile levigatura ostacolando lo slittamento" [Marmi Italiani, 1939]. This grey, fine-grained sandstone pertains to the "Formazione del Macigno" (mid to upper Oligocene) made of beds of sandstone (a few centimetres to 5 m thick) in alternating with thin beds of clay. The size of the sandstone elements was about 0.40 by 0.2 m and 0.2 m thick. An example of this kind of stone paving is present in Piazza dei Mercanti.

Since the last decades of the 20th century other kinds of stone were employed for paving, mainly in the refurbishment of streets and squares or in the new stations of the underground railways. Example: via Gorani (Ghiandone Sardo); Stations of Underground line 3 (Aurisina – Roman stone); Station Buonarroti Underground line 1 (Bargiolina, a Permian yellow to grey quartzite, coming from Monte Bracco, south of Barge, province of Cuneo, Piedmont).

Cubic stone paving

The most important stones for road paving are the porphyries (Permian) from Trentino and Alto Adige/Süd Tirol, featuring homogenous texture and regular fractures. Their use was developed in the 1930s, employing cubic elements (side about 0.08–0.1 m) set without mortar upon a bed of loose sand on the basis of a fan-shaped setup (about 1 m wide).

The process of preparation of porphyry cubes involves the use of a simple press device: despite its hardness, the porphyry is easily broken under pressure and a series of small cubes is obtained, starting from a big block. The joints among the cubes were traditionally filled with sand, but in more recent times, the joints among the cubes were filled with bitumen or mortar.

Cubic elements of granophyre (Permian hypabyssal red rock from Cuasso al Monte, province of Varese), were used in new setting up of central street devoted to pedestrian use (via Marino).

Other systems of road paving or surfacing are lacking dimension stone. The asphalt paving, coming down from the "Macadam" system (early 19th century), involves compressed layers of small broken stones bound together with asphalt. The concrete paving is made of Portland cement, coarse aggregate, sand and water. In Milan it was employed in the 1930s, and it is still in use in some urban streets (via Domodossola). The preparation involved multiple wide pieces of concrete (about 10 m long) separated by expansion joints. It is worth it to note the use of Granofiro di Cuasso (red granophyre) as an aggregate of the concrete (size of the fragments about 10–15 cm).

REFERENCES

Alberti L.B., De re aedificatoria – On the Art of Building in Ten Books (J. Rykwert, N. Leach, R. Tavernor eds.), MIT Press, Cambridge, 1988.

Aston B.G., Harrel J.A., Shaw I., Stone, *in* Ancient Egyptian Materials and Technology (P.T. Nicholson and I. Shaw eds.), Cambridge University Press, Cambridge, 2000.

Bergier N., Histoire des grands chemins de l'Empire romain, Leonard, Bruxelles, 1736.

Bugini R., Folli L., I litotipi usati nei pavimenti a mosaico e cementizi della Lombardia romana, *in* Atti XVIII Convegno AISCOM, Roma, 2013, 123–132.

Bugini R., Folli L., On Tesserae of Roman Mosaics in Lombardy, *in* The Study of Marble and Other Stones in Antiquity – Proceedings of ASMOSIA 7 (Y. Maniatis ed.), BCH Suppl., 51, 2009, 547–557.

Bugini R., Folli L., Pavimenti in opus sectile a Milano e in Lombardia: l'uso dei calcari neri, *in* Atti del V Congresso Nazionale di Archeometria, Siracusa, 2008, 265–278.

Bugini R., Folli L., Ceresa A., Features of two Roman mosaic floors from Milano, *in* Proceedings 12th AIEMA Colloquium, Scripta, Paris, 2015, 489–492.

Bugini R., Folli L., Portulano B., Roffia E, The Analytical Approach to the Roman Mosaics: A Case Study in Northern Italy, *in* Proceedings 11th AIEMA Colloquium, Zero Books, Istanbul, 2012, 163–170.

Claussen P.C., Marmi antichi nel Medioevo romano – L'arte dei Cosmati, *in* Marmi antichi (G. Borghini ed.), De Luca, Roma, 1998, 65–80.

Dunbabin K.M.D., Mosaics of the Greek and Roman World, Cambridge University Press, Cambridge, 1999.

Guidobaldi Guiglia A., Pavimenti marmorei a Roma e nel suburbio nei secoli IV-VII, *in* Materiali e tecniche dell'edilizia Paleocristiana a Roma (M. Cecchelli ed.), De Luca, Roma, 2001.

Kolbitz K., Signori G., Entryways of Milan, Ingressi di Milano, Taschen, 2016.

Latuada S., Descrizione di Milano, Cairoli, Milano, 1738.

Marmi Italiani (G. Peverelli, F. Squarzina eds.), Confederazione Fascista degli Industriali, Roma, 1939.

Milizia F., Principj di architettura civile, Remondini, Venezia, 1785.

Montaigne M., Journal de Voyage (F. Garavini ed.), Gallimard, Paris, 1983.

Moore R.E.M., A newly observed stratum in Roman floor mosaics, *in* American Journal of Archaeology, 72, 1968, 57–68.

Palladio A., I quattro libri dell'architettura: The Architecture of A. Palladio in Four Books (N. DuBois ed.), Watts, London, 1715.

Piranesi G.B., The Antiquities of Rome – Tomo Terzo, Rome, 1756.

Plinius, Natural History (D.E. Eichholz ed.), Loeb Classical Library, Harward University Press, Cambridge, 1962.

Plutarch, The Parallel Lives: The Life of Caius Gracchus (B. Perrin ed.), Loeb Classical Library, Harward University Press, Cambridge, 1921.

Previato C., Aquileia – Materiali, forme e sistemi costruttivi, Padova University Press, Padova, 2015.

Preziosi D., Hitchcock L.A., Aegean Art and Architecture, Oxford University Press, Oxford, 1999.

Rondelet J., Traité théorique et pratique de l'art de bâtir, Paris, 1802.

Scamozzi V., L'idea dell'architettura universale, Venezia, 1615.

Vasari G., Le vite de' più eccellenti Pittori, Scultori e Architettori – Introduzione (L. S. Maclehose ed., The Lives of the Most Excellent Painters, Sculptors and Architects – On Technique), J.M. Dent, London, 1907 (1st ed. 1568).

Vitruvius, De Architectura (F. Granger ed., On architecture), Harward University Press, Cambridge, 1931.

White J., Art and Architecture in Italy 1250–1400, Yale University Press, London, 1993.

Part VI

Special artefacts

Chapter 25

Church altars

INTRODUCTION

The altars of the Catholic churches became matter, especially from the 16th to early 19th centuries, of skilfully worked structures, more and more elaborated and complicated with columns, entablatures, pediments, statues and decorative and ornamental elements.

To understand this flourishing of ostentatious displays it is necessary to report some historical notes on the last decades of the 16th century and the confrontation between the Roman Catholic Church and the Reformation.

The Council of the Counter-Reformation was held in Trent in different sessions: 1545–47, 1551–52, 1562–63. Carlo Borromeo (1538–84), from 1560 onward archbishop of Milan and secretary of state of Pope Pius IV, promoted the re-assembling of the Council; the twenty-fifth and last session was held 3 and 4 December 1563 (Sarpi, 1619). The final decrees of the Council confirmed the Catholic response to the Reformation and made important changes in the liturgy and church building: the Counter-Reformers imposed the cross as a church plan and made the ceremonies more and more splendid as a reaction against the Protestant choice of unadorned ceremonies (Blunt, 1962). After the Council of Trent (1545–63) and the statement of the Counter-Reformation, new rules to build the churches and the altars were printed in 1577 by the Lombard cardinal Carlo Borromeo. These "Instructions" contain dispositions, even down to minute details, about the dimensions and proportions of churches, chapels, choirs, baptisteries, altars, tabernacles, staircases, etc. and simple rules about the materials, often prescribing stone or marble. Some examples are: the high chapel with "two marble or stone steps" (ch. 11), the tabernacle "made of gilded laminae or of precious marble" (ch. 13); the altars "made of stone or brick" (ch. 15); the canopy "supported by four marble or stone columns" (ch. 15); the baptistery font "made of a monolithic block of marble or stone material or of two marble or stone slabs" (ch. 25); the baptistery ciborium "made of marble or solid stone" (ch. 25) [Borromeo, 1577]. Actually, the architects and the craftsmen were self-ruling to plan the patterns in order to create very stunning and impactful decorations and ornamentations. The stone materials are represented by white marbles or black limestones and by coloured marbles. The first two were employed to form the framework; coloured marbles (moulded pieces, slabs or inlays) were added to improve the ornamentation. The provenance of these materials was disparate: from nearby territory (i.e. black limestone from Lake Como or Bergamo Prealps), from the Italian country (i.e. white marble from the Apuanian Alps) or from abroad (i.e. red limestone from Languedoc). The accuracy of the whole artefact is witnessed by the precision of the cuts and the perfection of the setting of each piece of stone or marble. In particular, curved surfaces were achieved using a single

DOI: 10.1201/9781003361008-31

piece with curvy cut or using different small pieces put together on a curved background; some slabs of veined materials (alabaster) were fabricated putting together irregular pieces, paying attention to join the pieces along the undulating veins; the combination of four slabs, carefully cut from the same block of veined marble, allowed to create unique and unrepeatable patterns (the "open-book style" or "macchia aperta"). Semi-precious gems, metal, alloy, wood, glass, plaster, terracotta and also textile were added to the marble in order to reach the apex of magnificence and virtuosity.

Together with sculpturing and carving marbles and stones, the "inlay" technique was also often employed. This technique consists of thin pieces of marble specifically cut and fit into carved-out recesses on other stone; each piece (about 1 cm thick) was normally fixed to the background by an adhesive made of colophony or turpentine or beeswax, etc. mixed with quartz sand or crushed marble; finally, the joints (often curvilinear) among the pieces were almost invisible thanks to the accuracy of the cut of each stone piece.

The inlay technique was called "*commesso alla fiorentina*" because of the high degree of skill reached by "Opificio delle Pietre Dure", a marble workshop established in Florence in 1588. The term "*commesso*" (Italian verb "*commettere*" meaning "put together two or more things") was employed by Vasari to describe the "opus sectile" featuring Roman pavements: "stone pavements diversified with various blending of porphyry, serpentine and granite. . . . They cut the marble into little pieces. . . . As we still see in the Baths of Caracalla in Rome" [Vasari, 1907; On technique, ch. 6]. The Florentine "Opificio" produced, for more than three centuries, thousands of "*commessi*", mainly for furniture (tabletops with geometric to figurate scenes, consoles, cabinets, boxes, etc.) or for religious items (reliquaries, tabernacles, etc.), together with the magnificent veneers and decorations of the former Tribuna degli Uffizi (since 1584) and of the Medicis' tombs (Cappella dei Prìncipi, San Lorenzo, Florence, since 1602). Different materials were used, materials coming from all around the known world: coloured ancient marbles (Egyptian red porphyry, alabasters, black limestone, etc.), jasper, mother of pearl, amethyst and precious gems.

The inlay technique, together with carving and sculpture, was applied in the construction of altars: altar pieces, altar frontals, balustrades (handrail and baluster), stairs, floors, etc. This is the case of Lombard skilled craftsmen, mostly unknown, operating for more than two centuries and makers of the one thousand altars located in Lombard churches.

Parallel to the "*commesso*" technique ran the "*scagliola*" technique, mainly used to make altar frontals: coloured stone pieces are replaced by pulverized and pigmented selenite (calcium sulphate) applied in appropriate recesses carved in a wet gypsum ground (normally black) and then set under heat and polished. This technique was brought to perfection and spread throughout Europe by the craftsmen of Valle d'Intelvi (province of Como), and it was precisely described by Nicola Cavalieri San Bertolo as an imitation of true natural marbles using a particular kind of plaster where the first coat is made of gypsum and glue and the second one is thinner and made of very fine-grained gypsum (scagliola or Gesso Speculare) and glued together with red and yellow ochres or other different colours. The dry surface of this plaster may be polished using sand, pumice, water or oil; in this way the plaster takes the aspect of a marble [Cavalieri, 1831; book 3, ch. 9, §699]. A masterful combination of colours allows a perfect imitation of veins and patches of natural stones (see Chapter 21, Volume 2).

Many surveys were made in order to list the stone materials used in some Milanese altars of which the name of the craftsmen and the construction date is known, because it was impossible to consider, one after the other, the altars present in all the churches.

MILAN – STONE MATERIALS OF THE CHURCH ALTARS (17TH–20TH CENTURIES)

The construction of an altar mainly involved coloured stones (marbles, white too; veined limestones, breccias) and other stone materials (plaster, stucco) often mixed with wood, metal, ceramic, glass, etc. Apart from the use of monolithic pieces (shaft), slabs (cladding), moulded or carved elements (cornice or bas-relief), it is worth it to note some artifices used by craftsmen in order to obtain exceptional achievements: shaft of coloured marble made using several thin and narrow slabs, of different shape, assembled together to improve the colour variegation of the stone or according to the spots and the veins; slabs of coloured marble made of several small slabs assembled together to create a more attractive spot or vein.

The framework of each altar was made of white marble, generally coming from the Apuanian Alps (Tuscany), or of black limestone coming from Triassic formations of the Lombard Prealps (i.e. Calcari di Perledo e Varenna on Lake Como; Argilliti di Riva di Solto on Lake Iseo).

The inlay includes different kinds of coloured Lombard stones (*Arabescato orobico, Grande antico d'Italia, Occhiadino, Alabaster*) and stones from abroad, such as *Rosso di Francia* and *Broccatello di Spagna.*

Balustrades, stairs and floors before and around the altars, were mainly made of coloured stones from Arzo (Rosso, Broccatello and Macchiavecchia).

Local materials (Lombardy)

*Stones from the Lombard Prealps: these stones match the features of ancient coloured marbles. These stones are also separately described (see Chapters 4–10, Volume 2).

- Arabescato orobico: streaked red and white or grey and white limestone, from Val Brembana (province of Bergamo); used for column (joining of small rectangular slabs), frieze of the rail. Matching: Fior di Pesco.
- Occhiadino: grey stromatolithic limestone with undulated white veins and spots, from Val Camonica (province of Brescia); used for baluster, cornice, monolithic column. Matching: Bigio antico.
- Calcare nero (black limestone): totally black fine-grained limestone from Varenna (Lake Como, province of Lecco) or from Cene and Gazzaniga (Val Seriana); used for the framework, column (a variety containing narrow white veins called "Grande antico d'Italia"), baluster. Matching: Bianco e nero antico.
- Alabastro: brownish calcite alabaster (travertine) from Induno Olona (province of Varese) or from Val Seriana (province of Bergamo); used for panel. Matching: Egyptian and Turkish alabasters.

*Stones from Arzo (Canton Ticino, Switzerland).

- Rosso: red limestone with occasional white veins or spots; used for floor slab, stair, baluster, rail.
- Broccatello: fossiliferous purple to red limestone; used for stair, baluster, etc.
- Macchiavecchia: multi-coloured breccia with irregular red, yellow, brown or grey patches; used for stair, baluster, rail, monolithic column, panel. Matching: Portasanta.

Materials from the Italian territory

- White marble: totally white or grey or veined marble from the Apuanian Alps (provinces of Carrara and Massa, Tuscany); used for framework, moulded elements, ornaments, statues.
- Rosso di Verona: red nodular limestone with undulated bedding and macrofossils from monti Lessini (province of Verona, Venetia); used for pavement.
- Giallo di Verona: yellow nodular limestone with undulated bedding and macrofossils from monti Lessini (province of Verona, Venetia); used for cornice. Matching: Giallo antico.
- Giallo di Siena: yellow marble with dark undulated veins from Montagnola Senese (province of Siena, Tuscany); used for panels. Matching: Giallo antico.
- Verde Alpi: green ophicalcite with white patches and veins from Piedmont or Valle d'Aosta; used for panels, monolithic columns.

Materials from abroad

Only two marbles were traded to Italy and used in the decoration of altars: Rosso di Francia and Broccatello di Spagna.

Each marble is already described in this text (see Chapters 4–18, Volume 2 and Chapter 20).

REFERENCES

Blunt A., Artistic Theory in Italy 1460–1600, Oxford University Press, Oxford, 1962.

Borromeo C., Institutionum fabricae et suppellectilis ecclesiasticae – Instructions for Ecclesiastical Buildings and Furnishings, Milano, 1577.

Cavalieri San Bertolo N., Istituzioni di architettura statica e idraulica, Negretti, Mantova, 1831.

Sarpi P. (P.S. Polano), Istoria del Concilio Tridentino, London, 1619.

Vasari G., The Lives of the Most Excellent Painters, Sculptors and Architects (L.S. Maclehose ed.), Dent and Company, London, 1907 (1st ed. 1568).

DESCRIPTION OF SOME ALTARS IN THE CHURCHES OF MILAN

Only a few well-documented altars are described in order to appreciate the use of each stone for different purposes and the skill of craftsmen using coloured stones to create impressive structures. It is worth it to note the lack of documents about the construction of altars and about the stones employed. In many cases the architect was also the craftsman and the entrepreneur of the construction, so the formal agreements report only the request of the use of "*marmi mischi*", meaning marbles with coloured veins and spots. The choice of different stone materials was left to the architect-craftsman-entrepreneur.

In particular, Lombard craftsmen of the 17th century were known for their special ability and expertise to perform the cutting and carving of coloured stones (marble, limestone, breccia, etc.) and of semi-precious ones (jasper, lapislazuli, rock crystal, etc.). The most significant examples of this craftsmanship are the altars of Santa Maria della Passione together with the Pulpit and the Confessionals of Sant'Alessandro.

Another example in the Lombard territory is present in the church of Certosa di Pavia: the altar frontal of several chapels shows excellent inlays of coloured marbles made by different craftsmen since the second half of the 17th century. In particular, the chapel of Santi Pietro e Paolo (sixth of the right aisle) shows an altar frontal made by Carlo Battista Sacchi (1688): the inlays reproduce a porch with twisted columns and arches together with flower decorations using white marble (ground), alabaster (column shafts), Giallo Antico (flowers), etc.

Finally, it is almost impossible to enumerate the countless elements that make up an altar, and it is very difficult to exactly identify the stones due to the height of the artefacts, the scarcity of light and the viewing distance. Moreover, the nomenclature of such complicated artefacts is very uncertain: only the most significant elements are described here.

Duomo

The Counter-Reformation and their architectural rules were quickly applied in new constructions, but also involved the ancient ones and the Duomo, the Milan cathedral, represented a test for these rules. Cardinal Borromeo promoted a drastic change in the cathedral, which at the time was still under construction and co-existed with some parts of the earlier cathedral. In particular, the ancient tombs of the dukes and the emblems of the noble families were completely dismantled and new altars were built along the aisles. The architect Pellegrino Tibaldi, called Pellegrini (1527–96), after some years in Rome and Bologna, was nominated (1567) architect of the cathedral and became the executer of Borromeo's ideas. His first work was the Baptistery, erected in the middle of the nave (then moved to the left aisle – 1685), involving four high pedestals made of Candoglia marble with column shafts made of Macchiavecchia; the shafts are connected by wide entablatures with pediments, and the entablatures are made by jack arches hidden with Macchiavecchia slabs; the agreement for the marble supply is dated 12 August 1567 [Buratti, 1990]. The font consists of a thermal basin made of Egyptian red porphyry (Porfido rosso antico), coming from Roman baths then reused in the former church of San Dionigi. Tibaldi also planned a circular crypt with the "Confession" altar separated from the ambulatory by a circle of eight columns, made of Macchiavecchia and ordered 19 May 1569 [Buratti, 1990]. In 1585 Tibaldi left the cathedral project when called to the Escorial by Philip II, king of Spain. The last work he planned for the cathedral was the floor, carried out in the 17th century, using black limestone from Varenna (Lake Como) and red limestone from Arzo in a ground made of Candoglia marble. It is worth it to note that black limestone matches *Bianco e nero antico – Grand antique* (from Pyrénées) and Macchiavecchia matches *Portasanta* (from Chios).

The new altars, three on the right aisle and three on the left, were planned by Pellegrino Tibaldi (circa 1567) following the new rules of the Counter-Reformation.

Reference – Latuada: "Gli ornamenti di tutte le Cappelle sono di marmi scolpiti con sue colonne, architravi, cornici, volute e statue al naturale, disegno di due peritissimi architetti, cioè Pellegrino Pellegrini e Martino Basso ed hanno buona parte di quadri o sculture bellissime" [Latuada, 1738; #2].

Bianconi: "Gli altari marmorei, quasi tutti ordinati dal Santo Pastore, sono disegnati da Pellegrino Pellegrini . . ., dal Cerani e da Martino Bassi, il quale di poi fu fatto architetto anch'esso di questo mirabile edificio" [Bianconi, 1787; 34].

*First pair of altars – sixth span (right: Sant'Agata; left: Crocefisso di San Carlo), showing a complicated structure:

- basement, stylobate, entablature, broken pediment: Candoglia.

*Second pair – seventh span (right: Sacro Cuore; left: San Giuseppe), showing another complicated structure:

- basement, stylobate, herms, entablature, broken pediment: Candoglia.
- rectangular slabs on stylobate, shafts, pilasters, frieze: black Prealpine limestone with white veins
- rectangular slabs on stylobate, shafts, pilasters, frieze: white veined black limestone.

*Third pair – eighth span (right: Virgo Potens; left: Sant'Ambrogio), showing a simple structure:

- pedestals, herms, entablature, curved pediments: Candoglia.
- shafts, pilasters for each side, frieze of the entablature: Macchiavecchia.

The contract for the first marble order from Arzo had already been signed on 17 December 1571; the marbles for the second altar (opposite in the same span) were ordered 9 October 1577 [Rocco, 1939].

The altars were built by Martino Bassi following Tibaldi's plans, after Tibaldi left Milan for Madrid (1585). The delay among the planning, the purchase of marbles and the construction of altars was probably due to the lasting disputes between Tibaldi and the Board of the cathedral, called *Fabbrica del Duomo.*

Two altars, located in the right transept of the cathedral and linked to the wishes of Pope Pius IV (Giovanni Angelo de' Medici, maternal uncle of the cardinal Carlo Borromeo.) are among the rare examples of the use of ancient marbles in the Mannerist period. The description of the first altar, built by Leone Leoni (1560–63), is reported by Vasari:

> The same Leone has made . . . the tomb of Signor Giovanni Jacopo Medici, Marquis de Marignano and brother of Pope Pius IV, which stands in the Duomo of Milan, about twenty-eight palms in length and forty in height. This tomb is all of Carrara marble, and adorned with four columns, two of them black and white [Bianco e nero antico], which were sent by the Pope as rare things from Rome to Milan, and two others, larger, which are of spotted stone similar to jasper [Macchiavecchia]; which are all accommodated under one and the same cornice, an unusual contrivance, by the desire of that Pope, who caused the whole work to be executed after the directions of Michelangelo, excepting only the five figures of bronze that are there, which are by the hand of Leone.
>
> [Vasari, 1912; Life of Leone Lioni].

Later, Latuada described the same altar:

> Entro a forti cancelli di ferro si ritrova la Cappella, chiamata di Gioanni Giacopo de' Medici zio materno di San Carlo e fratello del sommo Pontefice Pio IV. . . . Entro di esse v'è il maestoso suo Deposito lavorato di bianco marmo con la di lui statua al naturale in abito militare col bastone di comando in mano, quattro figure, ed altri bassi rilievi gettati in bronzo da Leone Aretino, avendone fatto il disegno, per quanto scrive il Vasari, Michelangelo Buonarroti
>
> [Latuada, 1738; Book 1, #2]

A slightly different description is reported by Bianconi:

> la cappella di Gian Giacomo Medici, zio di San Carlo, fratello di Pio IV e condottiero famoso d'armi. Il sepolcro fattogli erigere dal Pontefice fu disegnato da Michelangelo Buonarotti. Leone Leoni aretino vi fece in bronzo le statue ed i bassi rilievi; e le sei colonne di fini marmi orientali, che l'adornano, furono mandate dall'istesso Sommo Pastore unitamente ai marmi del ricco altare.
>
> [Bianconi, 1787; 35]

The second altar was the altar where Cardinal Borromeo worshipped; it was built in 1565 using "Portasanta" (altar-frontal), "Africano" (veneering), "Verde antico" (veneering, small columns) and "Giallo antico" (small columns) supplied from Rome. This altar is not mentioned by ancient authors.

Santa Maria della Passione (via Conservatorio)

HIGH ALTAR (circa 1620)
Architect: unknown
Reference – Latuada:

> Il grande Altare posto alla testa della Croce . . . vi si ascende per pochi gradini, entrando la balaustrata di marmi fini a macchiati, che lo chiude. . . . sostiene un Ciborio di squisita bellezza, e di gran prezzo, fabbricato in maniera di Tempio, tutto incastrato di Lapislazzuli, Diaspri, Agate, Corniole ed altre nobili pietre con ornamenti di metallo dorato.
>
> [Latuada, 1738; book 1, #29]

Bianconi: "Può osservarsi il bel ciborio ed altare di pietre dure, e metallo dorato" [Bianconi, 1787, 85].
*Balustrade: Rosso Arzo (stair); Rosso Verona (baluster); white veined black limestone (base and handrail).
*Floor: alabaster, Arabescato orobico, Portoro.
*Altar: Macchiavecchia (stair); black limestone (side volute), Macchiavecchia, Arabescato (inlay on side volute).
*Frontal: black limestone (structure), semiprecious stones (panel).
*Ciborium: black limestone (pedestal); wood (structure); silver (column); alabaster, lapislazuli, semiprecious stones (small panel); metal (cornice); mother of pearl (decoration of dome).
*Square-octagonal panels, on side piers (in front of the altar): Macchiavecchia and black limestone (concentric cornice); alabaster (centre).
EIGHT PANELS (Passion of Christ), one for each pier of the dome.

- structure: Boden (moulding); black limestone (inscription); alabaster, semiprecious stones (inlay on structure).

Madonna del Carmine (piazza del Carmine)

"Cappella della Madonna del Carmine" (right of the presbytery).
Architect: Girolamo Quadrio (1676) then partially transformed (1730).
Reference – Latuada:

> A mano dritta di chi entra, dopo la maggiore, si trova la Cappella di Maria Vergine titolare del sacro Scapolare, rappresentata in statua di bianco marmo, e riposta entro gran nicchia pure di marmo nell'anno 1676, intagliata dal celebre scultore Giovambattista Volpi. La stessa Cappella, che forma come un piccolo Oratorio, ella è in ogni parte adorna di marmi e pietre macchiate con stucchi dorati.
>
> [Latuada, 1738; Book 5, #198].

Bianconi: "La Cappella della Madonna del Carmine, con tanta copia d'ornati marmorei e stucchi così pesanti, ha la statua della Vergine ed Angioli in marmo del nostro Volpi" [Bianconi, 1787, 352].

The chapel is preceded by a vestibule, the altar area is surrounded by decorated walls, the altar includes a statue of the Virgin with two angels.

Vestibule

Staircase: Rosso Arzo (stair)
*Balustrade: black limestone (baseboard and rail), Rosso Arzo (baluster); Occhiadino, Verona yellow, Arabescato orobico, Breccia Seravezza (inlay on balusters)
*Floor: irregular slabs of Rosso Arzo, Rosso Verona, Botticino
*Walls: black limestone (pilaster and column with base and capital made of white marble, cornice of panel and cornice of painting); Macchiavecchia (panel); Rosso Arzo (background); white marble (lintel and cornice of entablature), Calcare Nero (frieze of entablature)
*Impost of the vault: black limestone (ground); Macchiavecchia (panel)
*Ceiling: stucco and painting

Around the altar

*Stairs: white veined black limestone
*Balustrade: black limestone (baseboard, baluster, rail); Rosso Arzo, Occhiadino (inlay on baluster)
*Floor: terracotta (squared tile), Rosso Arzo, black limestone (triangular tile)
*Walls: black limestone (pilaster and column with base and capital made of white marble, cornice of panel and painting); Macchiavecchia (panel, cornice of niche with statue, base of statue); Occhiadino (door jamb); Rosso Arzo (background); white marble (lintel and cornice of entablature); black limestone (frieze of entablature); white marble (statue of niches and conch close to the baseboard)
*Band above the entablature: Rosso Arzo (upper and lower part, panel), black limestone (cornice); plaster (coating of dome)
*Vault ceiling: stucco, paintings

Altar

*Pavement and stairs: Rosso Arzo
*Frontal: black limestone (baseboard); Macchiavecchia and Breccia Seravezza (background), Occhiadino (outer emblem); green ophicalcite (inner emblem); Rosso Verona (lower medallion); white marble (cherub statue); Macchiavecchia (background of sides)

*Altarpiece: black limestone (cornice); white marble (inner cornice), Macchiavecchia (panel)
*Tabernacle: metal, alabaster (small column); black limestone (cornice)
*Ciborium: Macchiavecchia (pilaster, honed); Occhiadino (column, twisted); white marble (base, capital, entablature); black limestone (inscription on entablature); white marble (statue)

Santa Maria dei Miracoli (corso Italia)

High altar
Architects: C. Garavaglia (17th century), L. Canonica (early 19th century)
Reference – Latuada:

> L'altare è dei più insigni della città non solamente per la ricchezza dei Sacri Arredi, ma ancora per essere egli ai giorni nostri (e ciò avvenne nel 1731) rifatto dal piano del pavimento sino ai gradini dei Candelieri, tutto di pietre preziose innestate, e farà perfetta mostra di tanti sui tesori, quando sarà finito ancora il Tabernacolo.
> [Latuada,1738; book 3, #85]

Bianconi: "L'Altar maggiore è copiosamente ornato di pietre dure incastrate in marmo con bronzi" [Bianconi, 1787, 143].
*Stairs: Rosso Arzo
*Balustrade: white Apuanian marble (base, little piers, handrail); Rosso Arzo (baluster); Pavonazzetto (inlay on baluster); Rosso Arzo and Varenna (twist decoration on little piers and handrail)

- Stairs: black limestone (structure); jasper, lapislazuli and other semiprecious stones (inlay)

*Altar: black limestone, Macchiavecchia (structure, lower part); Occhiadino (structure, upper part); alabaster, jasper, lapislazzuli and other semiprecious stones (inlay)
*Ciborium (architect L. Canonica)

- Base: Bardiglio (structure), semiprecious stones (inlay)
- Columns: veined white Apuanian marble; black limestone (cabled flute); metal (base, capital)
- Entablature: Bardiglio
- Dome: Bardiglio (structure), semiprecious stones (inlay)

Altare della Madonna (left of High Altar)

Architect: M. Bassi (late 16th century). Sculptor: A. Fontana (1586)
Reference – Latuada:

> L'altare poi è formato in ordine Corintio, con pilastri di fino marmo, e colonne coperte di argento massiccio, che hanno le basi, capitelli, fregi e rose di bronzo dorato. Nel mezzo di questi ornamenti . . . la statua di Maria Vergine Assunta . . . intagliata in marmo di Carrara . . . da Annibale Fontana. . . . Poggia questa statua sopra piedistallo di bronzo, che nel mezzo racchiude un pezzo quadrato di Diaspro Orientale.
> [Latuada, 1738; Book 3, #85].

Bianconi: "ne si pensò, come dopo si volle, a costruire quello [altare] per la miracolosa immagine nel sito presente, così incomodo per le funzioni e tanto nemico della necessaria euritmia" [Bianconi, 1787, 137].

"Nell'altare poi della Madonna che ha colonne d'argento oltre la bellissima statua della medesima, opera eccellente d'Annibale Fontana; evvi dello stesso bravissimo Autore una pietà in oro posta nella ricca base della marmorea statua" [Bianconi, 1787, 141].

- balustrade: white Apuanian marble (base, pier, handrail); Rosso Arzo (baluster); Macchiavecchia, Varenna (inlay on the base); Rosso Arzo and Varenna (geometric decoration on little piers and handrail)
- altar: Varenna (pedestal); metal (column base, shaft and capital); Rosso Verona (base of pedestal); Boden (entablature, pediment)
- statues: white Apuanian marble
- panels (on base and pedestal): Rosso Arzo (slab)

Sant'Alessandro (piazza Sant'Alessandro)

Pulpit

Architect: C. Garavaglia? (circa 1640)

Reference – Latuada: "Delle stesse pietre [Diaspri, Ametisti, e Sardoniche] è ricoperto il gran pulpito . . . con pezzi di rara grandezza" [Latuada, 1738; Book 3, #90].

Bianconi: "Il Marchese Alessandro Modrone . . . diede molta mano alla rara collezione di esse pietre [dure]; porzione delle quali rende osservabile il Pulpito" [Bianconi, 1787; 198].

*Upright structure: Macchiavecchia (base, lower part), Levanto red (base, upper part), *Pulpit: Broccatello Spagna and Levanto green (lower part), Macchiavecchia (upper part, handrail)

*Inlay: semiprecious stones (different jaspers, amethyst, agate, carnelian, chalcedony); metal (structure)

*Ceiling: metal, semiprecious stones, white marble

Confessionals

Reference – Latuada: "ed un confessionale che è situato . . . vicino all'Altar Maggiore nell'ingresso alla piccolo nave" [Latuada, 1738; book 3, #90].

Bianconi: "Il Marchese Alessandro Modrone . . . diede molta mano alla rara collezione di esse pietre [dure]; porzione delle quali rende osservabile . . . un confessio . . . vicino all'Altar Maggiore" [Bianconi, 1787; 198].

*Structure: Varenna (moulded element); Rosso Verona (double entablature)

*Cornices and contours of inlay: Varenna

*Inlay: semiprecious stones (jasper, amethyst, chalcedony, etc.), coloured marbles (Rosso Francia, Broccatello di Spagna, Pietra Paesina, Giallo Siena, alabaster).

High altar

Architect: Giovan Battista Riccardi (1741)

Reference – Latuada:

> Finora il riferito Altar Maggiore è di semplice lavoro d'intaglio in legno colorito, ma in breve sarà rifatto con tale preziosità e maestria, che senza iperbole non avrà forse pari

> in tutta l'Italia. Il disegno di esso fu ritrovamento dell'ingegnoso prontissimo Dipintore ed Architetto . . . Giambattista Riccardi di Milano, stimandosi dai periti la di lui invenzione per la sodezza e bella simmetria delle parti, ma singolarmente per la vaghezza e distribuzione delle pietre preziose, che quello compongono. Formeranno adunque li fianchi, gradini e tabernacolo le Gemme orientali più rare non tanto per la qualità, che è singolare, . . . quanto per la quantità, che ascende a molte migliaia, ed assai più per la strana grandezza di alcune fra di esse, come di Diaspri, Ametisti, e Sardoniche.
> [Latuada, 1738; Book 3, #90].

Bianconi: "l'Altar Maggiore ricco sommamente di pietre dure, stimabilissime per la qualità e grandezza loro, ma che potrebbe aver sortito miglior disegno" [Bianconi, 1787; 198].
*Stairs: Macchiavecchia
*Balustrade: Macchiavecchia (stair, baluster), Varenna (base, little pier, handrail), Arabescato orobico (panel on little pier)
*Altar: metal (structure, ornamentation), lapis lazuli, jasper, amethyst, chalcedony and other semiprecious stones (inlay)

Sant'Angelo (via della Moscova)

High altar
Architect: Giovanni Battista Dominioni, 1708
Reference – Latuada: "la Cappella Maggiore, ornata di un Altare di lisci marmi, sopra cui si venera una bellissima statua di marmo di Carrara, rappresentante l'Immacolata Concezione di Maria Vergine" [Latuada, 1738; book 5, #228].
Bianconi: "L'Altar maggiore resta decoroso per la sua elevatezza . . . ed evvi una ragionevole marmorea statua della Vergine Immacolata" [Bianconi, 1787; 370].
*Staircase: Macchiavecchia
*Floor of the chapel: Varenna, Carrara white, Rosso Arzo (rhombus)
*Floor of the altar: Varenna, Carrara white, Rosso Arzo (rhombus)
*Altar (frontal, table): Rosso Francia; Macchiavecchia (spiral)
*Altarpiece (from the bottom): Rosso Arzo (pedestal), Varenna (baseboard), Rosso Francia, Ophicalcite (cornice), Macchiavecchia (central panel), different marbles (decorated cornice), Varenna (upper cornice); Rosso Francia (under the statue)
*Support of Sculpture: Macchiavecchia
*Sculpture: white marble
Modern altar: Travertino, Macchiavecchia (table)

San Lorenzo

High altar
Architect: C. Garavaglia
Reference – Latuada: "Fu adunque questa immagine [della Madonna] collocata sopra l'Altare maggiore nel mezzo di vaghi ornamenti fatti di marmo e legno, secondo il disegno formato da Carlo Garavaglia ed approvato dall'architetto Francesco Richini" [Latuada, 1738; book 3, #122].
*Stairs: Macchiavecchia
*Balustrade: Macchiavecchia (baluster, handrail, stair)
*Floor: Varenna, veined white Apuanian marble (geometric pattern)

*Altar (stair): Macchiavecchia
*Altar (frontal): Ophicalcite (structure), Arabescato (panel); Occhialino (upper part)
*Altar (centre): Ophicacite (structure), Macchiavecchia (panel)
*Altarpiece: Occhiadino and Macchiavecchia (pier); Rosso Verona and Varenna (rear pier); Ophicalcite (column); metal (base, capital)
*Entablature: Macchiavecchia (structure); Ophicacite (panel)
*Pediment: Occhiadino (structure); Ophicalcite (lower cornice); metal (fronton) Arabescato (rear panel); wood with gilding (statue)

San Giorgio al Palazzo (via Torino)

High altar
Architect: Francesco Croce (1740)
Reference
Latuada: "L'Altare maggiore venne interamente rifatto fino ai gradini pei candelieri di lisci marmi con un bellissimo ciborio, attendendosi in breve di vederlo terminato con corrispondente Tabernacolo, disegnato dal mentovato Architetto Francesco Croce" [Latuada, 1738; book 3, #94].
*floor, staircase: Macchiavecchia
*balustrade: Arzo (baluster), Varenna (rail), Arabescato Orobico (panel)

- altar: Occhiadino, Arabescato Orobico, Varenna
- tabernacle: Varenna, Occhiadino
- canopy: Arabescato Orobico (shaft), Occhiadino

San Sebastiano (via Torino)

High altar
Architect: Carlo G. Merlo (1756)
*Balustrade: Macchiavecchia (staircase), Varenna (handrail), Rosso Arzo (baluster)
*Floor: Macchiavecchia (side), Chiampo and Botticino (lobate slab, centre)
*Altar: Macchiavecchia (staircase), Occhiadino (background), Varenna (cornice), Bardiglio (volute), black Breccia (panel)
*Canopy: Macchiavecchia (ciborium, six shafts)

San Giuseppe (via Verdi)

High altar
Architect: Isidoro M. Piana (1764)
*Balustrade: Rosso Arzo (stair, baluster, handrail)
*Floor: Candoglia and Nero (polygonal tile); Rosso Arzo (side edge)
*Staircase: Macchiavecchia
*Frontal: Verde and Arabescato
Rear: Macchiavecchia (pilaster, lower part); Alabaster (column and pilaster, upper part); Bardiglio (cimation); Carrara (statue); Giallo Siena and Bardiglio (around statue)

Side altar

*Balustrade: Rosso Arzo (stair); Rosso Arzo (handrail); Nero Varenna (baluster)
*Floor and stairs: Rosso Arzo
*Frontal: scagliola

San Marco

Chapel "Madonna della Cintura"
Architect: G. Quadrio (second quarter of the 18th century)
Reference – Latuada: "Nella seconda Cappella dopo l'ingresso del Tempio fu trasferita la Statua di Maria Vergine della Cintura . . .: fu ornato il di lei Altare con marmi lisci di vari colori, e Cupoletta al di sopra, sopra disegno di Giuseppe Quadrio" [Latuada, 1738; book 5, #222].
*Balustrade: black limestone (base, rail); Rosso Arzo (baluster); black limestone (central baluster with Rosso Francia); Macchiavecchia (stair)
*Staircase to altar: Macchiavecchia
*Frontal: black limestone (baseboard); Macchiavecchia (background); Rosso Arzo (curvilinear ornament); white marble (cartiglio); Bardiglio (central cornice); Macchiavecchia (small panel with metallic cross); Macchiavecchia (side element); Ophicalcite (panel on Macchiavecchia)
*Table: black limestone
*Altarpiece: Macchiavecchia (outer pilasters with reed moulding of Verde Ophicalcite), Verde Ophicalcite (inner pilasters with reed moulding of Rosso Francia); gilded stucco (capital); black limestone (lintel, cornice of entablature), Macchiavecchia (frieze of entablature)
*Central image: black limestone (pedestal); Arabescato orobico and Giallo Siena (lobate panel on the pedestal); black limestone (moulded cornice with lobated arch)
*Pediment with statue: wood, stucco, gilding

ALTARS OF THE 20TH CENTURY

The use of coloured marbles was also significant in churches built in the 20th century, despite the changes of liturgy. Some examples are reported.

Santa Croce (via Sidoli)

High altar
Architect: C. Arpesani, 1917
*Ciborium

- columns: Botticino (base), Breccia Medicea (shafts); with marble (capital)
- tabernacle: Macchiavecchia, Botticino (polished block)
- arches, pediment: stucco

Santa Maria Annunciata in Chiesa Rossa (via Montegani)

High altar
Architect: G. Muzio, 1932
*stairs: Zandobbio
*floor: Macchiavecchia, Occhiadino, Botticino, Rosso Verona
Ciborium

- stairs: Bardiglio, Macchiavecchia
- columns: Verde Alpi (pedestal), black limestone (base, capital), Cipollino Apuano (shaft)
- pediment: Galleggione

Altar

- cladding: Rosso Francia, Occhiadino, Arabescato Orobico

Altar left side

Architect: G. Muzio, 1932
*Balustrade: Occhiadino, Bardiglio (baluster)
*Side walls (cladding): Occhiadino, Arabescato (open-book style)
*Altar: Occhiadino, Arabescato

Sacro Cuore (Chapel of Università Cattolica, largo Gemelli 1)

High altar
Architect: G. Muzio, 1931–32

- stairs: Macchiavecchia
- floor: Macchiavecchia, white Apuanian marble, black Prealps limestone

*Altar: Occhiadino and white Apuanian marble; Rosso Francia
*Altar of the Crypt: Galleggione

San Vincenzo in Prato – Baptistery

Altar
Architect: P. Mezzanotte, 1932
*Base: serpentinite
*Column shafts: Portasanta Caldana
*Capital: serpentinite
*Front: Alabastro a Pecorella
*Table: serpentinite
*Font: Breccia corallina (reused column shaft)

REFERENCES

Bianconi C., Nuova Guida di Milano per gli Amanti delle Belle Arti, Sirtori, Milano, 1787.
Buratti Mazzotta A., Introduzione e note, *in* Tibaldi (Pellegrini) P., L'architettura, Il Polifilo, Milano, 1990.
Latuada S., Descrizione di Milano, Cairoli, Milano, 1738.
Rocco G., Pellegrino Tibaldi (Pellegrini) – L'architetto di S. Carlo e le sue opere nel Duomo di Milano, Milano, 1939.
Vasari G., Vita di Leone Lioni aretino, *in* Le vite de' più eccellenti Pittori, Scultori e Architettori (G. du C. De Vere ed., Lives of the Most Eminent Painters, Sculptors & Architects), MacMillan and Co., London, 1912.

Chapter 26

Monuments

Despite the famous lines of Horace, the Roman poet: "I have raised a monument, more durable than bronze" [Horace, 1882; book 3, 30, lines 1–2], in Milan, in Italy and abroad dozens of metallic statues were erected in commemoration of a person or an event. These monuments are subject, unlike the poetry, to: "devouring rain, or fierce northerly gale . . . or the immeasurable succession of years, and the swift passage of time" [Horace, 1882; book 3, 30, lines 3–5]. Really, the use to erect statues or buildings to gods or kings appeared with religious worship and with the authority of the power: i.e. the marble statue of Hermes of Praxiteles, from the Temple of Hera at Olympia (now in the local museum) or the bronze equestrian statue of the emperor Marcus Aurelius on the capitol hill (now in the Museo Capitolino). Later on, monuments were erected to commemorate saints or military leaders or distinguished personages or historical events. Alberti described different thoughts about the erection of statues: "Others especially were for setting up to public view in consecrated places, the effigies of such as had deserved well of mankind . . . believing it must inspire posterity, when they came to worship them, with a love of glory, and an emulation of their virtue" [Alberti, 1988; book 7, ch. 17].

The most important monument ever of Milan was the bronze equestrian statue of Francesco Sforza, duke of Milan, but this monument, planned by Leonardo da Vinci, was left unachieved, and the enormous clay model, put together by the artist (1493), was soon destroyed. This model was admired by Vasari:

> [Leonardo] proposed to the Duke [of Milan] to make a horse in bronze, of a marvellous greatness, in order to place upon it, as a memorial, the image of the Duke. And on so vast a scale did he begin and continue it, that it could never be completed . . . because of being of so great a size, an incredible difficulty was encountered in seeking to cast it in one piece. . . . And, indeed, those who saw the great model that Leonardo made in clay vow that they have never seen a more beautiful thing, or a more superb; and it was preserved until the French came to Milano with King Louis of France, and broke it all to pieces.
>
> [Vasari, 1912; Life of Leonardo da Vinci]

Few ancient monuments have survived over the centuries; in some cases they were knocked down during troubles and disorders: i.e. the statue of Felipe II, king of Spain, located in the central niche of the tower of palazzo dei Giureconsulti (now via Mercanti), destroyed in 1799 (Repubblica Cisalpina, in the mood of the French Revolution) and then

DOI: 10.1201/9781003361008-32

replaced by a statue (made of Candoglia marble) of St Ambrose's blessing, manufactured by L. Scorzini and unveiled in 1833.

The foundation of the Regno d'Italia (1861), the first unitary state encompassing the whole Italian territory, was crucial to start a long period of erection of monuments dedicated to contemporary or ancient Italian personages (soldiers, martyrs, scientists, poets, painters, etc.). Several sculptors and architects were involved in contests and competitions, and the winners were chosen by specific commissions, starting, almost always, long disputes, controversies and quarrels. The second half of the 20th century, after World War II, featured a decrease of monuments dedicated to a single personage and an increase of monuments dedicated to historical events and symbolized by abstract forms. Finally, in some cases the artefact is installed as an embellishment of the urban landscape without celebratory purpose [Pietrantoni, 1997].

The list, in chronological order, includes the monuments now existing in Milan. Most of the monuments include a statue of the honoured personage and a high pedestal; the use of stone material concerns both the statue and the pedestal or the stone is combined with other materials, such as bronze, steel, etc. Together with monuments are listed some fountains and columns including the surviving ones, called "Crocette", once erected by the archbishop Carlo Borromeo.

REFERENCES

Alberti L.B., De re aedificatoria: On the Art of Building in Ten Books (J. Rykwert, N. Leach, R. Tavernor eds.), MIT Press, Cambridge, 1988.

Horace, The Odes and Carmen Saeculare of Horace (J. Conington ed.), Bell and Sons, London, 1882.

Pietrantoni M. (ed.), Memorie nel bronzo e nel marmo, Motta, Milano, 1997.

Vasari G., Vita di Lionardo da Vinci – Pittore e scultore fiorentino, *in* Le vite de' più eccellenti Pittori, Scultori e Architettori (G. du C. De Vere ed., Lives of the Most Eminent Painters, Sculptors & Architects), MacMillan and Co., London, 1912.

MONUMENTS OF THE 16TH CENTURY

Crocetta di San Carlo

Address: piazza De Angeli
Background: one of the votive columns erected by the Cardinal Carlo Borromeo in the late 16th century (1576–77) in memory of the epidemic pestilence
Materials
Column: Baveno (basement, pedestal, base, monolithic shaft, capital, entablature)
Statue: Ceppo del Brembo

Croce di San Calimero

Address: largo della Crocetta, corso di Porta Romana
Background: erected by the Cardinal Carlo Borromeo (1581); transformed in the early 18th century
Materials
Pedestal: Montorfano (lower part), Baveno (upper part)

Shaft (quadrangular): Baveno (moulding)
Statue: Ceppo Brembo

Colonna di Sant'Eufemia

Address: corso Italia, via Sant'Eufemia
Background: erected by the Cardinal Carlo Borromeo (1581); transformed in 1614–16; moved from piazza Bertarelli (1955)
Materials
Pedestal: Baveno (squared block)
Support: Boden (thick slab); marble (decorative element)
Column: Oira (base, capital), Baveno (shaft)
Entablature: Candoglia (lintel), Boden (frieze), Candoglia (cornice)
Statue (Sant'Elena): white marble

MONUMENTS OF THE 17TH CENTURY

Obelisco di San Glicerio

Address: Boschetti, via Marina
Background: made in the early 17th century (1607) in the town quarter called Bottonuto (now via Larga – via Albricci); moved by Piermarini (1787) as part of the new configuration of the area around villa Belgioioso
Materials
Pedestal: Baveno (moulded element)
Spheres: Baveno
Obelisk: Baveno (monolithic element)

San Carlo Borromeo

Address: piazza Borromeo (near church of Santa Maria Podone)
Background: 25 August 1624 (unveiling); 1786 (new location)
Architect: D. Bussola (drawing)
Materials
Statue: bronze
Pedestal: Baveno (lower and upper part), white marble (intermediate part and epigraphs)

Colonna del Leone

Address: piazza San Babila (in front of San Babila)
Background: built in 1626 to mark the border of a city quarter (lion statue, 12th century)
Architect: G. Robecco
Materials
Pedestal: Ceppo del Brembo (squared block); Candoglia (panels)
Cap of pedestal: Angera (moulded element)
Column shaft: Ceppo del Brembo (superimposed drums)
Decorative elements: Molera (moulded element)
Lion figure: Oira

Colonna del Verziere

Address: Verziere, largo Augusto
Background: 1673 (erection), but planned earlier (1580). Transformed as "Colonna della Vittoria" after Italian unification (1860)
Architects: G.B. Lonato (1580), A. Trezzi (1611), G. and G.B. Vismara (1673)
Materials
Base: Montorfano
Pedestal: Viggiù
Column: Baveno (shaft), Oira (capital)
Entablature: Baveno (lintel, cornice); Viggiù (frieze)
Statue (the Redeemer): bronze

Colonna di San Pietro Martire

Address: piazza Sant'Eustorgio (in front of Sant'Eustorgio)
Background: built around 1680
Materials
Pedestal: Montorfano (squared block)
Lower part: Viggiù
Column: Baveno (shaft), Viggiù (base), Angera (capital and entablature)
Statue (San Pietro Martire): white marble

Sant'Arderico Vescovo

Address: piazzetta San Nazaro
Background: unknown
Materials
Circular base (1985): Montorfano
Pedestal: Beola (lower part), Baveno (upper part)
Shaft (quadrangular): Baveno (moulding)
Statue: Ceppo Brembo

MONUMENTS OF THE 18TH CENTURY

San Giovanni Nepomuceno

Address: Castello Sforzesco (main court)
Background: 1727 or 1729
Author: G. Dugnani
Materials
Statue: white marble (Apuanian Alps)
Pedestal: Candoglia
Baseboard: Baveno, Montorfano

Colonna di San Lazzaro

Address: piazza della Vetra (near San Lorenzo)
Background: 1728 (erection)

Materials
Pedestal: Baveno
Support: Ceppo del Brembo (coarse grained)
Statue: Ceppo del Brembo (fine grained)

Fontana del Piermarini

Address: piazza Fontana.
Background: 1782 (setting)
Authors: G. Piermarini (architect), G. Franchi (sculptor)
Materials
Statues (women, dolphins): white marble (Apuanian Alps)
Basins (three, decreasing in diameter): Baveno (moulded element)

MONUMENTS OF THE 19TH CENTURY

Napoleone I

Address: Accademia di Brera (main court), via Brera 28
Background: 1809 (manufacture); 15 August 1859 (unveiling)
Author: A. Canova
Materials
Statue: bronze
Pedestal: Baveno, white marble

Cesare Beccaria

Address: Accademia di Brera (main staircase, left), via Brera 28
Background: 1838 (unveiling)
Author: P. Marchesi
Materials
Statue: white Apuanian marble
Pedestal: Baveno

Giuseppe Parini

Address: Accademia di Brera (main staircase, right), via Brera 28
Background: 1838 (unveiling)
Author: G. Monti
Materials
Statue: white Apuanian marble
Pedestal: Baveno

Italia

Address: Giardini Pubblici (monte Merlo)
Background: 1861 (unveiling); 1850 (manufacture)
Author: A. Puttinati

Materials
Statue: white marble (Apuanian Alps)
Pedestal: Ceppo del Brembo

Carlo Porta

Address: via Brolo, via Larga, piazza Fontana
Background: 4 October 1862 (unveiling); 1943 (destruction by aerial bombings); 18 September 1966 (unveiling of the reproduction)
Authors: A. Puttinati (1862); I. Soli (1966, reproduction)
Materials
Statue: marble (1862); bronze (1966)
Pedestal: Montorfano

Camillo Benso di Cavour

Address: piazza Cavour
Background: 5 June 1865 (unveiling)
Authors: O. Tabacchi (statue of Cavour); A. Tantardini (statue of Italia)
Materials
Statues: bronze
Pedestal: Baveno

Federico Borromeo

Address: piazza San Sepolcro (near Biblioteca Ambrosiana)
Background: 16 July 1865 (unveiling)
Author: C. Corti
Materials
Statue: white marble (Apuanian Alps)
Pedestal: Baveno; Botticino (slabs)

Mosè and Aronne

Address: via dell'Arcivescovado 1 (main court of palazzo Arcivescovile)
Background: 1870 (unveiling); 1865 (manufacture)
Authors: A. Tantardini (Mosè), G. Strazza (Aronne)
Materials
Statues: Candoglia
Pedestal: Ceppo Brembo

Cesare Beccaria

Address: piazza Beccaria
Background: 19 March 1871 (unveiling)
Author: G. Grandi
Materials
Statue: bronze
Pedestal: Montorfano; white marble (epigraphs)

Leonardo da Vinci and four followers

Address: piazza della Scala
Background: 4 September 1872 (unveiling)
Author: P. Magni
Materials
Statues: white marble (Apuanian Alps)
Pedestal: Baveno

Caduti di Mentana

Address: piazza Mentana
Background: 3 November 1880 (unveiling)
Author: L. Belli
Materials
Statue: white marble (Apuanian Alps)
Pedestal: Baveno
Stairs: Baveno

Alessandro Manzoni

Address: piazza San Fedele
Background: 22 May 1883 (unveiling)
Author: F. Barzaghi
Materials
Statue: bronze
Pedestal: Montorfano (lower part), Baveno

Agostino Bertani

Address: piazza Fratelli Bandiera
Background: 30 April 1888 (unveiling)
Author: V. Vela
Materials
Statue: bronze
Pedestal: Baveno
New fountain: Baveno

Francesco Hayez

Address: piazzetta Brera
Background: 10 February 1890 (unveiling)
Author: F. Barzaghi
Materials
Statue: bronze
Pedestal: Balma, Baveno

Giuseppe Sirtori

Address: Giardini Pubblici

Background: 5 June 1892 (unveiling)
Author: E. Butti
Materials
Statue: bronze
Pedestal: Verrucano

Luciano Manara

Address: Giardini Pubblici (Bastioni Porta Venezia)
Background: 8 July 1894 (unveiling)
Author: F. Barzaghi
Materials
Statue: bronze
Pedestal: Chiampo, Viggiù, Baveno

Giovan Battista Piatti

Address: largo la Foppa
Background: 18 November 1894 (unveiling)
Author: S. Pisani
Materials
Statue: bronze
Pedestal: Botticino, Baveno

Cinque Giornate di Milano

Address: piazza Cinque Giornate
Background: 18 March 1895 (unveiling), 1894 (manufacture)
Author: G. Grandi
Materials
Obelisk and statues: bronze
Basement: red granite from Sweden (upper); Balma (lower)

Giuseppe Garibaldi

Address: largo Cairoli
Background: 3 November 1895 (unveiling)
Author: E. Ximenes
Materials
Statues: bronze
Pedestal: Baveno; Botticino (epigraphs)

Vittorio Emanuele II, king of Italy

Address: piazza del Duomo
Background: 24 June 1896 (unveiling)
Author: E. Rosa
Materials
Equestrian statue: bronze

Reliefs: bronze
Lions: veined marble
Pedestal: Porfido Monumentale (lower part), Baveno (upper part)

Antonio Rosmini

Address: Giardini Pubblici (in front of palazzo Dugnani)
Background: 12 July 1896 (unveiling)
Author: F. Confalonieri
Materials
Statue: bronze
Pedestal: Botticino, Balma

Antonio Stoppani

Address: Giardini Pubblici
Background: 9 June 1898 (unveiling)
Author: F. Confalonieri
Materials
Statue: bronze
Pedestal: Baveno

Giuseppe Parini

Address: piazza Cordusio
Background: 26 November 1899 (unveiling)
Author: L. Secchi
Materials
Statue: bronze
Pedestal: Botticino
Hemicycle: Ghiandone

MONUMENTS OF THE 20TH CENTURY

Carlo Cattaneo

Address: via Santa Margherita
Background: 27 June 1901 (unveiling)
Author: E. Ferrari
Materials
Statue: bronze
Pedestal: Baveno, Lumachella

Cesare Correnti

Address: piazza Resistenza Partigiana
Background: 24 June 1901 (unveiling)
Author: L. Secchi
Materials

Base: Ghiandone
Pedestal: Botticino
Statue: bronze

Giuseppe Dezza

Address: via Marina, via Palestro
Background: 2 November 1902 (unveiling)
Author: E. Cassi
Materials
Statue: bronze
Pedestal: Baveno

Andrea Verga

Address: largo Richini
Background: 1903 (manufacture)
Author: G. Branca
Materials
Statue: white marble (Apuanian Alps)
Pedestal: Baveno

Felice Cavallotti

Address: via Marina, via Senato
Background: 15 July 1906 (unveiling)
Author: E. Bazzaro
Materials
Statue: white marble (Apuanian Alps)
Pedestal: Ghiandone

Eugenio Villoresi

Address: piazza Leonardo da Vinci
Background: 1 December 1907 (unveiling); 1966 (new location)
Author: L. Panzeri
Materials
Statue: bronze
Pedestal: Zandobbio

Gaetano Negri

Address: Giardini Pubblici
Background: 11 October 1908 (unveiling)
Author: L. Secchi
Materials
Statue: bronze
Pedestal: Balma

Giuseppe Giacosa

Address: Giardini Pubblici
Background: 21 May 1910 (unveiling)
Author: L. Secchi
Materials
Statue: bronze
Pedestal: Balma

Giuseppe Verdi

Address: piazza Buonarroti
Background: 10 October 1913 (unveiling)
Author: E. Butti
Materials
Statue: bronze
Pedestal: Granito Bellinzona

Alpini d'Italia

Address: piazza Giovanni XXIII
Background: 1915 (manufacture); 1928 (unveiling, via Pagano); 21 November 1948 (moved to piazza Cadorna); 23 September 1963 (moved to piazza Giovanni XXIII)
Author: E. Bisi
Materials
Statue: bronze
Pedestal: Montorfano (1948)
Stairs: Serizzo Ossola (1948)

Giuseppe Missori

Address: piazza Missori
Background: 7 May 1916 (unveiling)
Author: R. Ripamonti
Materials
Statue: bronze
Pedestal: Ceppo Brembo (rough block, coarse and fine grained)
Baseboard: Ceppo Brembo (saw finish, coarse grained)

Filippo Carcano

Address: Giardini Pubblici
Background: 8 September 1916 (unveiling)
Author: E. Boninsegna
Materials
Statue: white marble (Apuanian Alps)
Pedestal: Ghiandone

Caduti di Porta Romana

Address: via Tiraboschi, via Muratori, via Lattuada
Background: 24 June 1923 (unveiling)
Author: E. Saroldi
Materials
Statues: bronze
Pedestal: Ghiandone; Botticino (epigraphs)
Base of the statues: Ghiandone

Leone XIII

Address: Cortile della Canonica (Sant'Ambrogio)
Background: 1924 (unveiling)
Author: F. Confalonieri
Materials
Statue: bronze
Pedestal: Molera

Ernesto Teodoro Moneta

Address: Giardini Pubblici
Background: 29 June 1924 (unveiling); 1920 (manufacture)
Author: T. Brianzi
Materials
Statue: white marble (Carrara)
Pedestal: Baveno

San Francesco d'Assisi

Address: piazza Sant'Angelo
Background: 1926 (setting)
Author: G. Castiglioni
Materials
Statue: bronze
Fountain: Serizzo Ossola

Aviatore

Address: piazza Novelli
Background: 20 September 1926 (unveiling at Cascina Malpensa); 30 March 2005 (moved to piazza Novelli)
Author: unknown
Materials
Structure: Ghiandone
Relief: bronze
Inscription: veined marble

Napoleone III

Address: Parco Sempione (Montetordo)
Background: 27 February 1927 (unveiling); 1886 (manufacture)
Author: F. Barzaghi
Materials
Statue: bronze
Pedestal: Balma
Baseboard: Balma
Staircase: Ceppo Brembo

San Francesco d'Assisi

Address: piazza Risorgimento
Background: 28 October 1927 (unveiling)
Author: D. Trentacoste
Materials
Statue: bronze
Pedestal with reliefs: Travertino
Stairs: Baveno

Fontana delle Quattro stagioni

Address: piazza Giulio Cesare
Background: 12 April 1927 (unveiling); 1943 (aerial bombing, partial destruction)
Author: R. Gerla
Materials
Statues: Vicenza (Berici)
Obelisks: Sarnico
Basin with decorations: Sarnico

Fontana dei Tritoni

Address: via Romagnosi – via Andegari
Background: 1930
Authors: A. Minali (architecture), S. Saponaro (sculptures)
Materials
Statues: white marble (Apuanian Alps)
Architecture: Baveno

Francesco Baracca

Address: piazzale Baracca
Background: 27 September 1931 (unveiling)
Author: S. Monfrini
Materials
Statue: bronze
Pedestal: Valcundria

Sant'Antonio da Padova

Address: via Farini – via Tazzoli
Background: 1932
Author: G. Maretto
Materials
Statue: bronze
Fountain: Ghiandone

Giuseppe Grandi

Address: piazza Grandi
Background: 30 November 1936 (unveiling)
Authors: W. Sevèr, E. Noël Winderling
Materials
Statue: bronze
Fountain: Alzo
Fountain (central part): travertine

Costantino Imperatore

Address: piazza san Lorenzo (in front of the basilica)
Background: 1937 (unveiling)
Author: from the statue (4th century) located in San Giovanni in Laterano (Rome)
Materials
Statue: bronze
Pedestal: Ceppo Brembo
Another statue of "Costantino Imperatore" is located in front of Santa Croce (via Sidoli)

Caduti nei campi di sterminio nazisti

Address: Cimitero Monumentale (piazzale interno)
Background: 1945, 1950, 1955, 1961 (seven slabs)
Authors: Studio BBPR
Materials
Pedestal: Serizzo Ossola (rough block)
Structure (cube): bronze (the centre of the structure holds a vase with the ashes from the concentration camp at Mauthausen, Oberoesterreich – Austria); Serizzo Ossola (pedestal)
Epigraphs (on the structure): Candoglia, black granite (slab)
Lying slabs (reporting the names of the fallen): Serizzo Ossola

Piccoli Martiri di Gorla

Address: piazza Piccoli Martiri
Background: 20 October 1947 (unveiling)
Author: R. Brioschi
Materials
Statues: bronze
Stele: Montorfano
Stairs: Diorite

Fontana Montecatini

Address: largo Donegani – via Turati
Background: 1951 (unveiling), 1992 (rebuilding)
Author: G. Ponti
Materials
Basin: Bardiglio

Caduti di piazzale Loreto

Address: piazzale Loreto, via Doria
Background: 1960 (manufacture)
Author: G. Castiglioni
Materials
Stele: Simona
Pedestal: Balma
Floor: Trachyte

Cavallo impennato

Address: via Brera, garden L. Romano
Background: 1960 (manufacture)
Author: A. Sassu
Materials
Statue: bronze
Pedestal: Serizzo Ossola

Marinai d'Italia

Address: largo Marinari d'Italia, corso XXII Marzo
Background: 10 September 1967 (unveiling)
Author: F. Somaini
Materials
Statue: bronze
Cylindrical elements: concrete

Finestra nel Cielo (Ai Caduti per la libertà)

Address: piazza Miani
Background: 1968 (unveiling)
Author: C. Ramous
Materials
Figure: steel
Pedestal: reinforced concrete

Caduti del quartiere Isola

Address: piazzale Segrino (moved from via Sassetti)
Background: 25 April 1972 (unveiling)
Author: C. Ramous

Materials
Statue: steel
Pedestal: reinforced concrete

Fontana Bagni Misteriosi

Address: Parco Sempione (near Palazzo dell'Arte)
Background: 1973 (XV Triennale d'Arte); 2009 (restoration)
Author: G. De Chirico
Materials
Statues: Vicenza

Bersagliere

Address: Verziere
Background: 23 September 1973 (unveiling)
Author: M. Robaudi
Materials
Statues: bronze
Wall: reinforced concrete

Giuseppe Mazzini

Address: piazza della Repubblica
Background: 1 June 1974 (unveiling)
Author: P. Cascella
Materials
Statue: bronze
Sculptural and architectural elements: white marble (Apuanian Alps)

Colonna a grandi fogli

Address: Gruppo Mondadori, Segrate
Background: 1975 (unveiling); 1972–75 (manufacture)
Author: A. Pomodoro
Materials
Structure: bronze, steel

Mère Ubu

Address: via Senato (in front of palazzo del Senato)
Background: 1976 (unveiling); 1975 (manufacture)
Author: J. Mirò
Materials
Statue: bronze

Roberto Franceschi

Address: via Bocconi
Background: 16 April 1977 (unveiling)

Author: collective
Materials
Statue: steel (former industrial hammer)

Grande disco

Address: piazza Meda
Background: 11 November 1980 (unveiling); 1972 (manufacture); 2006 (moved to via Lanza, in front of Piccolo Teatro della Città di Milano)
Author: A. Pomodoro
Materials
Statue: bronze

Emilio Alessandrini

Address: parco Alessandrini
Background: 25 April 1980 (partial unveiling); 3 October 1982 (unveiling)
Author: A. Cascella
Materials
Pedestal: reinforced concrete
Epigraph: Zandobbio (slab)
Spherical element: Montorfano

Carabiniere

Address: piazza Diaz
Background: 3 December 1981 (unveiling)
Author: L. Minguzzi
Materials
Emblem of military corps: steel

Gesto per la Libertà

Address: piazza Conciliazione
Background: 1981 (unveiling in the present location); 1974 (first location: piazzetta Reale); 1972 (manufacture)
Author: C. Ramous
Materials
Structure: steel

Condizione umana

Address: viale della Liberazione
Background: 1983 (unveiling, via Melchiorre Gioia); 2019 (moved to viale Liberazione)
Author: A. Fabbri
Materials
Statue: bronze
Beams: steel
Pedestal: Aurisina (slab)

Finanziere (Cielo, Terra e Mare)

Address: piazza Tricolore
Background: 25 April 1985 (unveiling)
Author: A. Sassu
Materials
Statue: bronze
Cylindres: steel
Pedestal: Montorfano

Fontana della Libertà

Address: piazza Leonardo Vinci
Background: 1985
Author: A. Cascella
Materials
Sculptural elements: granite of Sardinia (Rosa Nule)

Dedalo

Address: viale Forlanini (Linate Airport)
Background: 13 September 1985 (unveiling)
Authors: A. Cascella (sculpture), F. Tartaglia (architecture)
Materials
Sculpture: white marble (Apuanian Alps)
Basin: Botticino

Cometa di Halley

Address: viale Forlanini (Linate Airport)
Background: 28 October 1985
Author: C. Mo
Materials
Sculpture: steel
Basement: gneiss

Grande Toscano

Address: piazza del Carmine
Background: 1986 (unveiling); 1981 (manufacture)
Author: I. Mitoraj
Materials
Statue: bronze
Pedestal: Serizzo Ossola (slab)

Cerchi in movimento

Address: piazza Sei Febbraio
Background: 1987 (unveiling)

Author: C. Cappello
Materials
Circles: steel
Pedestal: Africa black, Serizzo Ossola (slab)

Sandro Pertini

Address: via Croce Rossa, via Manzoni
Background: 1 May 1990 (unveiling)
Author: A. Rossi
Materials
Architecture: Candoglia – Ornavasso (slab)
Rear fountain: metal

Luci nel Bosco (Alle vittime del dovere in tempo di pace)

Address: largo Grassi (near Teatro Strehler)
Background: April 1990 (unveiling)
Author: L. Minguzzi
Materials
Statue: bronze
Pedestal: Serizzo Ossola (slab)

I Monti, i laghi e i fiumi di Lombardia

Address: piazza San Babila
Background: 1996 (unveiling)
Author: L. Caccia Dominioni
Materials
Basin and pyramid: Serizzo Masino, Serizzo Dubino, Baveno, Montorfano

Cavallo di Leonardo da Vinci

Address: piazzale dello Sport 16 (Ippodromo del Galoppo, San Siro)
Background: September 1999 (unveiling)
Author: N. Akamu
Materials
Statue: bronze
Pedestal: veined white marble (Apuanian Alps)

MONUMENTS OF THE 21ST CENTURY

Ago, filo e nodo

Address: piazzale Cadorna
Background: 2000 (unveiling)
Authors: C. Oldenburg and C. van Bruggen
Materials
Statues: steel, fiberglass

Gian Giacomo Mora

Address: via Mora – corso Porta Ticinese
Background: June 2005 (unveiling)
Author: R. Menergon
Materials
Image: bronze

Indro Montanelli

Address: Giardini Pubblici
Background: 22 May 2006 (unveiling)
Author: V. Tongiani
Materials
Statue: bronze
Pedestal: serpentine, white marble
Seat: white marble
Enclosure: serpentine

L'uomo della Luce (Alle vittime del terrorismo)

Address: via Vivaio 1, largo 11 September 2001
Background: 9 May 2008 (unveiling)
Author: B. Roig
Materials
Statue: bronze
Structure: steel (beam)

Torre a spirale

Address: largo Greppi (near Teatro Strehler)
Background: May 2010 (unveiling)
Author: A. Pomodoro
Materials
Structure: bronze

L.O.V.E. (Libertà – Odio – Vendetta – Eternità)

Address: piazza Affari
Background: 24 September 2010 (unveiling)
Author: M. Cattelan
Materials
Statue: white marble (Apuanian Alps)
Pedestal: travertine

Il Dio Pan è morto

Address: piazza Piemonte
Background: 22 May 2012 (unveiling); 1995 (manufacture)
Author: A. Sassu

Materials
Statue: bronze
Pedestal: Botticino

Grande Ciclista

Address: piazza Piemonte
Background: 22 May 2012 (unveiling); 1995 (manufacture)
Author: A. Sassu
Materials
Statue: bronze
Pedestal: Porfido Monumentale (slab)
Floor: Porfido Trentino (slab)

Cavallo del Mare

Address: piazza Piemonte
Background: 22 May 2012 (unveiling); 1995 (manufacture)
Author: A. Sassu
Materials
Statue: bronze
Pedestal: Porfido Monumentale

Le Tre Grazie

Address: piazza Piemonte
Background: 12 October 2013 (unveiling); 1994 (manufacture)
Author: S. Fiume
Materials
Statues: bronze
Pedestal: gneiss

Seme dell'Altissimo

Address: via Paleocapa
Background: 2015 (unveiling)
Author: E. Isgrò
Materials
Object: Arabescato marble (Apuanian Alps)

Nelson Mandela

Address: vicolo San Giovanni sul Muro
Background: October 2015 (unveiling)
Author: P. Scampini
Materials
Statue: bronze

Giulio Ricordi

Address: largo Ghiringhelli, piazza della Scala

Background: 25 November 2016 (unveiling, moved from Ricordi publishing house, via Berchet); 1922 (manufacture)
Author: L. Secchi
Materials
Statue: white marble (Apuanian Alps)
Pedestal: Botticino

Ruggero Boscovich

Address: Giardini Pubblici
Background: 13 February 2017 (unveiling); 1992 (donation)
Author: I. Mestrovic
Materials
Statue: bronze
Pedestal: Negro Africa

Pencho Slaveykov

Address: Verziere, via del Brolo
Background: 30 October 2019 (unveiling)
Author: A. Novakov
Materials
Statue: bronze

Vittime delle Foibe

Address: piazza Repubblica
Background: 10 October 2020 (unveiling)
Author: P. Tatricchio
Materials
Stele: grey granite
Basement: Porfido Monumentale

Partigiani

Address: via Mercanti, piazza Mercanti (porch of Palazzo della Ragione)
Background: 13 May 2021 (unveiling)
Architect: C. Boeri (2010)
Materials
Stelae: Beola
Sheets: glass

Chapter 27

Funeral chapels of Cimitero Monumentale

A particular field of stone employ, such as building material, concerns the architecture and the sculpture of burial sites; this aspect is excellently symbolized in Milan at Cimitero Monumentale.

The *Cimitero Monumentale* was built on the plan of the architect Carlo Maciachini (1818–99) and opened in 1866, gathering the corpses from five urban cemeteries no longer used according to the new burial rules. These rules made the first move in Bonaparte's period with the abolition of church cemeteries and the creation of the common graves ("Titre premier. Des Sépulture, et des lieux qui leur sont consacrés. Article premier. Aucune inhumation n'aura lieu dans les églises, temples, synagogues, hôpitaux, chapelles publiques et générallement dans aucune des édifices clos et fermés où les citoyens ce réunissent pour la célébration de leurs cultes, ni dans les enceinte des villes et bourgs" [Décret imperial, 1804].) The decree was applied in the Reign of Italy since 5 September 1806, taking into account the menaces to the public health caused by careless burials. These rules were condemned by Ugo Foscolo in his poem "On sepulchres": "And yet a law now places sepulchres/beyond the gaze of pity and denies/the dead a name" [Foscolo, 2010; lines 51–53].

The *Cimitero Monumentale* grew rapidly, occupying the planned spaces; at the turn of the century, it was supported with a larger cemetery, called Cimitero Maggiore (1895) and located in the north-western outskirts of Milano (Musocco). A crematory was built (1876) in the farthest part of the burial field, following scientific studies on the cremation [Gorini, 1876]; the building was planned by Maciachini, featuring a Doric colonnade; the industrialist Alberto Keller (1800–74) provided the necessary funds for the construction.

The Cimitero Monumentale features a complicated building on the front, with a central domed chapel (called Famedio, designated to the distinguished citizens) together with covered galleries, domes, arches, staircases; a wide area, on the rear, lined with trees and split in square partitions (called "Riparto") was dedicated to the burial. Since its inauguration this area was filled by hundreds of chapels, sepulchres, tombs and graves. In particular, the chapels, owned by wealthy families of Milan, were almost always built like an edifice with walls, roofs, different architectural elements (columns, entablatures, cornices, arches, doors, pediments, etc.) and different decorative elements (reliefs, bas-reliefs, statues, etc.) [Ginex, Selvafolta, 1996].

A selection of funeral chapels of Cimitero Monumentale was made, taking into consideration the works of the most significant architects, makers of remarkable attempts of Milan architecture in the late 19th and early 20th centuries.

Each funeral monument is listed according to the address based on the partition (Riparto) of the cemetery area, and it is identified by the name of the architect and by the name of the owner. Several maps are present in different sites of the Cimitero Monumentale.

DOI: 10.1201/9781003361008-33

REFERENCES

Bertrand R., Groud, G., Décret impérial sur les sépultures – 23 prairial an XII (1804, june 12) *in* Cimetières et tombeaux. Editions du Patrimoine, Paris, 2016.

Foscolo U., Dei Sepolcri (J.G. Nichols ed., Sepulchres), Oneworld Classics, London, 2010.

Ginex G., Selvafolta O., Il Cimitero Monumentale di Milano – Guida storico-artistica, Silvana, Milano, 1996.

Gorini P., Sulla purificazione dei morti per mezzo del fuoco, considerazioni, sperimenti e proposte, Battezzati, Milano, 1876.

RIPARTO I AREA CENTRALE – EMICICLO – PIAZZALE INTERNO

*Caduti nei Campi di sterminio nazisti
Background: 1945, 1950, 1955, 1961 (seven slabs)
Authors: Studio BBPR
Pedestal: Serizzo Ossola (rough block, slab)
Structure (cube): bronze (the centre of the structure holds a vase with the ashes from the concentration camp at Mauthausen, Oberoesterreich – Austria); Serizzo Ossola (pedestal)
Epigraphs (on the structure): Candoglia, black granite (slab)
Lying slabs (reporting the names of the fallen): Serizzo Ossola

RIPARTO I
*Riparto I, #173
Bocconi (owner), G. Boni (architect), O. Grossoni (sculptor) – 1901/14
A high chapel with four columns (Montorfano) standing on a squared basement (Montorfano) together with figures and the Crucifixion (white Carrara marble)

*Riparto I, #175
Erba (owner), G.B. Borsani and A. Savoldi (architect) – 1898/1910
A high pyramidal chapel featuring three sepulchres above the openings and a loggia. A selected choice of stone materials includes Balma (base); Rosso Verona (openings, sepulchres, loggia) and Candoglia (pyramid, cladding around the main entrance).

*Crossing of Riparto I, II, III and IV
Falck (owner), M. Fiocchi (architect), G. Castiglioni (sculptor) – 1939–42
A pyramidal chapel made of squared blocks (Dubino). A white marble group (Pietà con Angelo) is set on the front (G. Castiglioni, 1942); a bronze relief is set on the staircase to the crypt (A. Minerbi, 1955).

RIPARTO II
*Riparto II, #13
Chinelli (owner), R. Zavanella (architect), L. Fontana (sculptor) – 1949
A monument with some thick slabs of stone (Serizzo and Beola) around a figure (Angel) made of glazed ceramic

*Riparto II, #51–54
Reggiori (owner), F. Reggiori (architect) – 1901.
A family tomb featuring a stone enclosure (Serizzo Ossola) with arches and gravestones made of Candoglia. The base is made of Serizzo Ossola.

*Riparto II, 166
Sommaruga Biffi (owner), G. Sommaruga (architect) – 1906/38
A small chapel made of Ghiandone with an insertion of serpentinite

*Riparto II, #177
Feltrinelli (owner), G. Casarini (architect) – 1914–21
A huge chapel with columns, pediment, dome and ornaments; all the elements are made of Botticino

*Riparto II, #183
Broggi (owner), L. Broggi (architect) – 1909/10
A neo-Romanesque church-like chapel featuring brick masonry with stone blocks (marble, gneiss, sandstone, etc.). The portal is made of Majolica with a base made of Mazzano.

Riparto II, #192
Baj (owner), C. Nava (architect), A. Carminati (sculptor) – 1905/09
A huge sepulchre (white Apuanian marble) set on a base (Balma) and a structure (Simona), including the portrait of the owner, and standing on six columns (four made of Verona and two of Oira, with bases and capitals made of Oira)

*Riparto II, 204
Verga (owner), E. Pirovano (architect) – 1906/07
A pyramidal chapel made of Serizzo Ossola

RIPARTO III
*Riparto III, #143
Pomelli Cimbardi (owner), F. Pestagalli (architect), E. Longoni (painter) – 1897/99
A Neo-Renaissance chapel made of ashlars of Botticino; a mosaic, featuring "L'Angelo della Speranza", stands over the portal

*Riparto III, #129
Gritti (owner), L. Beltrami (architect); A. Mazzucchelli (sculptor) – 1909
A sarcophagus with sculptures, made of Candoglia, standing on a base of finely grained gneiss and of well-squared blocks of Anzola

RIPARTO IV
*Riparto IV, #84
Crespi (owner), E. Bazzaro (sculptor) – 1909/12
A sculptural group (white Carrara marble) with four figures standing on a support made of big blocks and piers (Serizzo Ossola)

*Riparto IV, #85
Gabba, Frua (owner), P. Cesa Bianchi (architect) – 1890/91
A high chapel, made of Serizzo Ossola (base) and Viggiù (ashlar, moulded elements), with a colonnaded portal made of Balma (shaft) and Oira (base and capital)

*Riparto IV, #88
Cottini (owner), L. Beltrami (architect) – 1901
A pyramidal chapel made of Ghiandone (base) and Mazzano (block)

*Riparto IV, #89
Beaux (owner), U. Stacchini (architect) – 1904
A pyramidal chapel made of Baveno (ashlar with curvilinear edges) together with bronze branches

RIPARTO V
*Riparto V, #82
Branca (owner), G. Boni (architect), M. Vedani (sculptor) – 1909/12
A massive chapel, transformed during the times, featuring a lower part made of Anzola and an upper part made of Balma. A bronze group (Pietà) is set on the top.

*Riparto V, #83
Civico Mausoleo, M. Palanti (architect) – 1928–30
A gigantic sepulchre (big blocks of Baveno) standing on very short columns (Baveno). Some vases (Porfido Monumentale) are present between the columns.

*Riparto V, #84
Squadrelli (owner), E. Bazzaro (sculptor) – 1911
A sepulchre (Serizzo Ossola) with some bronze figures, standing on a squared base made of a very big block of Serizzo Ossola

*Riparto V, #85
Frova (owner), L. Beltrami (architect) – 1888/89
A tall chapel with a dome, made of Viggiù (base), Mazzano (ashlar, portal) and Oira (columns)

*Riparto V, #86
Casati (owner), Author: E. Butti (sculptor) – 1884
A simple chapel with a lying down figure (bronze) in a structure made of Ceppo Iseo

Riparto V, #92
Pinardi (owner), U. Stacchini (architect) – 1903/05
A prismatic chapel featuring a huge portal and four crosses on the top. Baveno is the sole stone employed, and sometimes it is worked in curvilinear shapes.

*Riparto V, #95
Molina (owner), G. Moretti (architect), U. Zanoni (sculptor) – 1892
A chapel featuring the base of Serizzo Ossola, the walls of Baveno together with a white marble statue

*Crossing of Riparto V, VI, VII and VIII

Brambilla (owner), C. Maciachini (architect), P. Magni (sculptor) – 1870–71

A centrally planned chapel featuring a dome and four arches with pediments. The masonry is made of alternating bands of Mazzano with different shades, the dark base is made of Balma. The inner part shows a cladding of white or grey Apuanian marbles and an altar made of Bardiglio together with a white marble statue.

RIPARTO VI

*Riparto VI, #127

Besenzanica (owner), E. Butti (sculptor) – 1912

A ploughing scene ("Il Lavoro") featuring two countrymen and two cows (bronze) in a harsh land (Simona). A gigantic statue (Simona), representing the "Nature", stands in the rear, from the ground to the top.

*Riparto VI, #128

Pierd Houy (owner), P. Giudici (sculptor) – 1900/01

A pyramidal chapel with a bronze angel on the top, featuring complicatedly carved ornaments (Simona) on each side and a central part with Ceppo Brembo framing bronze bas-reliefs

*Riparto VI, #130

Galimberti (owner), A. Manfredini (architect), L. Brivio (sculptor) – 1902

A chapel made of Ghiandone with some bronze decorations

*Riparto VI, #135

Caminada (owner), P. Portaluppi (architect) – 1927/30

A polyhedric chapel made of Travertino together with a portal made of white Apuanian marble. The floor of the entrance includes Rouge Languedoc, Bardiglio, black limestone, Rosa Garda, Botticino, etc.

*Riparto VI, #142

Casati (owner), G. Moretti (architect), A. Carminati (sculptor) – 1900/01

A pyramidal chapel (Serizzo Ossola) with four sepulchres (Serizzo Ossola) surmounted by bronze figures supporting another sepulchre (white Apuanian marble)

RIPARTO VII

*Riparto VII, #174

Dompé di Mondarco (owner), G. Clerici di Cavenago and S. Lo Bianco (architects); N. Conti (sculptor) – 1959/63

A chapel consisting in a vault made of two arches (ashlars of Montorfano with a particular cut and copper slabs) over an ancient sepulchre (white marble). The pavement is made of irregular slabs of red granite from Finland. The staircase is made of red granite (riser) and Serizzo Ossola (tread).

*Riparto VII, #176

Marelli (owner), C. Arpesani (architect) – 1920

A church-like chapel with a squared plan made of Serizzo Ossola (base), veined white marble and veined grey marble in superimposed courses (ashlar masonry); white marble (semi-columns). Small discs of coloured marbles decorate the upper part.

*Riparto VII, #179
Vanoni, Pasi (owner), A. Menni (architect), E. Boninsegna (sculptor) – 1907/08
A chapel made of Serizzo Ossola with some insertion of a darker variety of Serizzo and reliefs made of gilded bronze. The staircase and the upper cross are made of Anzola.

*Riparto VII, #184
Toscanini (owner), Leonardo Bistolfi (sculptor) – 1909/11
A cubic chapel with bas-reliefs upon a plain base (white veined marble of Apuanian Alps)

*Riparto VII, #190
Giudici (owner), P. Mezzanotte (architect) – 1904/05
A squared chapel with base and walls made of Baveno (ashlar). A mosaic is set over the portal and foliage decorations are set on the edges.

*Riparto VII, #191
Pavesi (owner), F. Solmi (architect) – 1903
An arched chapel featuring a rose window with coloured glass. The base and the columns are made of Serizzo Ossola, the walls are made of Urago.

RIPARTO VIII
*Riparto VIII, #182
Biffi (owner), G.B. Bossi (architect) – 1900/03
A tall chapel with colonnaded portal, arches and hexagonal spire. The whole chapel is made of Mazzano.

RIPARTO IX
*Riparto IX, #541
Pirelli (owner), L. Beltrami (architect) – 1919/21
A chapel with colonnaded entrance made of brick masonry, a portal made of Candoglia (frame) together with two columns (Candoglia), bases and two capitals (Oira)

RIPARTO X
*Riparto X, #246
Bandelli (owner), Gaetano Moretti (architect) – 1916
A chapel featuring a high base (Ghiandone), some stairs (Candoglia), a sepulchre (Candoglia) surmounted by a canopy (Candoglia). The canopy stands on four columns made of red granite (Finland) with bronze bases and capitals.

*Riparto X, #250
Piazza (owner), A. Manfredini (architect) – 1909
A cubic chapel made of Ghiandone with rough finishing together with an insertion of a foliated gneiss

*Riparto X, #253
Bonomi (former Galbiati, owner), M. Fiocchi (architect), F. Messina (sculptor) – 1950
A prismatic chapel made of "Granito di Montjovet" (as reported in archive documents) with a white marble sculpture set in a niche

RIPARTO XI
*Between Riparto XI and Riparto XII
Chiesa (owner), R. Ferrini (architect) – 1925 (reconstruction 1949)
A chapel with a masonry made of Baveno (ashlar) and a dome with circular windows (alabaster). The entrance is clad with slabs of Cipollino.

RIPARTO XII
*Riparto XII, #391–92
Achille (owner), L. Figini and G. Pollini (architects); F. Melotti (sculptor) – 1950/52
A cubic volume made of blocks and slabs of Candoglia surrounded by thin walls of Candoglia with blocks set to leave cross-shaped voids. A bronze statue stands on the left.

NECROPOLI
*Necropoli, #1
Bernocchi Antonio (owner), A. Minali (architect), G. Castiglioni (sculptor) – 1936
A conical chapel with a spiral structure (Musso), including a series of sculptures (Musso) of the *Via Crucis* or the Passion of Christ. The floor of the crypt is made of Musso and Larvikite (slab).

*Necropoli, #4
Turati (owner), C. Maciachini (architect), L.G. Buzzi (sculptor) – 1872/73
A centrally planned chapel with a base made of Balma; the main body (Botticino) includes columns, pillars, arches, etc. The side windows are made of Vicenza.

*Necropoli, #10
Melzi d'Eril, Branca (owner), L. Broggi and C. Nava (architects), U. Nono (sculptor) – 1913/14
A cavern-like chapel with the stone set to create a perspective in front of a Jesus statue. Irregular blocks of Ceppo del Brembo, Simona, Mazzano are spread on the floor or they form the masonry of the chapel.

*Necropoli, #11
Calegari (owner), C. Maciachini (architect) – 1878/79
A Neo-Gothic chapel with spires made of Botticino, the base board in chromatic alternace is made of Simona.

*Necropoli, #15
Bernocchi Andrea (owner), G. Baroggi (architect), G. Castiglioni (sculptor) – 1931/33
A squared chapel featuring a three-sided carved frieze made of three monolithic slabs. Both ashlars and carved slabs are made of Musso.

*Necropoli, #17
Crosti (owner), G. Greppi (architect), E. Butti (sculptor) – 1922
The simple chapel includes a base (Balma) and a pyramidal body (Simona). The sculpture represents "La Maddalena al Sepolcro".

*Necropoli, #28
Conti (owner), P. Portaluppi (architect) – 1921/22

A brick chapel with a dome featuring rhombic windows and terracotta decorations. A huge, sculpted portal, made of Rosso Verona, stands on the stairs and a base made of Serizzo Ossola.

*Necropoli, #24
Biraghi (owner), C. Maciachini (architect) – 1875/80
A squared chapel made of Baveno (base) and Botticino (structure and ornaments) together with, on the sides, two slabs of white marble

*Necropoli. #34
Spedizione Porro (owner), C. Borromeo (architect) – 1888
A rocky base (irregular blocks of Ceppo del Brembo) supporting four bronze lions; a monolithic obelisk (Baveno) stands upon the lions

*Necropoli, #38
Cesa Bianchi (owner), E.A. Griffini (architect) – 1938
A chapel showing a base made of Balma, a body made of Verona and brick. The ornamentation is also made of Verona.

*Necropoli, #81
Bernocchi Michele (owner), A. Minali (architect) – 1940
A huge chapel reproducing the "Monument of the Julii" (*Glanum*, Saint Rémy-de-Provence, Bouches du Rhône, France; 1st century BCE). Above the base, made of Ghiandone, the structure and the ornaments are made of a white marble with grey veins (Apuanian Alps).

*Necropoli, #33
Moretti (owner), G. Sommaruga (architect) – 1913/14
A huge prismatic chapel featuring a base and stairs made of Serizzo Ossola; the lower part is made of big ashlars (Ceppo del Brembo), the upper part is made of a foliated gneiss coming from erratic blocks (Valganna, province of Varese) with rough surfaces

*Necropoli, #89
Bertarelli (owner), P. Portaluppi (architect) 1915/18; A. Mazzucchelli 1930
The present-day chapel is the result of different works; the stone employed is dark syenite (Balma). A bronze sculture (P. Miglioretti 1880) stands over the entrance.

*Necropoli, #130
Pigni (owner), O. Tabacchi (sculptor) – 1877
A sepulchre made of Montorfano together with a statue (white Carrara marble) representing "L'angelo custode del sepolcro"

*Necropoli, #142
Borghi (owner), E. Butti (sculptor) – 1884
A great sepulchre (Baveno) with a crucifix (Baveno) in the background and a winged figure (white Carrara marble) representing the "Il Tempo". A wing of the figure is now broken into three fragments lying behind the sepulchre.

*Necropoli, #146
Bruni (owner), A. Colla (architect), G. Monteverde (sculptor) – 1876
A perfect replica of an Egyptian pyramid made of Viggiù (ashlar). The statues (a female figure and a lion) near the moulded Portal, made of Viggiù, are made of white Carrara marble.

*Necropoli, #147
Mongeri (owner), C. Maciachini (architect) – 1888
A squared chapel with a base made of Simona, a structure made of Botticino, two marble slab of the sides and two columns on the portal made of Oira (shaft) and of white marble (base and capital).

*Necropoli, #168
Belinzaghi (owner), G. Landriani (architect) – 1892/93
A squared chapel featuring the blocks of Serizzo Ossola (rustication), the portal (jambs and lintel) and two side slabs of Montorfano

*Necropoli, #169
Dell'Acqua (owner), C. Maciachini (architect) – 1887
A structure made of Viggiù on a base of Montorfano and includes four columns, without base, made of Viggiù

*Necropoli, #178
Borletti (owner), G. Ponti (architect), L. Andreotti (sculptor) – 1931
A cubic chapel with a cladding made of wide slabs of Valle Strona; the base and the eaves cornice are made of Anzola as the door frames and the cross in the centre. Slabs of alabaster are set in the side windows.

*Necropoli, #184
Ghiotti, Ligresti (owner), C. Maciachini (architect) – 1875/80 (restored in 2003)
A squared chapel made of Balma (base), Botticino (structure and ornaments) and Molera (four shafts on the corners). The name of the owner is made of mosaic.

*Necropoli, #185
Chierichetti (owner), A. Wildt (architect and sculptor) – 1921
An unfinished chapel showing only the first step of the original plan: sixteen crosses made of white Carrara marble on a bronze base

RIPARTO XIII
*Riparto XIII, 405
Nathan (owner), M. Piacentini (architect), A. Mazzucchelli (sculptor) – 1918
A classical sarcophagus (Travertino) standing on small columns (Travertino) and on a rectangular base (Travertino)

RIPARTO XIV
*Riparto XIV, 442
Malugani (owner), E. Pirovano (architect) – 1925
A stepped pyramid made of Ghiandone

RIPARTO XV

*Riparto XV, 371

Koerner (owner), G.U. Arata (architect), A. Wildt (sculptor) – 1929

A cylindrical chapel made of ashlars (various size and shape) of Porfido Monumentale and Verrucano; the rear windows show thin Alabaster slabs instead of glass. Two bronze figures stand near the entrance.

RIPARTO XVI

*Riparto XVI, 376–78

Valente (owner), P. Portaluppi (architect) – 1923

A finely carved sarcophage (Rosso Verona) set on a base made of travertine. A high wall (travertine) stands behind.

RIPARTO XVII

*Riparto XVII, L

Porcile (owner), E.A. Griffini (architect) – 1934

A huge chapel with two parts: a lower one with base and portal of Anzola and ashlars of Serizzo Màsino showing in alternating polished and rough surface; an upper one with pillars, arches and pediment made of Serizzo Antigorio, two columns made of Serizzo Màsino and small windows made of Ornavasso

*Riparto XVII, M

Collini (owner), G. Greppi (architect), R. Del Bo (sculptor) – 1932/35

A symmetric chapel with pediment showing alternating bands of Ghiandone and Dubino. The portal, the pediment and the rear window are made of Dubino; two sculpture reliefs are made of Candoglia.

*Riparto XVII, Z

Rizzoli (owner), A. Castiglioni and P.G. Castiglioni (architects), G. Castiglioni (sculptor) – 1952/53

A hexagonal chapel with a loggia on the top, supported by angel statues. The sole stone is Musso (ashlar) and some reliefs on the rear are cut in the ashlars of the masonry.

*Riparto XVII, 259

Castellotti (owner), R. Zavanella (architect), L. Fontana (sculptor) – 1935

A bronze sculpture (Christ) placed on a horizontal slab (Larvikite) and standing before a vertical slab of Valle Strona

RIPARTO XVIII

*Riparto XVIII B

Motta (owner), M. Bega (architect), Manzù (sculptor) – 1939/52

A conical chapel with rough surfaces (Balma) and six bronze sculptures on the portal (Balma, polished surfaces).

*Riparto XVIII, D

Belloni (owner), C. Chierichetti (architect) – 1941/55

A small-sized facsimile of "Cappella dei Pazzi" (1430, F. Brunelleschi, Florence). The portico is supported by four columns (Verona, shaft; Botticino, base and capital) with arches

(Giallo Verona). The portal shows two pilasters made of Oira (shaft), Botticino (base, capital) and Occhiadino (pedestal). The two-order edifice is made of Botticino (ashlar) with a base of Occhiadino; some pilasters (Rosso Verona, shaft; Botticino, base and capital) mark the masonry of both orders.

RIPARTO XIX
*Riparto XIX, 309
Cella (owner), M. Baciocchi (architect), Manzù (sculptor) – 1941/43
A squared chapel with bronze reliefs, a staircase on the side and a sepulchre on the top. The stone material is unique (gabbro).

*Riparto XIX, 314
Conegliano (owner): A. Cassi Ramelli (architect) – 1941/42
A squared chapel (Granito Mergozzo) with rising arches (Granito Mergozzo) above the sarcophagus (Verde Aosta)

*Riparto XIX, 315
Manusardi (owner), L. Figini and G. Pollini (architects); F. Melotti (sculptor) – 1940
A prismatic chapel made of Baveno ashlars of two different sizes, set to leave squared voids and framed by two vertical monolithic supports (Baveno) and a monolithic lintel (Baveno). The sculpture on the right side is made of single piece of Montorfano.

*Riparto XIX, 326
Girola (owner), P. Portaluppi (architect), G. Castiglioni (sculptor) – 1941
A prismatic chapel with high windows and reliefs (workers) on the side ashlars (Arabescato veined marble from the Apuanian Alps). The base and the short staircase are made of Anzola.

RIPARTO XX
*Riparto XX Esterno Ponente 61
Birolli (owner), M. Zanuso (architect), L. Viani (sculptor) – 1961
A simple enclosure made of Granite of Sardinia with a spiral bronze figure

GALLERIE
*Galleria D Ponente superiore 179
Carcano di Bregnano (owner), C. Nava (architect), A. Carminati (sculptor) – 1904
A monument featuring a basement made of Oira and Ghiandone, a sepulchre made of Botticino and two bronze figures. The niche on the rear is coated with a mosaic (gilt glass tesserae).

*Galleria G Levante superiore 7–9
Introini (owner), P. Mezzanotte (architect) – 1925
A monument made of a sepulchre (Verde Levanto), with small niches (glass mosaic), supporting a bronze female figure

*Galleria BG Ponente Superiore
Sommaruga (owner), G. Castiglioni (sculptor) – 1935
A sepulchre (Ghiandone) standing in a niche coated with plaster (graffito)

RIALZATI

*Rialzato AB Ponente
Campari (owner), G. Castiglioni (sculptor) – 1935
The "Last Supper" with thirteen figures larger than life standing on a base made of a purple conglomerate (Verrucano)

*Rialzato AB Levante
Cusini (owner), A. Zacchi (architect), A. Minerbi (sculptor) – 1925
A simple monument featuring a base made of Porfido Monumentale, with different sculptures of birds, and a statue of Saint Francis d'Assisi supported by Oira blocks

*Rialzato B Ponente, 1489
Omodeo (owner), G. Sartirana (architect), O. Tabacchi (sculptor) – 1876/1930
A monument featuring a base made of Balma, a finely carved white Carrara marble figure (Memore omaggio). On the rear was added a monolithic pierced screen made of veined white marble.

CIRCONDANTI

*Circondante Ponente, 8 ½
Nava (owner), C. Nava (architect), A. Sassi (sculptor) – 1894
A brick-made chapel featuring a sculpted sepulchre (Viggiù) supported by columns and pillars (Serizzo Ossola shaft, Viggiù base and capital)

*Circondante Ponente, 13
Bazzaro (owner), L. Broggi (architect) – 1883
A neo-Romanesque church-like chapel featuring brick masonry with a base made of Ghiandone. The same stone was also used for the door jambs; the gravestone is made of a white fine-grained marble

*Circondante Ponente, 15 2/3
Occa (owner), C. Boito (architect) – 1889
A nearly pyramidal chapel with a façade made of Botticino and featuring a relief on the portal made of Apuanian white marble. The rear is made of brick masonry.

*Circondante Ponente, 701–03
Righini (owner), A. Cassi Ramelli (architect) – 1934/36
A squared chapel (Serizzo Màsino) with rising arches (Serizzo Màsino) above the sarcophagus (Porfido Monumentale)

*Circondante Ponente 715–16
Wildt (owner), G. Muzio (architect), A. Wildt (sculptor) – 1931
Two simple T-shaped stelae, each made of two blocks of Ghiandone and each supporting a bronze mask-portrait, set in a green grass

GIARDINI

*Giardini Levante, 102
Fratelli Praga (owner), P. Portaluppi (architect) – 1929
A simple memorial stone made of granitic gneiss standing on a base made of Anzola

ACATTOLICI

*Acattolici I, 1

Keller (owner), C. Maciachini (architect); G. Argenti (sculptor) – 1875

A squared chapel with high base (Simona), pillars and arches (Botticino), ornaments (Verona). The statue (La Protezione) is made of white Apuanian marble.

ISRAELITI

*Israeliti 1, 1

Leonino Davide (owner), C. Maciachini (architect) – 1875

A chapel featuring a high basement made of Balma and a "portico" made of four columns of white Carrara marble

*Israeliti 2, 14

Segre (owner), L. Conconi (architect) – 1900

A double-order chapel with columns, dome and four rose windows featuring the "menorah". The sole stone employed is Botticino (Mazzano).

*Israeliti 2, 9–10

Pisa (owner), C. Maciachini (architect) – 1885

A tall and compact chapel (Viggiù, ashlar) with a highly decorated rose window. The base and the stairs are made of Baveno.

*Israeliti 37 campo 1

Consolo Zanini (owner), G. Zanini (architect) – 1933

A small edicule made of Cardoso stone

*Israeliti 102–04

De Benedetti (owner), G.B. Bossi (architect) – 1904

A small monument featuring a base (Ghiandone) and a structure made of stone (Rosso Verona) with bronze ornaments

Atlas of stones and their textures

The aim of this atlas is to supply a guide to recognize the stones employed for building in Milan and Lombardy. The pictures at actual size with different surface finishing are coupled with the microscopic observation on thin section.

The appearance of a stone is not unique, but it may be also very different. It is impossible to report all the different varieties of each stone, so we only consider the most significant one. This concerns more the sedimentary rocks and metamorphic (massive or foliated) than igneous ones.

For each stone a picture will be presented which allows it to be recognized at actual size, without the aid of specific investigation methods.

To complete the textural description, a micro-graph on thin section is also provided, mainly in cross-polarized light. The scientific equipment used is a petrographic microscope, also called a polarizing light microscope (PLM), based on the nature of the minerals crossed by transmitted light (thin section): the results are the "interference colours". These colours are not the real ones of the mineral but are due to the interference of polarized rays that have passed through it. The "interference colours", characteristic of each mineral, depend on its optical properties, in particular on the birefringence, but also on the orientation and thickness of the specimen. The identification of the minerals allows the petrographic classification.

The stones are listed according to petro-genetic criterion, based on their geological origin (igneous, sedimentary, metamorphic).

For each stone, the first picture (actual size) includes the commercial name, the origin of the sample (quarry/work), the surface (rough/polished) and the quarry site. The second picture (PLM microscopy) includes the texture and the minerals, together with the kind of transmitted light (PPL plane-polarized/XPL cross-polarized light).

The stones are numbered in relation of their use: Roman times to late 19th century (Figures 28.1–28.41); 20th century up to the present (Figures 28.42–28.73).

A detailed description, of both macroscopic and microscopic appearances, of each stone is reported in Chapters 4–19, Volume 2.

DOI: 10.1201/9781003361008-34

Figure 28.1a Trachite dei Colli Euganei (trachyte); Euganean Hills (Colli Euganei) Zovon di Vò (PD). Quarry sample, honed surface.

Figure 28.1b Trachite dei Colli Euganei – thin section, XPL. Porphyritic texture, phenocrysts of sanidine, amphiboles and sodic plagioclase; very fine groundmass made of potash feldspar and plagioclase.

Figure 28.2a Granito di Baveno (granite); Feriolo (Baveno, VCO). Quarry sample, polished surface.

Figure 28.2b Granito di Baveno – thin section, XPL. Granular texture; crystals of quartz, potash feldspar, plagioclase and biotite with tabular habit.

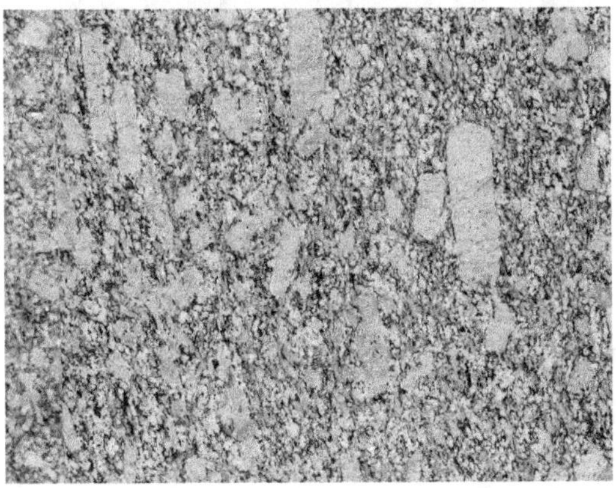

Figure 28.3a Ghiandone (granodiorite); val Màsino (SO). Quarry sample, polished surface.

Figure 28.3b Ghiandone – thin section, XPL. Granular texture; crystals of potash feldspar, plagioclase, quartz and biotite.

Figure 28.4a Granito di Montorfano (granite); Montorfano (VCO). Quarry sample, polished surface.

Figure 28.4b Granito di Montorfano – thin section, XPL. Granular texture; crystals of potash feldspar, plagioclase and biotite with tabular habit.

Figure 28.5a Granito di San Fedelino (granite); Novate Mezzola (lower Val Chiavenna, SO). Quarry sample, rough surface.

Figure 28.5b Granito di San Fedelino – thin section, XPL. Granular texture; crystals of quartz, plagioclase (oligoclase), potash feldspar, muscovite and biotite tabular crystals.

Figure 28.6a Serizzo (diorite); val Màsino (SO). Quarry sample, rough surface.

Figure 28.6b Serizzo – thin section, XPL. Granular texture; crystals of plagioclase, hornblende and quartz.

Figure 28.7a Arenaria dell'Appennino pavese (sandstone); Santa Giuletta (PV). Work sample, decayed surface.

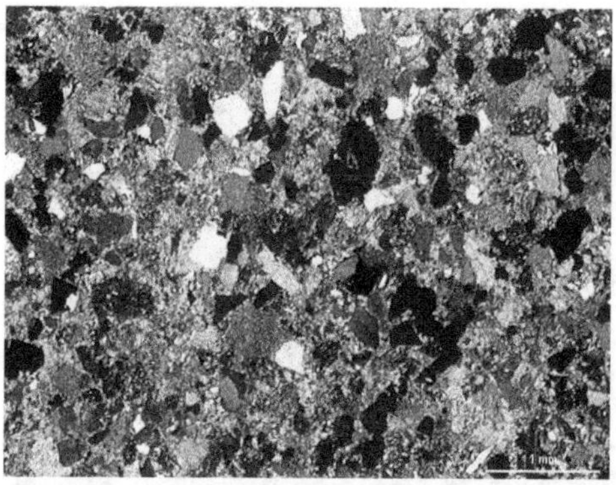

Figure 28.7b Arenaria dell'Appennino pavese – thin section, XPL. Clasts of quartz and muscovite, cement made of calcite spar.

Figure 28.8a Ceppo del Brembo (conglomerate, sandstone); Brembate (BG). Work sample, rough surface.

Figure 28.8b Ceppo del Brembo – thin section, XPL. Clasts of limestones, chert and quartz (single or polycrystalline); cement made of calcite spar.

Figure 28.9a Macchiavecchia (breccia); Arzo (Canton Ticino, Switzerland). Quarry sample, polished surface.

Figure 28.9b Macchiavecchia – thin section, XPL. Rhombohedral crystals of dolomite, fine-grained microcrystalline matrix made of calcite.

Figure 28.10a Pietra di Malnate (sandstone); Malnate (VA). Work sample, decayed surface.

Figure 28.10b Pietra di Malnate – thin section, XPL. Quartz (single crystals or composite), plagioclase, potash feldspar, muscovite, biotite and metamorphic rock clasts, argillaceous cement fills the spaces among clasts.

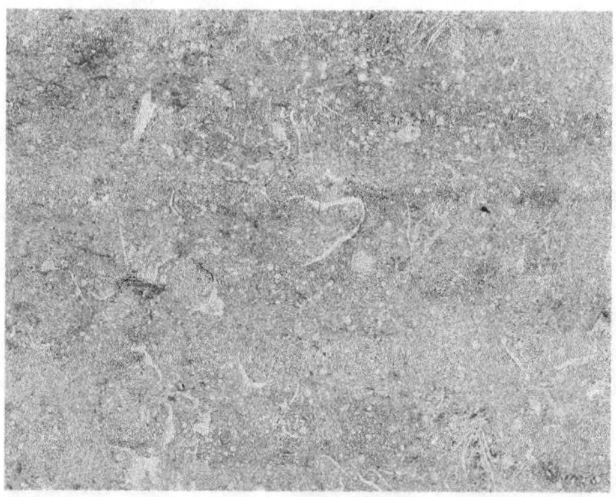

Figure 28.11a Pietra di Vicenza (bioclastic limestone); Zovencedo (Monti Berici, VI). Work sample, honed surface.

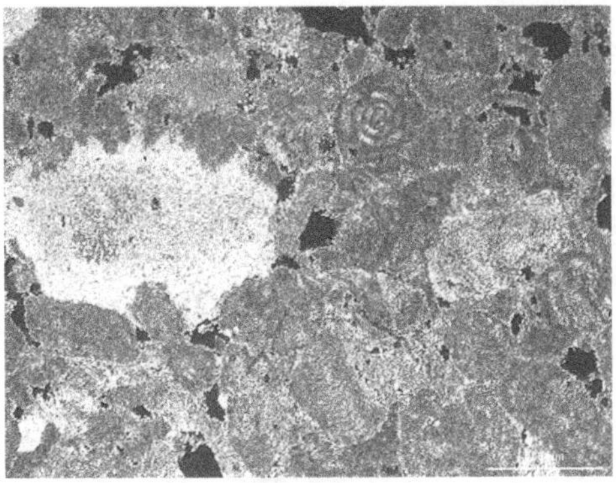

Figure 28.11b Pietra di Vicenza – thin section, XPL. Bioclasts (Foraminifera, Nullipores), mosaic cement of calcite and syntaxial overgrowth.

Figure 28.12a Matòn (bioclastic limestone); Montorio Veronese (Monti Lessini, VR). Work sample, honed surface.

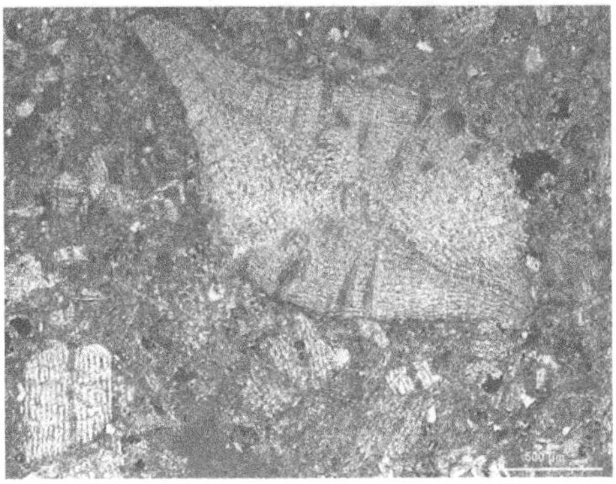

Figure 28.12b Matòn – thin section, XPL. Bioclasts (Foraminifera), mosaic cement of calcite.

Figure 28.13a Pietra Simona (sandstone); Darfo (BS). Quarry sample, honed surface.

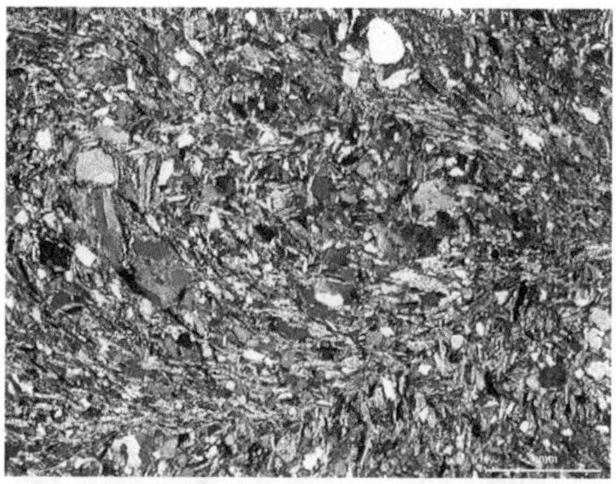

Figure 28.13b Pietra Simona – thin section, PPL. Thin crystals of muscovite with curvilinear alignment and quartz (single or composite).

Figure 28.14a Abbazia (limestone); Abbazia (Albino, BG). Quarry sample, polished surface.

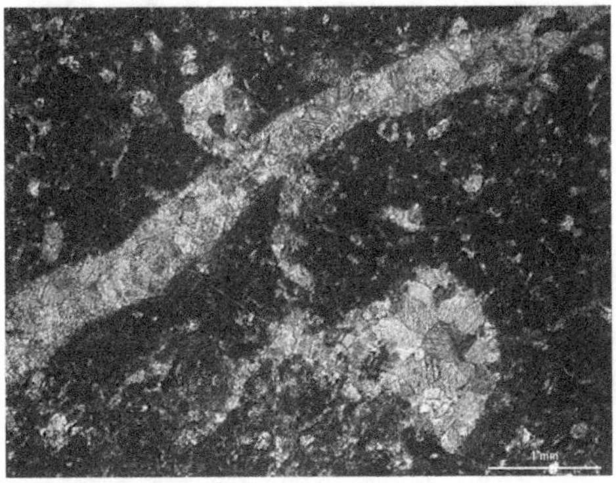

Figure 28.14b Abbazia – thin section, XPL. Microcrystalline texture, matrix-supported with bioclasts, veins filled by calcite spar.

Figure 28.15a Pietra di Angera (dolostone); Angera (VA). Work sample, honed surface.

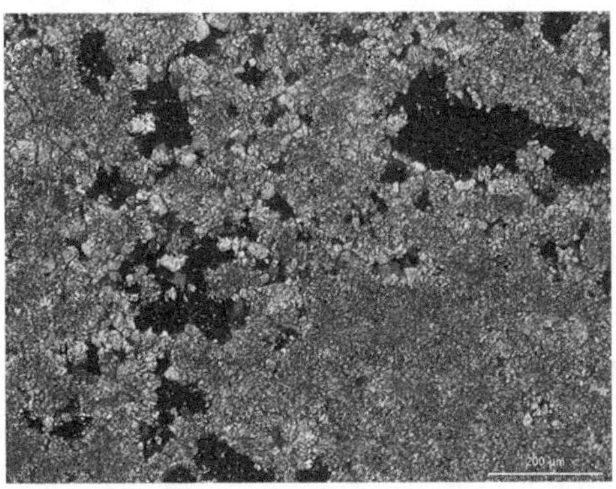

Figure 28.15b Pietra di Angera – thin section, XPL. Microcrystalline texture of dolomite, irregular cavities.

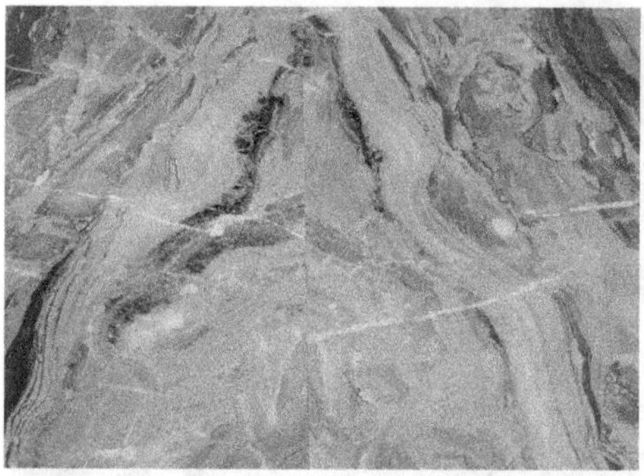

Figure 28.16a Arabescato orobico (limestone); Camerata Cornello (BG). Quarry sample, polished surface.

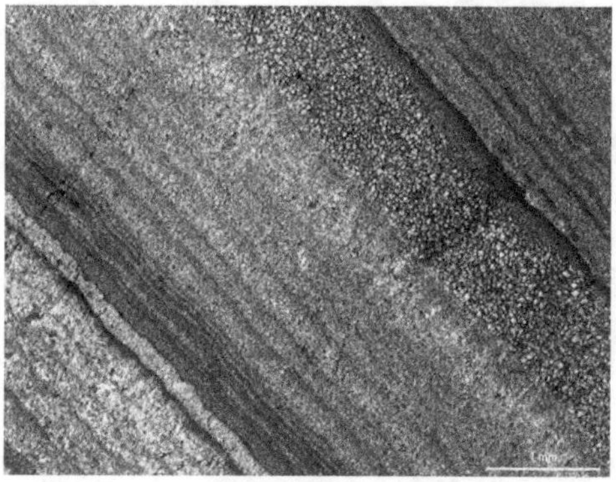

Figure 28.16b Arabescato orobico – thin section, PPL. Thin banded microcrystalline texture of calcite.

Figure 28.17a Pietra di Arzo (limestone); Arzo (Canton Ticino, Switzerland). Quarry sample, polished surface.

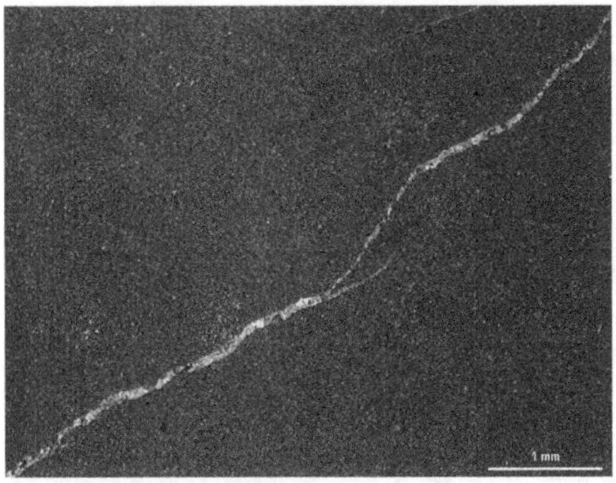

Figure 28.17b Pietra di Arzo – thin section, PPL. Microcrystalline texture of calcite.

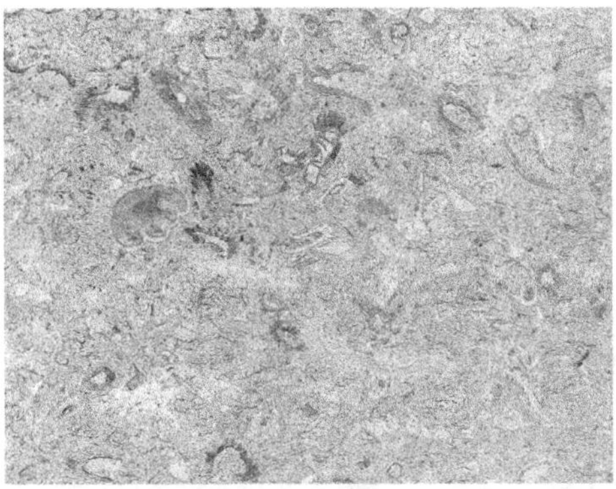

Figure 28.18a Pietra di Aurisina (bioclastic limestone); Aurisina (TS). Quarry sample, polished surface.

Figure 28.18b Pietra di Aurisina – thin section, PPL. Skeletal components (bioclasts featuring prismatic calcite crystals), micritic sediment and cement made of calcite spar.

Figure 28.19a Pietra di Botticino (dolomitic limestone); Botticino Mattina (BS). Quarry sample, polished surface.

Figure 28.19b Pietra di Botticino – thin section, XPL. Microcrystalline calcite together with euhedral rhomb-shaped crystals of dolomite.

Figure 28.20a Rosso di Entratico (nodular limestone); Entratico (BG). Quarry sample, rough surface.

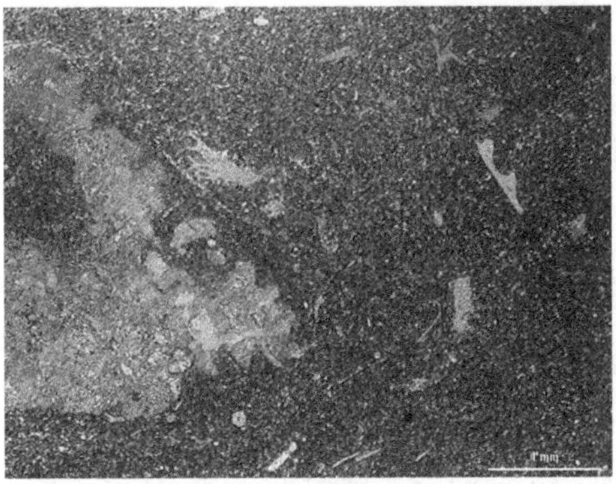

Figure 28.20b Rosso di Entratico – thin section, XPL. Microcrystalline calcite with bioclasts.

Figure 28.21a Nero di Bergamo (limestone); Cene (BG). Work sample, decayed surface (discolouration).

Figure 28.21b Nero di Bergamo – thin section, XPL. Microcrystalline calcite, irregular veins and pores.

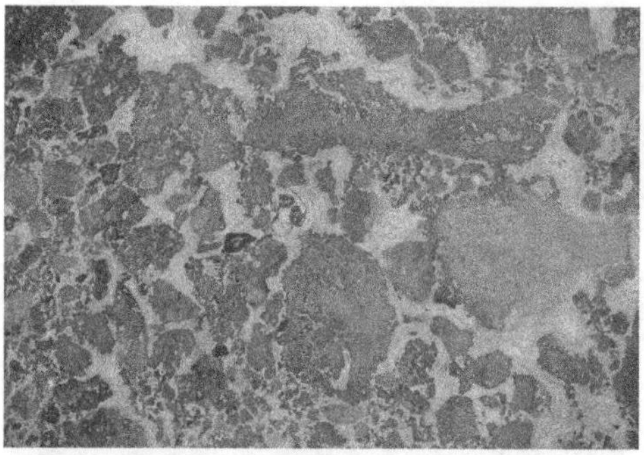

Figure 28.22a Occhiadino (limestone); Lozio (BS). Quarry sample, polished surface.

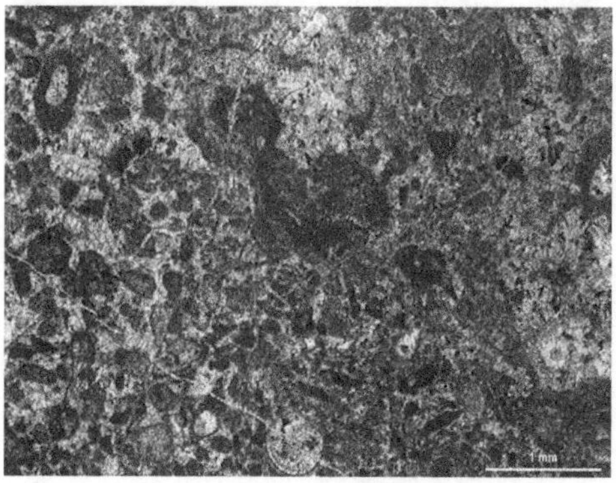

Figure 28.22b Occhiadino – thin section, XPL. Microcrystalline calcite, bioclasts, thin calcite veins.

Figure 28.23a Repen (bioclastic limestone); Monrupino/Repentabor (TS). Quarry sample, polished surface.

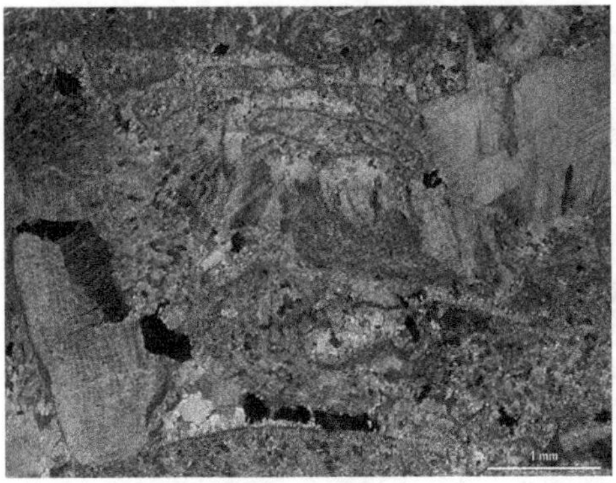

Figure 28.23b Repen – thin section, XPL. Skeletal components, micritic sediment, cement made of calcite spar.

Figure 28.24a Pietra di Saltrio (clastic limestone); Saltrio (VA). Work sample, honed surface.

Figure 28.24b Pietra di Saltrio – thin section, XPL. Intraclasts of uniform grain-size with interspersed nodules of chert.

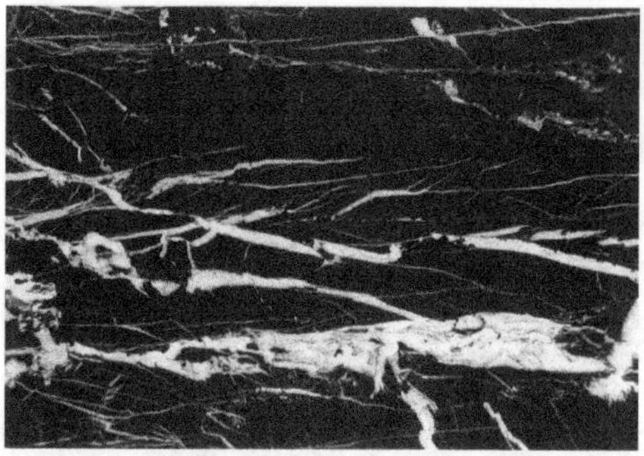

Figure 28.25a Pietra di Varenna (limestone); Varenna (CO). Quarry sample, polished surface.

Figure 28.25b Pietra di Varenna – thin section, PPL. Homogeneous microcrystalline calcite, veins made of calcite spar.

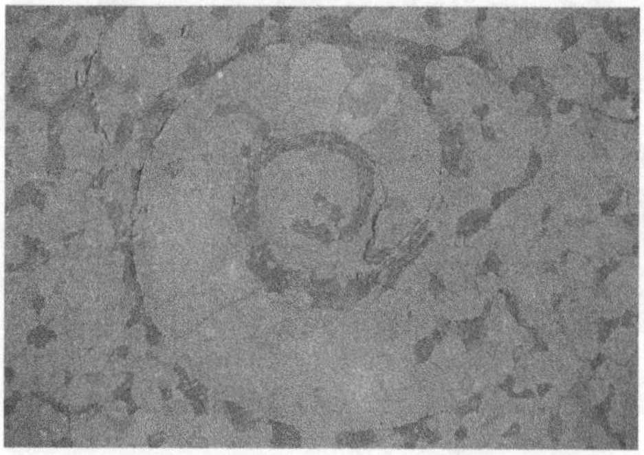

Figure 28.26a Rosso di Verona (nodular limestone); Sant'Ambrogio di Valpolicella (VR). Quarry sample, polished surface.

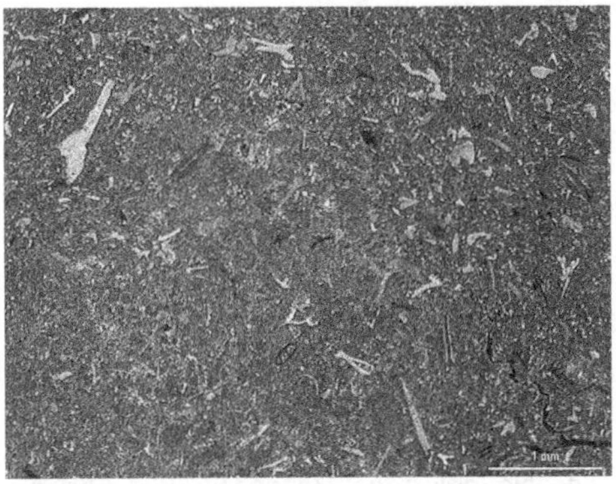

Figure 28.26b Rosso di Verona – thin section, XPL. Microcrystalline calcite matrix, bioclasts, irregular cavities.

Figure 28.27a Pietra di Viggiù (clastic limestone); Piamo (Viggiù, VA). Quarry sample, honed surface.

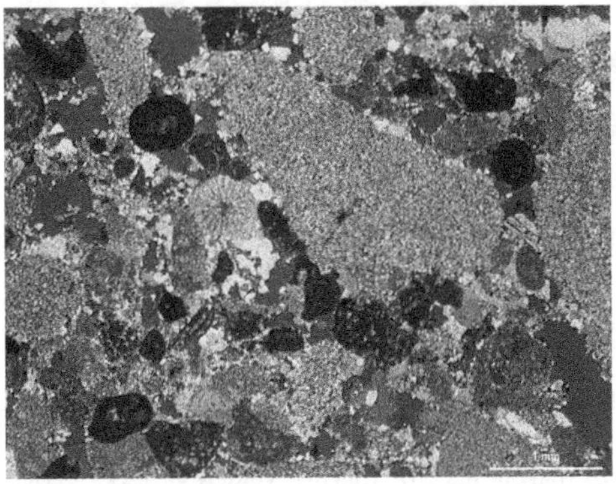

Figure 28.27b Pietra di Viggiù – thin section, PPL. Medium-grained intraclasts and oolites, cement made of calcite spar.

Figure 28.28a Marmo di Zandobbio (dolostone); Zandobbio (BG). Quarry sample, honed surface.

Figure 28.28b Marmo di Zandobbio – thin section, XPL. Mosaic fabric made of subhedral crystals of dolomite showing rhombohedral cleavage.

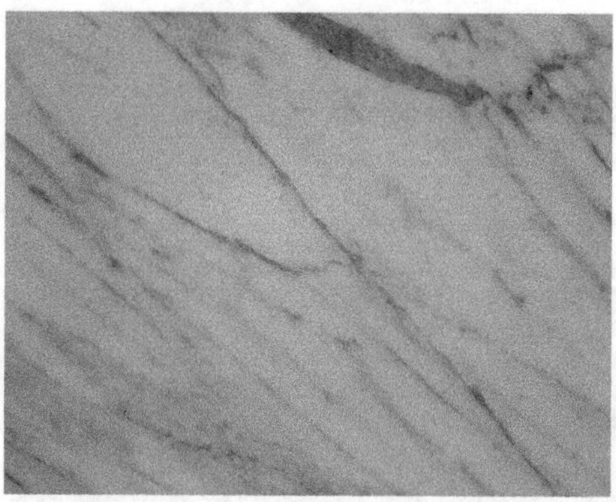

Figure 28.29a Apuanian Alps marble (calcitic marble, veined variety); Gioia (Colonnata, MS). Quarry sample, polished surface.

Figure 28.29b Apuanian Alps marble – thin section, XPL. Homeoblastic texture with polygonal crystals of calcite (straight to curved corners).

Figure 28.30a Bardiglio (calcitic marble); Carrara area (MS). Work sample, honed surface.

Figure 28.30b Bardiglio – thin section, XPL. Homeoblastic texture; very fine calcite crystals with polygonal shape.

Figure 28.31a Beola (gneiss); Bèura (Domodossola, VCO). Quarry sample, rough face.

Figure 28.31b Beola – thin section, XPL. Polycrystalline quartz and feldspar crystals, alignments of muscovite and biotite tabular crystals.

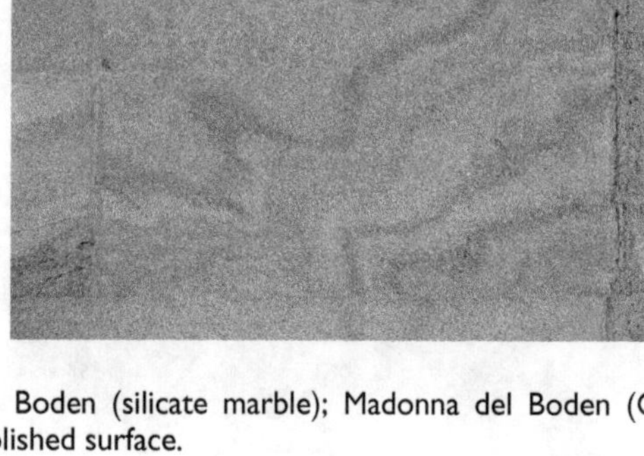

Figure 28.32a Grigio del Boden (silicate marble); Madonna del Boden (Ornavasso, VCO). Work sample, polished surface.

Figure 28.32b Grigio del Boden – thin section, XPL. Calcite crystals in alternating with quartz; phlogopite and polygonal crystals of opaque minerals.

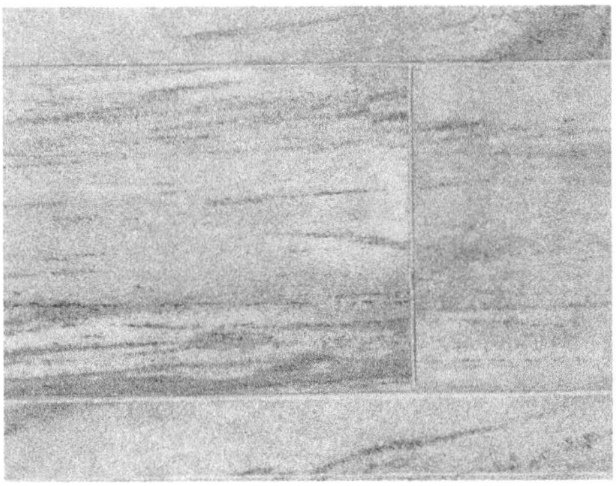

Figure 28.33a Marmo di Candoglia (calcitic marble); Candoglia (VCO). Work sample, polished surface.

Figure 28.33b Marmo di Candoglia – thin section, XPL. Granular calcite crystals with irregular boundaries and polysynthetic twinning.

Figure 28.34a Marmo di Crevola (dolomitic marble); Crevoladossola (VCO). Quarry sample, polished surface.

Figure 28.34b Marmo di Crevola– thin section, XPL. Euhedral crystals of dolomite with thin tabular crystals of phlogopite.

Figure 28.35a Marmo di Musso (calcitic marble); Musso (CO). Work sample, polished surface.

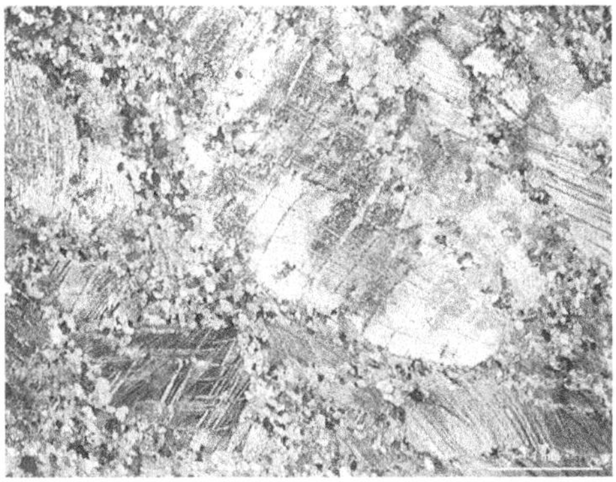

Figure 28.35b Marmo di Musso – thin section, XPL. "Mortar" texture with few large anhedral calcite crystals with curved twin lamellae, accompanied by numerous finer calcite crystals.

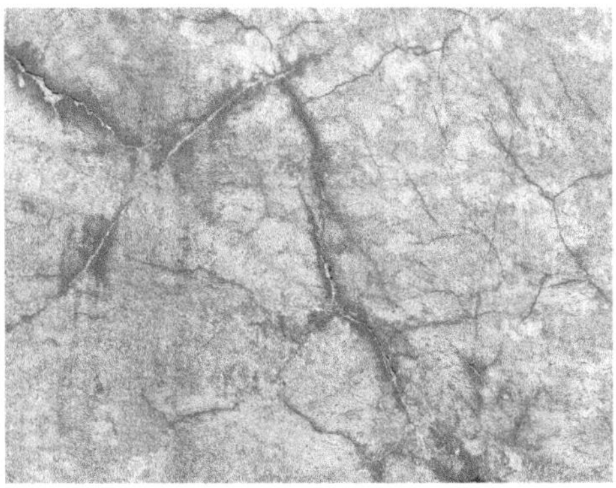

Figure 28.36a Pietra d'Oira (serpentinized peridotite); Oira (Nònio, VCO). Work sample, decayed surface (discolouration).

Figure 28.36b Pietra d'Oira – thin section, XPL. Large olivine crystals with fibrous serpentine.

Figure 28.37a Marmo di Ornavasso (calcitic marble); Ornavasso (VCO). Quarry sample, polished surface.

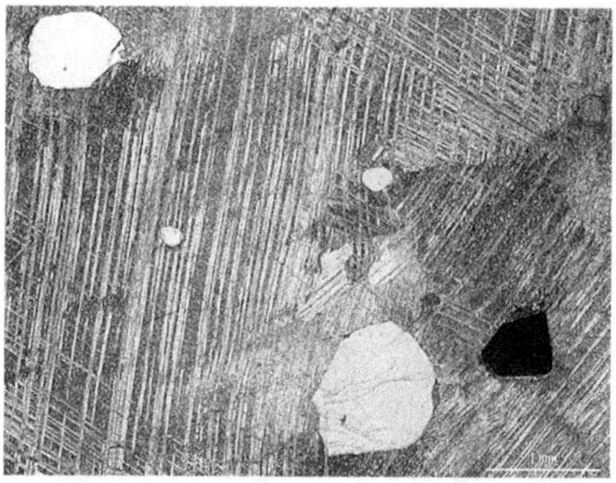

Figure 28.37b Marmo di Ornavasso – thin section, XPL. Heteroblastic calcite crystals with polysynthetic twinning, rounded crystals of quartz, triangular pyrite.

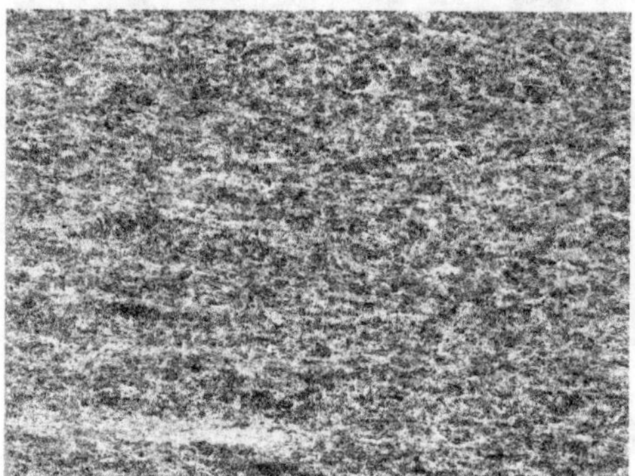

Figure 28.38a Serizzo dell'Ossola (gneiss); Valle Antigorio (VCO). Quarry sample, rough surface.

Figure 28.38b Serizzo dell'Ossola – thin section, XPL. Gneissic texture, quartz, potash feldspar and muscovite.

Figure 28.39a Serpentinite (serpentineschist); Chiareggio (Val Malenco, SO). Quarry sample, rough surface.

Figure 28.39b Serpentinite – thin section, XPL. Rodlike fine crystals of serpentine with preferred orientation (foliation).

Figure 28.40a Marmo di Valle Strona (calcitic marble); Sambughetto (VCO). Quarry sample, honed surface.

Figure 28.40b Marmo di Valle Strona – thin section, XPL. Heteroblastic calcite crystals and quartz (single or polycrystalline).

Figure 28.41a Marmo di Vezza d'Oglio (calcitic marble); Vezza d'Oglio (BS). Quarry sample, rough surface.

Figure 28.41b Marmo di Vezza d'Oglio – thin section, XPL. Heteroblastic calcite crystals with polysynthetic twinning and rhombohedral cleavage; rounded quartz crystals.

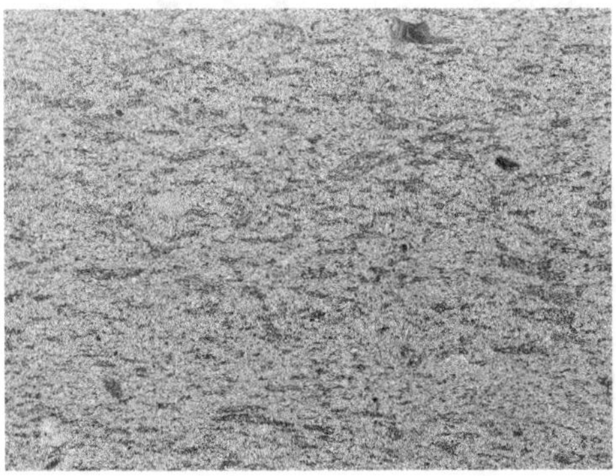

Figure 28.42a Peperino/Lavagrigia (pyroclastic flow); Vitorchiano (VT). Work sample, honed surface.

Figure 28.42b Peperino/Lavagrigia – thin section, XPL. Porphyritic texture; feldspar and pyroxene crystals.

Figure 28.43a Porfido Monumentale (rhyolite); Croce Domini Pass (Bienno, BS). Quarry sample, split surface.

Figure 28.43b Porfido Monumentale – thin section, XPL. Porphyritic texture; tabular phenocrysts of plagioclase, quartz and biotite; micro to cryptocrystalline groundmass.

Figure 28.44a Porfido del Trentino (rhyodacite); Val Cembra (TN). Quarry sample, split face.

Figure 28.44b Porfido del Trentino – thin section, XPL. Porphyritic texture; quartz, potash feldspar, biotite crystals, groundmass mainly made of glass.

Figure 28.45a Granito di Alzo (granite); Alzo (Pella, VCO). Quarry sample, rough surface.

Figure 28.45b Granito di Alzo – thin section, XPL. Granular texture; crystals of quartz, potash feldspar, zoned plagioclase, tabular crystals of biotite.

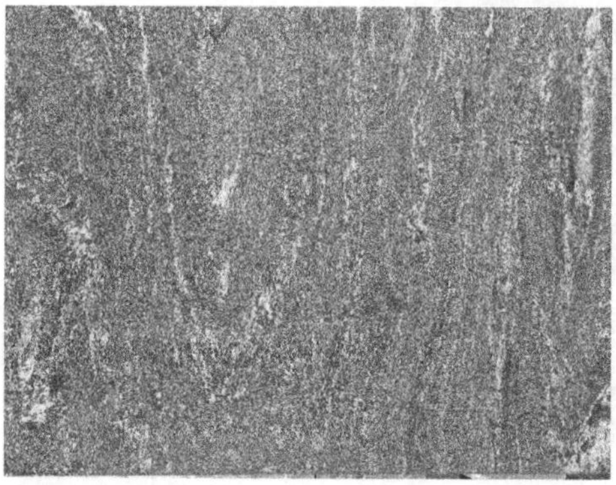

Figure 28.46a Granito nero di Anzola (gabbro); Anzola d'Ossola (VCO). Quarry sample, polished surface.

Figure 28.46b Granito nero di Anzola – thin section, XPL. Granular texture; crystals of hornblende, plagioclase (labradorite), olivine.

Figure 28.47a Sienite di Balma (syenite); San Paolo Cervo (BI). Quarry sample, polished surface.

Figure 28.47b Sienite di Balma – thin section, XPL. Granular texture; alkaline feldspar, zoned plagioclase and amphibole crystals.

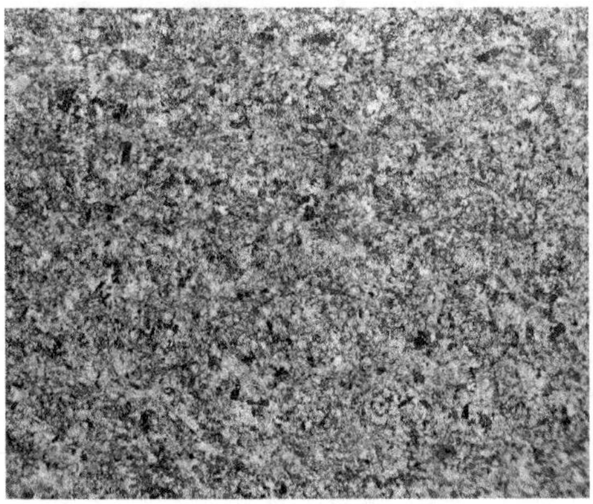

Figure 28.48a Granito di Dubino (quartzdiorite); Dubino (SO). Quarry sample, polished surface.

Figure 28.48b Granito di Dubino – thin section, XPL. Granular texture, irregular-sized plagioclase crystals together with hornblende and quartz.

Figure 28.49a Granito di Mergozzo (granite); Mergozzo (VCO). Quarry sample, polished surface.

Figure 28.49b Granito di Mergozzo – thin section, XPL. Granular texture; potash feldspar and plagioclase crystals, biotite mainly transformed in chlorite.

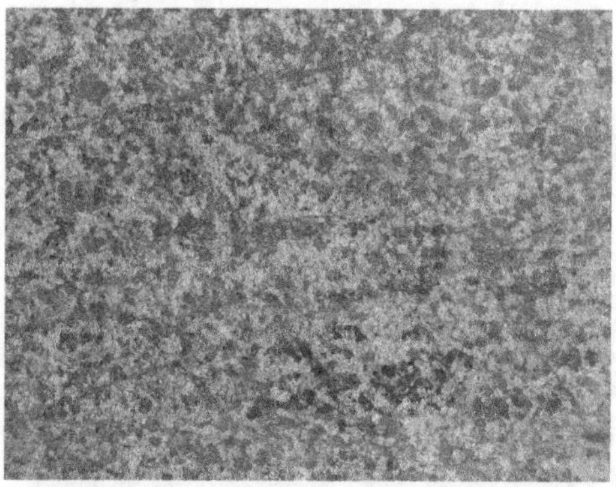

Figure 28.50a Rosso Pantheon (granite); San Colombano Belmonte (TO). Quarry sample, polished surface.

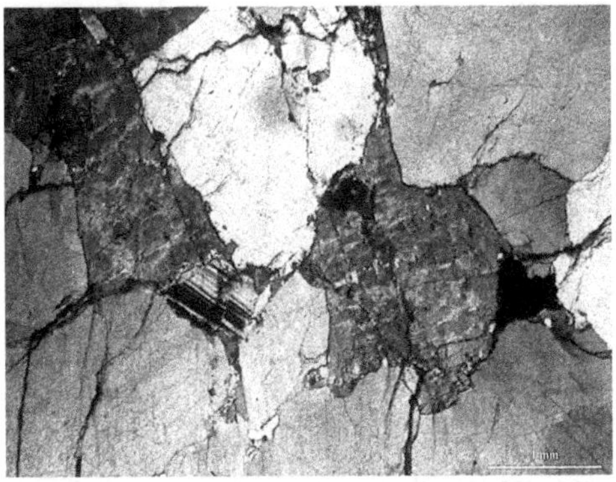

Figure 28.50b Rosso Pantheon – thin section, XPL. Granular texture; quartz and K-feldspar crystals, sodium plagioclase; some thin cracks.

Figure 28.51a Diorite di Vico Canavese (diorite); Vico Canavese (TO). Quarry sample, polished surface.

Figure 28.51b Diorite di Vico Canavese – thin section, XPL. Granular texture; plagioclase, hornblende, quartz and biotite crystals.

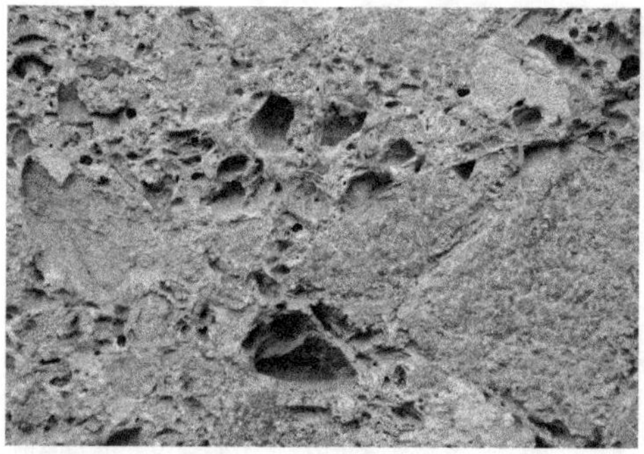

Figure 28.52a Ceppo del lago d'Iseo (conglomerate); Grè (Castro, BG). Quarry sample, sawn surface.

Figure 28.52b Ceppo del lago d'Iseo – thin section, XPL. Clasts of dolomite joined by a cement made of calcite spar.

Figure 28.53a Pietra di Finale (bioclastic limestone); Finale Ligure (SV). Work sample, honed surface.

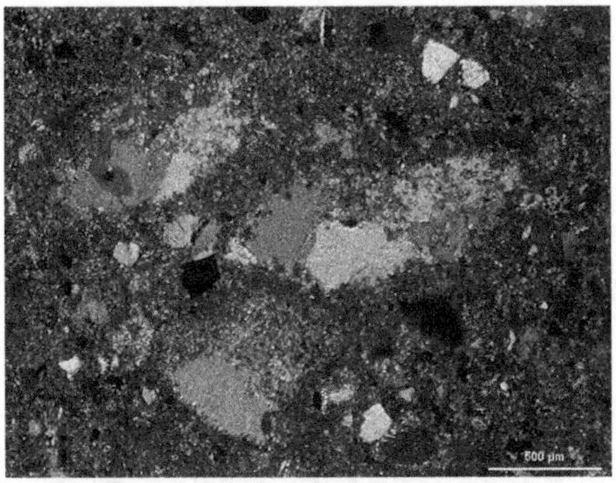

Figure 28.53b Pietra di Finale – thin section, XPL. Coarse calcite crystals together with angular quartz.

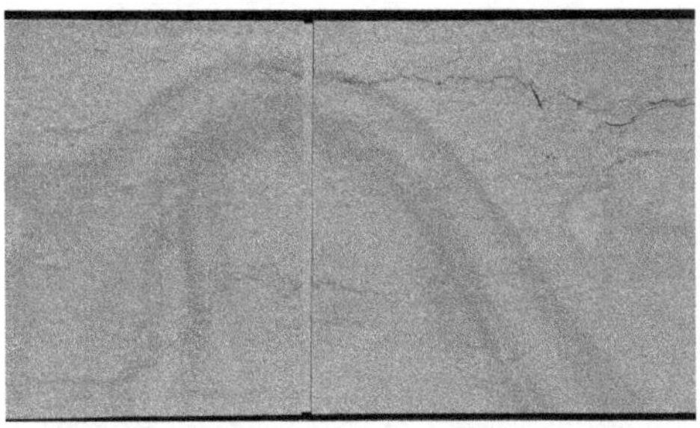

Figure 28.54a Pietra dorata (sandstone); Poggio Sanopia (Manciano, GR). Work surface, honed surface.

Figure 28.54b Pietra dorata – thin section, XPL. Angular crystals of quartz and K-feldspar, limestone fragments, cement of calcite spar.

Figure 28.55a Breccia di Urago (conglomerate); Urago (Montorfano, CO). Quarry sample, sawn surface.

Figure 28.55b Breccia di Urago – thin section, XPL. Clasts of gneiss, limestones (oolitites, bioclasts), chert; calcite spar cement fills the gaps among clasts.

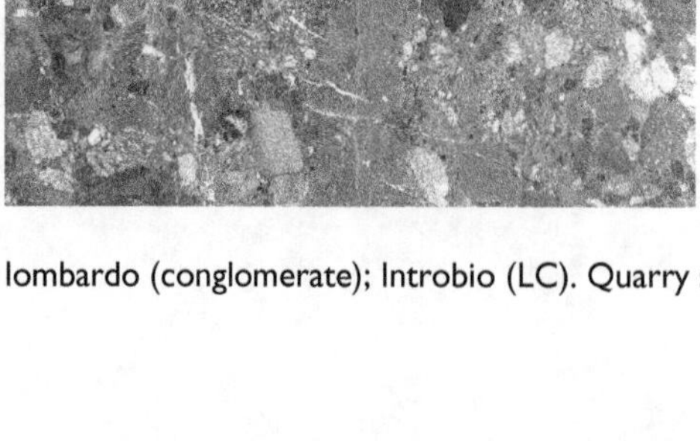

Figure 28.56a Verrucano lombardo (conglomerate); Introbio (LC). Quarry sample, polished surface.

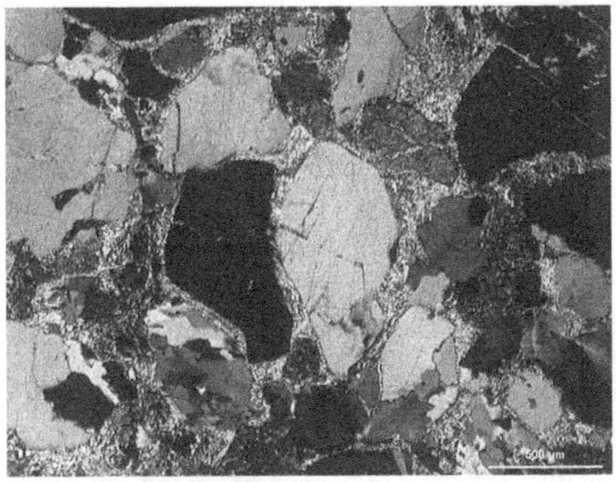

Figure 28.56b Verrucano lombardo – thin section, XPL. Fragments of volcanic rocks (quartz rhyolites) or igneous rocks (diorite) or schists; matrix containing quartz, feldspar, micas.

Figure 28.57a Pietra di Chiampo (limestone); Chiampo (VI). Quarry sample, polished surface.

Figure 28.57b Pietra di Chiampo – thin section, XPL. Skeletal components (Foraminifera), cement made of calcite spar.

Figure 28.58a Giallo Mori (limestone); Mori (TN). Work sample (mosaic), polished surface.

Figure 28.58b Giallo Mori – thin section, XPL. Microcrystalline calcite, rhombohedral dolomite; interstitial iron oxides.

Figure 28.59a Grigio carnico (limestone); Timau (Paluzza, UD). Quarry sample, polished surface.

Figure 28.59b Grigio carnico – thin section, XPL. Microcrystalline calcite with irregular cavities together with thick veins filled by large crystals of calcite.

Figure 28.60a Pietra d'Istria (limestone); Montižana (Vrsar/Orsera, Croatia). Work sample, honed surface.

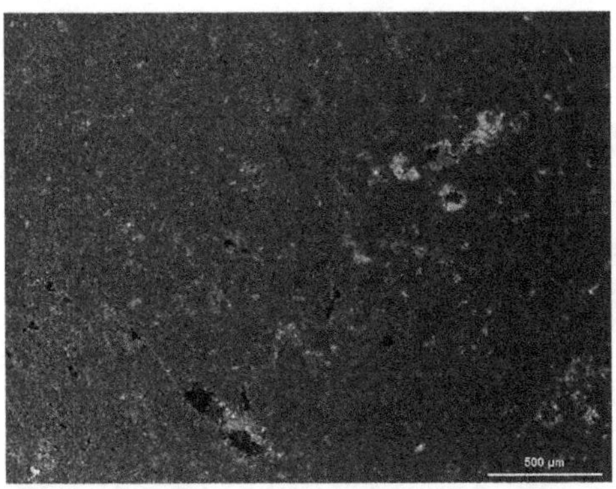

Figure 28.60b Pietra d'Istria – thin section, XPL. Microcrystalline calcite with few round cavities.

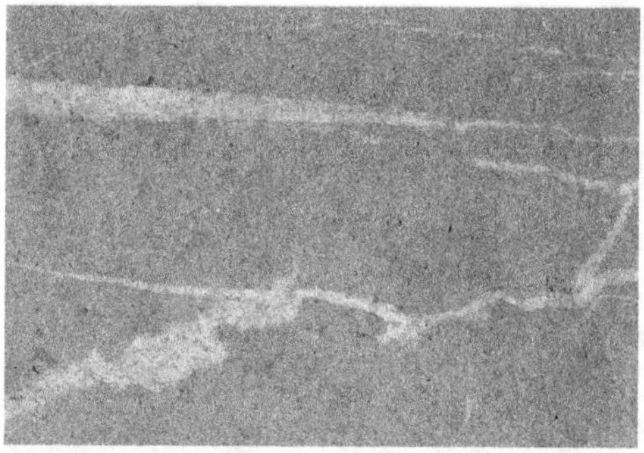

Figure 28.61a Pietra piasentina (clastic limestone); Torreano (UD). Work sample, honed surface.

Figure 28.61b Pietra piasentina – thin section, XPL. Intraclasts and skeletal components; cement of calcite spar.

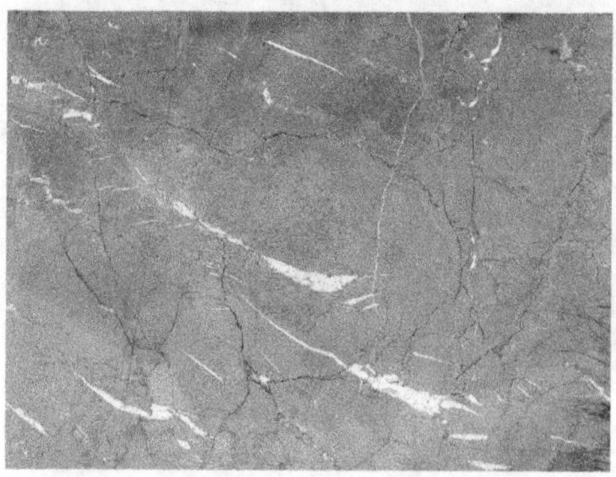

Figure 28.62a Rosso Collemandina (limestone); Villa Collemandina (Castelnuovo Garfagnana, LU). Quarry sample, honed surface.

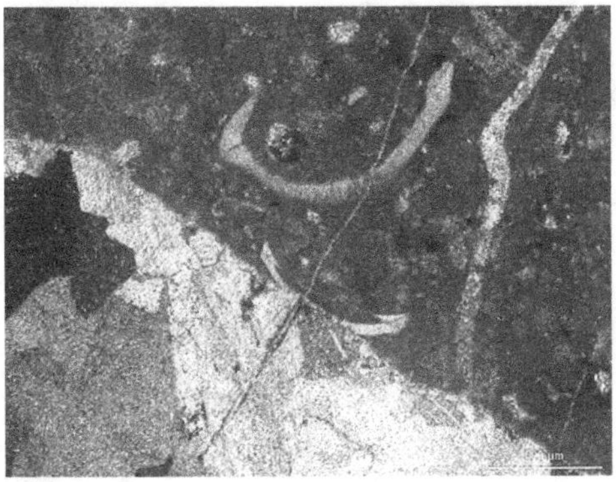

Figure 28.62b Rosso Collemandina – thin section, XPL. Microcrystalline calcite together with skeletal components and straight veins of calcite spar.

Figure 28.63a Rosso di Trento (nodular limestone); Trento. Quarry sample, polished surface.

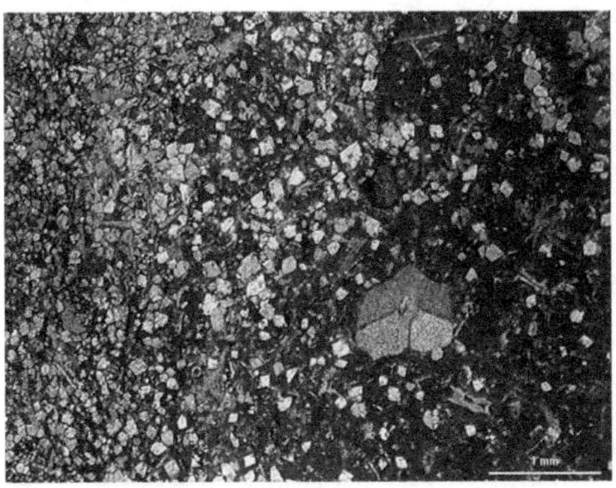

Figure 28.63b Rosso di Trento – thin section, XPL. Microcrystalline calcite; dolomite (rhombohedral crystals), skeletal components.

Figure 28.64a Travertino romano – *Lapis Tiburtinus* (non-clastic deposit); Bagni di Tivoli (Rome). Work sample, honed surface.

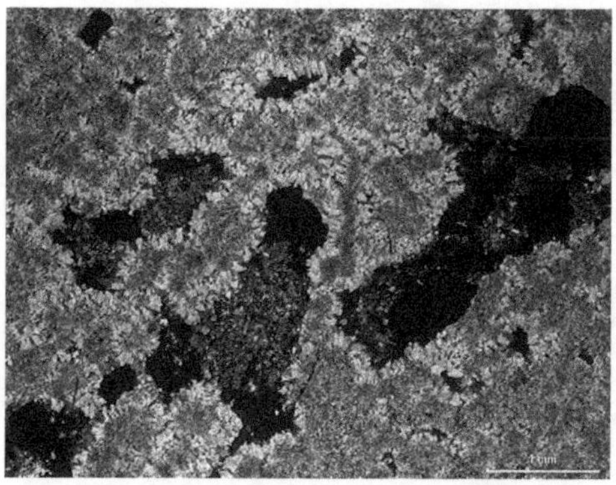

Figure 28.64b Travertino romano – thin section, XPL. Microcrystalline calcite, coarse rod-shaped calcite crystals arranged around irregular cavities.

Figure 28.65a Travertino of Tuscany (non-clastic deposit); Rapolano Terme (SI). Work sample, honed surface.

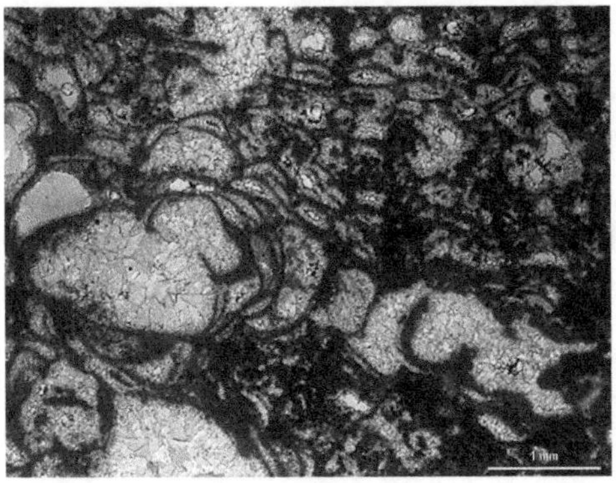

Figure 28.65b Travertino of Tuscany – thin section, XPL. Microcrystalline calcite, coarse calcite crystals filling roundish cavities.

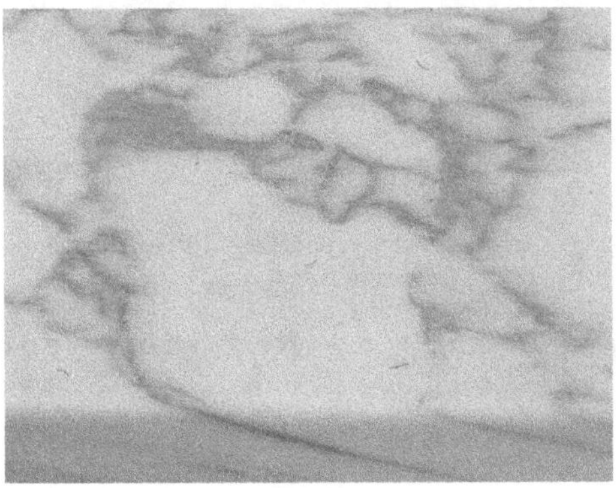

Figure 28.66a Calacata (calcitic marble); Calacata (Carrara, MS). Work sample, polished surface.

Figure 28.66b Calacata – thin section, XPL. Polygonal calcite crystals that increase in size within the veins.

Figure 28.67a Cipollino apuano (calcitic marble); Cardoso (Stazzema, LU). Work sample, polished surface.

Figure 28.67b Cipollino apuano – thin section, XPL. Anhedral calcite and thin tabular crystals of chlorite.

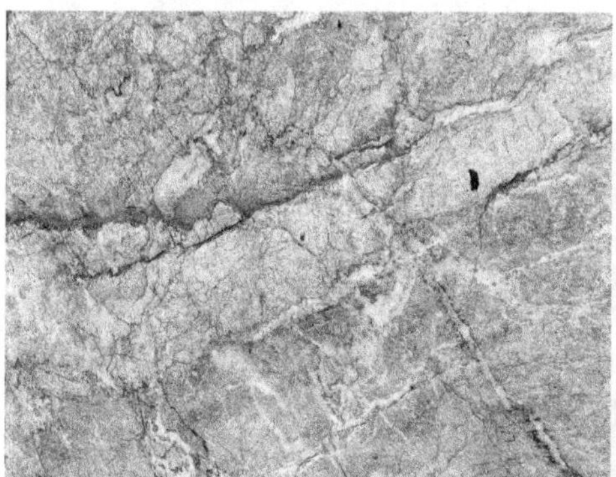

Figure 28.68a Fior di pesco carnico (calcitic marble); Forni Avoltri (UD). Quarry sample, honed surface.

Figure 28.68b Fior di pesco carnico – thin section, XPL. Anhedral calcite crystals with polysynthetic twinning.

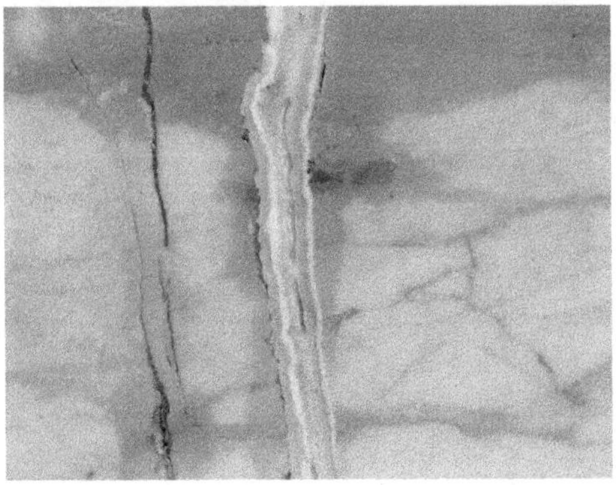

Figure 28.69a Marmo giallo di Siena (calcitic marble); Sovicille (Montagnola Senese, SI). Quarry sample, polished surface.

Figure 28.69b Marmo giallo di Siena – thin section, XPL. Anhedral fine calcite crystals, pyrite (polygonal crystals).

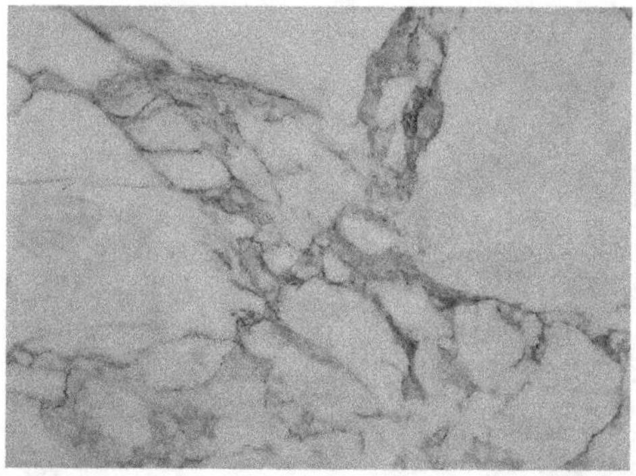

Figure 28.70a Marmo di Lasa (calcitic marble); Lasa/Laas, (BZ). Quarry sample, honed surface.

Figure 28.70b Marmo di Lasa – thin section, XPL. Anhedral crystals of calcite (polysynthetic twinning and rhombohedral cleavage), rounded crystals of quartz.

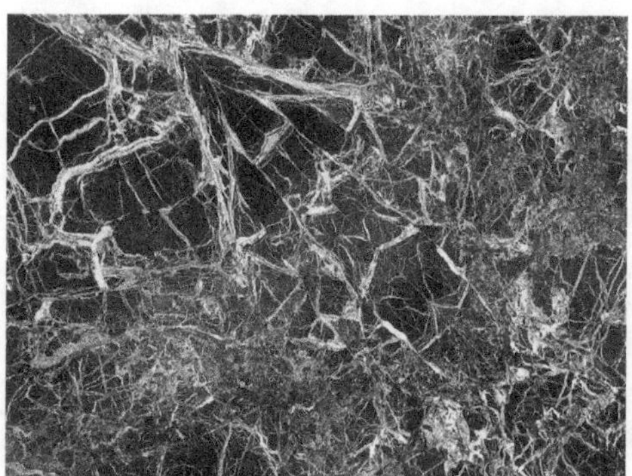

Figure 28.71a Rosso/Verde di Levanto (ophicalcite); Bonassola (SP). Quarry sample, polished surface.

Figure 28.71b Rosso/Verde di Levanto – thin section, XPL. Fibrous aggregates of serpentine.

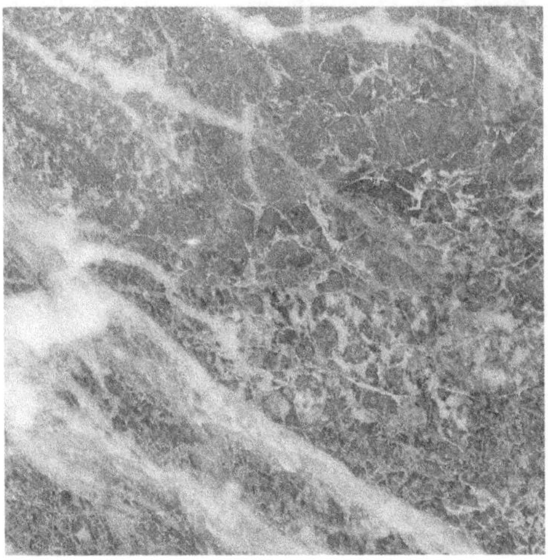

Figure 28.72a Moncervetto (marble); Monastero di Vasco (Mondovì, CN). Quarry sample, polished surface.

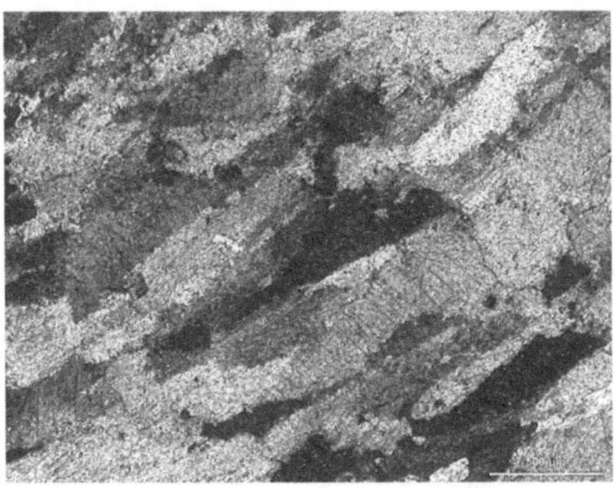

Figure 28.72b Moncervetto – thin section, XPL. Stretched calcite crystals showing fairly orientation.

Figure 28.73a Verde Roja (slate); San Dalmazzo di Tenda (France). Work sample, honed surface.

Figure 28.73b Verde Roja – thin section, XPL. Cryptocrystalline texture of clay minerals, very fine crystals of quartz.

Index

The entries concern:
(a) architects and other artists;
(b) buildings (apartment and office buildings, theatres, villas, hospitals, churches etc.; the term "Milan" is omitted, the names of other cities or towns are given in brackets);
(f) geological formations, units etc.;
(m) commemorative monuments (personal names and titles);
(o) owners of the funeral chapels of Cimitero Monumentale;
(q) quarry sites;
(r) raw materials (lime, gypsum, cast stone etc.);
(s) stones (terms as pietra, granito, marmo etc. are generally omitted);
(t) artistic techniques.
Urban addresses are reported letter by letter together with the Italian terms: via, piazza, corso, viale etc. The stone entries (s) apply only to chapters 2 to 20; stones are not indexed in the other chapters.

For Product Safety Concerns and Information please contact our EU
representative GPSR@taylorandfrancis.com
Taylor & Francis Verlag GmbH, Kaufingerstraße 24, 80331 München, Germany

www.ingramcontent.com/pod-product-compliance
Lightning Source LLC
LaVergne TN
LVHW080852300926
849162LV00003B/405
* 9 7 8 1 0 3 2 4 2 0 5 6 1 *